Methods in Molecular Biology

Series Editor
John M. Walker
School of Life and Medical Sciences
University of Hertfordshire
Hatfield, Hertfordshire, UK

For further volumes:
http://www.springer.com/series/7651

For over 35 years, biological scientists have come to rely on the research protocols and methodologies in the critically acclaimed *Methods in Molecular Biology series.* The series was the first to introduce the step-by-step protocols approach that has become the standard in all biomedical protocol publishing. Each protocol is provided in readily-reproducible step-by step fashion, opening with an introductory overview, a list of the materials and reagents needed to complete the experiment, and followed by a detailed procedure that is supported with a helpful notes section offering tips and tricks of the trade as well as troubleshooting advice. These hallmark features were introduced by series editor Dr. John Walker and constitute the key ingredient in each and every volume of the Methods in *Molecular Biology series.* Tested and trusted, comprehensive and reliable, all protocols from the series are indexed in PubMed.

Cartilage Tissue Engineering

Edited by

Martin J. Stoddart and Elena Della Bella

AO Research Institute Davos, Davos Platz, Switzerland

Angela R. Armiento

UCB Pharma, Slough, UK

Humana Press

Editors
Martin J. Stoddart
AO Research Institute Davos
Davos Platz, Switzerland

Elena Della Bella
AO Research Institute Davos
Davos Platz, Switzerland

Angela R. Armiento
UCB Pharma
Slough, UK

ISSN 1064-3745 ISSN 1940-6029 (electronic)
Methods in Molecular Biology
ISBN 978-1-0716-2838-6 ISBN 978-1-0716-2839-3 (eBook)
https://doi.org/10.1007/978-1-0716-2839-3

Cover Illustration Caption: Scaffold-filled osteochondral defect.

This Humana imprint is published by the registered company Springer Science+Business Media, LLC, part of Springer Nature.
The registered company address is: 1 New York Plaza, New York, NY 10004, U.S.A.

Preface

The repair of traumatic cartilage injuries is one of medicine's remaining dilemmas, with few clinical treatment options available. Current treatments include debridement, microfracture, and various implementations of autologous chondrocyte implantation, yet the results are largely unsatisfactory over the medium term (5 years onwards). For this reason, there is still significant research being performed on novel therapies for cartilage regeneration. Within this methods book, we attempt to bring together a collection of methodologies, from the most basic to more complex, that provide researchers with a platform they can use to embark on a cartilage research career. One of the factors delaying research on cartilage is likely the fact that 3D cartilage models are more challenging to work with compared to cells in monolayer. Such challenges are associated with RNA and protein extraction, imaging, gene transfer, and for some laboratories even access to such tissues. Stable differentiation and variations in cell phenotype from different tissue origins adds to the complexity. As technology evolves (omics, bioinformatics, data mining, organ-on-a-chip, to cite but a few) and scientists have a better appreciation of the relevant *in vitro* and *in vivo* models, the research on cartilage is more exciting than ever.

Davos Platz, Switzerland *Martin J. Stoddart*
Davos Platz, Switzerland *Elena Della Bella*
Slough, UK *Angela R. Armiento*

Contents

Contributors

MAURO ALINI • *AO Research Institute Davos, Davos Platz, Switzerland*
ANGELA R. ARMIENTO • *AO Research Institute Davos, Davos Platz, Switzerland; UCB Pharma, Slough, UK*
VICKY BATCHELOR • *Centre for Osteoarthritis Pathogenesis, Kennedy Institute of Rheumatology, University of Oxford, Oxford, UK*
AUSTIN BELL-HENSLEY • *Department of Biomedical Engineering, Washington University, St Louis, MO, USA*
ANGELO BOFFA • *Applied and Translational Research Center, IRCCS Istituto Ortopedico Rizzoli, Bologna, Italy*
PRIYANKA P. BRAHMACHARY • *Department of Mechanical & Industrial Engineering, Montana State University, Bozeman, MT, USA*
TERESA Z. BROSE • *Medical Center-Albert-Ludwigs-University, Department of Orthopedics and Trauma Surgery, Faculty of Medicine, Albert-Ludwigs-University of Freiburg, Freiburg, Germany*
JUSTYNA M. BUCHERT • *Research Centre for Experimental Orthopaedics, Heidelberg University Hospital, Heidelberg, Germany*
ROCÍO CASTRO-VIÑUELAS • *Human Movement Biomechanics Research Group, Department of Movement Sciences, KU Leuven, Leuven, Belgium; Tissue homeostasis and Disease Group, Skeletal Biology and Engineering Research Center, Department of Development and Regeneration, KU Leuven, Leuven, Belgium*
DEVA D. CHAN • *Purdue University, Weldon School of Biomedical Engineering, West Lafayette, IN, USA*
BLAINE A. CHRISTIANSEN • *University of California Davis Health, Department of Orthopaedic Surgery, Sacramento, CA, USA*
CLARISSA R. COVENEY • *Kennedy Institute of Rheumatology, Medical Sciences Division, University of Oxford, Oxford, UK; Department of Human Evolutionary Biology, Harvard University, Cambridge, MA, USA*
SALIM E. DARWICHE • *Musculoskeletal Research Unit (MSRU), Vetsuisse Faculty, University of Zürich, Zürich, Switzerland; Center for Applied Biotechnology and Molecular Medicine (CABMM), Vetsuisse Faculty, University of Zürich, Zürich, Switzerland*
RODOLFO E. DE LA VEGA • *Musculoskeletal Gene Therapy Laboratory, Mayo Clinic, Rochester, MN, USA*
FRANCESCO DELL'ACCIO • *Centre for Experimental Medicine and Rheumatology, William Harvey Research Institute, Barts and the London School of Medicine and Dentistry, Queen Mary University of London, London, UK*
ELENA DELLA BELLA • *AO Research Institute Davos, Davos Platz, Switzerland*
ANDREA DE LUNA • *Center for Regenerative Medicine, University For Continuing Education, Krems, Austria*
NICHOLAS N. DEPHILLIPO • *University of Pennsylvania, Department of Orthopaedic Surgery, McKay Orthopaedic Research Laboratory, Philadelphia, PA, USA; Mechano-Therapeutics LLC, Philadelphia, PA, USA*
AMANDA R. DICKS • *Department of Orthopaedic Surgery, Washington University, St. Louis, MO, USA; Shriners Hospitals for Children – St. Louis, St. Louis, MO, USA; Department of*

Biomedical Engineering, Washington University, St. Louis, MO, USA; Center of Regenerative Medicine, Washington University, St. Louis, MO, USA

SOLVIG DIEDERICHS • *Research Centre for Experimental Orthopaedics, Heidelberg University Hospital, Heidelberg, Germany*

DENITSA DOCHEVA • *Experimental Trauma Surgery, Department of Trauma Surgery, University Medical Center of Regensburg, Regensburg, Germany; Department of Musculoskeletal Tissue Regeneration, Orthopaedic Hospital König-Ludwig-Haus, University of Wurzburg, Wurzburg, Germany*

GEORGE R. DODGE • *University of Pennsylvania, Department of Orthopaedic Surgery, McKay Orthopaedic Research Laboratory, Philadelphia, PA, USA; Mechano-Therapeutics LLC, Philadelphia, PA, USA*

SEYED ALI ELAHI • *Human Movement Biomechanics Research Group, Department of Movement Sciences, KU Leuven, Leuven, Belgium; Soft Tissue Biomechanics Group, Biomechanics Division, Mechanical Engineering Department, KU Leuven, Leuven, Belgium*

SUZANNE E. ELDRIDGE • *Centre for Experimental Medicine and Rheumatology, William Harvey Research Institute, Barts and the London School of Medicine and Dentistry, Queen Mary University of London, London, UK*

CHRISTOPHER H. EVANS • *Musculoskeletal Gene Therapy Laboratory, Mayo Clinic, Rochester, MN, USA; Rehabilitation Medicine Research Center, Mayo Clinic, Rochester, MN, USA*

NELE FAMAEY • *Soft Tissue Biomechanics Group, Biomechanics Division, Mechanical Engineering Department, KU Leuven, Leuven, Belgium*

GIUSEPPE FILARDO • *Applied and Translational Research Center, IRCCS Istituto Ortopedico Rizzoli, Bologna, Italy; Service of Orthopaedics and Traumatology, Department of Surgery, EOC, Lugano, Switzerland; Faculty of Biomedical Sciences, Università della Svizzera Italiana, Lugano, Switzerland*

ANKE GOVAERTS • *Human Movement Biomechanics Research Group, Department of Movement Sciences, KU Leuven, Leuven, Belgium; Tissue homeostasis and Disease Group, Skeletal Biology and Engineering Research Center, Department of Development and Regeneration, KU Leuven, Leuven, Belgium*

SIBYLLE GRAD • *AO Research Institute Davos, Davos Platz, Switzerland*

FARSHID GUILAK • *Department of Orthopaedic Surgery, Washington University, St. Louis, MO, USA; Shriners Hospitals for Children – St. Louis, St. Louis, MO, USA; Department of Biomedical Engineering, Washington University, St. Louis, MO, USA; Center of Regenerative Medicine, Washington University, St. Louis, MO, USA*

MAKENNA J. HARDY • *Department of Biological Sciences, Biomolecular Research Center, Boise State University, Boise, ID, USA*

PAUL HARRISON • *RJAH Orthopaedic Hospital NHS Foundation Trust, Oswestry, Shropshire, UK*

TIMOTHY HOPKINS • *RJAH Orthopaedic Hospital NHS Foundation Trust, Oswestry, Shropshire, UK; School of Pharmacy and Bioengineering, Keele University, Staffordshire, UK*

CHARLOTTE HULME • *RJAH Orthopaedic Hospital NHS Foundation Trust, Oswestry, Shropshire, UK; School of Pharmacy and Bioengineering, Keele University, Staffordshire, UK*

NICHOLAS R. HUM • *Physical and Life Science Directorate, Lawrence Livermore National Laboratory, Livermore, CA, USA*

BRIAN JOHNSTONE • *Department of Orthopaedics & Rehabilitation, Oregon Health & Science University, Portland, OR, USA*

ILSE JONKERS • *Human Movement Biomechanics Research Group, Department of Movement Sciences, KU Leuven, Leuven, Belgium*

RONALD K. JUNE • *Department of Mechanical & Industrial Engineering, Montana State University, Bozeman, MT, USA; Department of Microbiology & Cell Biology, Montana State University, Bozeman, MT, USA; Department of Orthopedics & Sports Medicine, University of Washington, Seattle, WA, USA*

ILYAS M. KHAN • *Faculty of Medicine, Health & Life Science, Swansea University, Swansea, UK*

MARTIN M. KNIGHT • *Centre for Bioengineering, School of Engineering and Materials Science, Queen Mary University of London, London, UK; Centre for Predictive in vitro Models, Queen Mary University of London, London, UK*

JASMIJN V. KORPERSHOEK • *Department of Orthopaedics, University Medical Center Utrecht, Utrecht University, Utrecht, The Netherlands*

E. JOHANNA KUBOSCH • *Medical Center-Albert-Ludwigs-University, Department of Orthopedics and Trauma Surgery, Faculty of Medicine, Albert-Ludwigs-University of Freiburg, Freiburg, Germany*

BODO KURZ • *Department of Anatomy, Christian-Albrechts-University, Kiel, Germany*

YANN D. LADNER • *AO Research Institute Davos, Davos Platz, Switzerland; Institute for Biomechanics, ETH Zürich, Zürich, Switzerland*

THOMAS LANGE • *Medical Physics Department of Radiology, Faculty of Medicine, Medical Center—Albert-Ludwigs-University of Freiburg, Freiburg im Breisgau, Germany*

GABRIELA G. LOOTS • *Physical and Life Science Directorate, Lawrence Livermore National Laboratory, Livermore, CA, USA; School of Natural Sciences, University of California Merced, Merced, CA, USA; Department of Orthopaedic Surgery, University of California Davis Health, Sacramento, USA*

RIK LORIES • *Tissue homeostasis and Disease Group, Skeletal Biology and Engineering Research Center, Department of Development and Regeneration, KU Leuven, Leuven, Belgium*

BENEDICT LOTZ • *Clinic for Orthopaedics and Trauma Surgery, Heidelberg University Hospital, Heidelberg, Germany*

TRISTAN MAERZ • *University of Michigan, Departments of Orthopaedic Surgery and Biomedical Engineering, Ann Arbor, MI, USA*

BRANDON D. MARKWAY • *Department of Orthopaedics & Rehabilitation, Oregon Health & Science University, Portland, OR, USA*

JERAHME MARTINEZ • *University of Pennsylvania, Department of Orthopaedic Surgery, McKay Orthopaedic Research Laboratory, Philadelphia, PA, USA*

AUDREY MCALINDEN • *Department of Orthopaedic Surgery, Washington University School of Medicine, St Louis, MO, USA; Department of Cell Biology & Physiology, Washington University School of Medicine, St Louis, MO, USA; Shriners Hospitals for Children – St Louis, St Louis, MO, USA*

HELEN MCCARTHY • *RJAH Orthopaedic Hospital NHS Foundation Trust, Oswestry, Shropshire, UK; School of Pharmacy and Bioengineering, Keele University, Staffordshire, UK*

JILLIAN L. MCCOOL • *Physical and Life Science Directorate, Lawrence Livermore National Laboratory, Livermore, CA, USA; School of Natural Sciences, University of California Merced, Merced, CA, USA*

JOSHUA MCKENNA • *Faculty of Medicine, Health & Life Science, Swansea University, Swansea, UK*

HUAN MENG • *Centre for Bioengineering, School of Engineering and Materials Science, Queen Mary University of London, London, UK*

JADWIGA MIOTLA-ZAREBSKA • *Centre for Osteoarthritis Pathogenesis, Kennedy Institute of Rheumatology, University of Oxford, Oxford, UK*

ALI MOBASHERI • *Department of Regenerative Medicine, State Research Institute Centre for Innovative Medicine, Vilnius, Lithuania; Research Unit of Medical Imaging, Physics and Technology, Faculty of Medicine, University of Oulu, Oulu, Finland; Departments of Orthopedics, Rheumatology and Clinical Immunology, University Medical Center Utrecht, Utrecht, The Netherlands; Department of Joint Surgery, First Affiliated Hospital of Sun Yat-sen University, Guangzhou, Guangdong, China; World Health Organization Collaborating Center for Public Health Aspects of Musculoskeletal Health and Aging, Université de Liège, Liège, Belgium*

CHRISTOPHER V. NAGELLI • *Musculoskeletal Gene Therapy Laboratory, Mayo Clinic, Rochester, MN, USA*

STEFAN NEHRER • *Center for Regenerative Medicine, University For Continuing Education, Krems, Austria*

SYLVIA NÜRNBERGER • *Ludwig Boltzmann Institute for Traumatology, The Research Center in Cooperation with AUVA, Austrian Cluster for Tissue Regeneration, Vienna, Austria; Medical University of Vienna, Department of Orthopedics and Trauma Surgery, Division of Trauma Surgery, Vienna, Austria*

KATJA NUSS • *Musculoskeletal Research Unit (MSRU), Vetsuisse Faculty, University of Zürich, Zürich, Switzerland; Center for Applied Biotechnology and Molecular Medicine (CABMM), Vetsuisse Faculty, University of Zürich, Zürich, Switzerland*

ALEXANDER OTAHAL • *Center for Regenerative Medicine, University For Continuing Education, Krems, Austria*

JULIA THOM OXFORD • *Department of Biological Sciences, Biomolecular Research Center, Boise State University, Boise, ID, USA*

IDA PARISI • *Centre for Osteoarthritis Pathogenesis, Kennedy Institute of Rheumatology, University of Oxford, Oxford, UK*

GIRISH PATTAPPA • *Experimental Trauma Surgery, Department of Trauma Surgery, University Medical Center of Regensburg, Regensburg, Germany*

PAUL K. POTTER • *Department of Biological and Medical Sciences, Faculty of Health and Life Sciences, Oxford Brookes University, Oxford, UK*

XINZHU PU • *Department of Biological Sciences, Biomolecular Research Center, Boise State University, Boise, ID, USA*

WILTRUD RICHTER • *Research Centre for Experimental Orthopaedics, Heidelberg University Hospital, Heidelberg, Germany*

MARGOT RIKKERS • *Department of Orthopaedics, University Medical Center Utrecht, Utrecht University, Utrecht, The Netherlands*

BERND ROLAUFFS • *G.E.R.N. Research Center for Tissue Replacement, Regeneration & Neogenesis, Department of Orthopedics and Trauma Surgery, Faculty of Medicine, Medical Center—Albert-Ludwigs-University of Freiburg, Freiburg im Breisgau, Germany*

CORNELIA SCHNEIDER • *Ludwig Boltzmann Institute for Traumatology, The Research Center in Cooperation with AUVA, Austrian Cluster for Tissue Regeneration, Vienna, Austria*

AIMY SEBASTIAN • *Physical and Life Science Directorate, Lawrence Livermore National Laboratory, Livermore, CA, USA*

NANCY STEWARD • *Department of Orthopaedic Surgery, Washington University, St. Louis, MO, USA; Shriners Hospitals for Children – St. Louis, St. Louis, MO, USA; Center of Regenerative Medicine, Washington University, St. Louis, MO, USA*
MARTIN J. STODDART • *AO Research Institute Davos, Davos Platz, Switzerland*
MILENA TEGELKAMP • *Musculoskeletal Research Unit (MSRU), Vetsuisse Faculty, University of Zürich, Zürich, Switzerland*
CLARE L. THOMPSON • *Centre for Bioengineering, School of Engineering and Materials Science, Queen Mary University of London, London, UK; Centre for Predictive in vitro Models, Queen Mary University of London, London, UK*
ANNE-SOPHIE THORUP • *Centre for Experimental Medicine and Rheumatology, William Harvey Research Institute, Barts and the London School of Medicine and Dentistry, Queen Mary University of London, London, UK*
M. LETIZIA VAINIERI • *AO Research Institute Davos, Davos Platz, Switzerland*
PETER M. VAN DER KRAAN • *Experimental Rheumatology, Department of Rheumatology, Interdisciplinary Consortium for Clinical Movement Sciences and Technology (ICMS), Radboud Medical Centre, Nijmegen, The Netherlands*
MARJOLEIN C. H. VAN DER MEULEN • *Cornell University, Meinig School of Biomedical Engineering and Sibley School of Mechanical & Aerospace Engineering, Ithaca, NY, USA*
TONIA L. VINCENT • *Centre for Osteoarthritis Pathogenesis, Kennedy Institute of Rheumatology, University of Oxford, Oxford, UK*
MARITA VOELKER • *G.E.R.N. Research Center for Tissue Replacement, Regeneration & Neogenesis, Department of Orthopedics and Trauma Surgery, Faculty of Medicine, Medical Center—Albert-Ludwigs-University of Freiburg, Freiburg im Breisgau, Germany*
LUCIENNE A. VONK • *Department of Orthopaedics, University Medical Center Utrecht, Utrecht University, Utrecht, The Netherlands; CO.DON AG, Teltow, Germany; Xintela AB, Lund, Sweden*
BRIGITTE VON RECHENBERG • *Musculoskeletal Research Unit (MSRU), Vetsuisse Faculty, University of Zürich, Zürich, Switzerland; Center for Applied Biotechnology and Molecular Medicine (CABMM), Vetsuisse Faculty, University of Zürich, Zürich, Switzerland*
ANGUS K. WANN • *Kennedy Institute of Rheumatology, Medical Sciences Division, University of Oxford, Oxford, UK*
HOPE D. WELHAVEN • *Molecular Biosciences Program, Montana State University, Bozeman, MT, USA; Department of Chemistry & Biochemistry, Montana State University, Bozeman, MT, USA*
KARINA WRIGHT • *RJAH Orthopaedic Hospital NHS Foundation Trust, Oswestry, Shropshire, UK; School of Pharmacy and Bioengineering, Keele University, Staffordshire, UK*
CHIA-LUNG WU • *Department of Orthopaedic Surgery and Rehabilitation, Center for Musculoskeletal Research, University of Rochester, Rochester, NY, USA*
YADAN ZHANG • *Faculty of Medicine, Health & Life Science, Swansea University, Swansea, UK*
HONGJUN ZHENG • *Department of Orthopaedic Surgery, Washington University School of Medicine, St Louis, MO, USA*

Chapter 1

Cartilage Tissue Engineering: An Introduction

Martin J. Stoddart, Elena Della Bella, and Angela R. Armiento

Abstract

Once damaged, cartilage has limited healing capability. This has led to a huge body of research that aims to repair or regenerate this important tissue. Despite the progress made, significant hurdles still need to be overcome. This chapter highlights some of the progress made, while elaborating on areas that need further research. The concept of translation and the route to clinical translation must be kept in mind if some of the promising preclinical research is to make it to routine clinical application.

Key words Chondral, Cell source, Regenerative medicine, Cartilage, Preclinical models

1 Introduction

Ever since the original observations by Hunter in 1743 (*see* ref. 1), generations of researchers have been attempting to find ways to regenerate damaged articular cartilage. Focal traumatic injuries not only can lead to pain and mobility problems, but they can also lead to post-traumatic osteoarthritis with more severe long-term consequences [2]. Many reasons are suggested as to why cartilage tissue has such a problematic healing ability. Proposals such as lack of reparative cells due to the absence of vasculature, the mismatch of mechanical properties between the de novo regenerating tissue and the surrounding intact articular cartilage, and the role of chronic inflammation in preventing an adequate cell differentiation are among many of those discussed. The likelihood is that a combination of these factors leads to the issues routinely observed within clinical practice.

While progress in cartilage repair has been made with marrow stimulation techniques such as microfracture [3], and more traditional cell-based therapies such as autologous chondrocyte implantation and its variations [4, 5], there are still significant challenges in developing treatments for long-term repair of traumatic articular cartilage defects. Tissue engineering (TE) as a concept is generally

Martin J. Stoddart et al. (eds.), *Cartilage Tissue Engineering*,
Methods in Molecular Biology, vol. 2598, https://doi.org/10.1007/978-1-0716-2839-3_1,

considered to be born with the seminal paper of Vacanti and Langer, introducing the so-called tissue engineering triad of cells, a suitable scaffold material and the correct biological and biophysical stimuli [6]. This has led to great interest in using TE for cartilage repair.

To overcome some of the challenges in repairing cartilage, various experimental directions have been taken. Cell-based therapies have been extensively studied to overcome the lack of reparative cells within the defect, with investigations into various different types of cells from different tissue sources and how they may be utilized to attempt a repair [7]. The clinical relevance of each cell type should always be balanced with the practical issues surrounding their isolation. There is also still great debate regarding the use of autologous or allogeneic cells, with both having their own advantages and disadvantages [8]. Typically, autologous cells have the advantage of a lower risk profile when considering disease transfer and immunological responses; however, patient-specific cell expansion leads to challenges with both regulatory aspects and cost. Allogeneic cells have the potential to offer an off-the-shelf approach, but the potential for immunological rejection still has to be determined [8, 9]. While progress is being made, it is clear that the in vitro cell expansion conditions need to be further optimized and standardized. Furthermore, the issue of donor variation when using autologous cells needs to be more adequately addressed. Commonly used markers are often not predictive of cellular function, which is a major hurdle to clinical translation. This has led to increased effort to establish functionally predictive markers of chondrogenesis [10, 11]. It should also be considered that, due to regulatory approval, the use of monolayer expanded cells will be a more challenging pathway than intraoperative methods adopting a more regenerative medicine approach. The desire for an intraoperative or off-the-shelf solution has led to interest in the use of induced pluripotent stem cells (iPSCs), which would allow a cell bank of unlimited supply that could undergo safety and efficacy screening prior to implantation [12–14]. Also, considering the numerous attempts to repair cartilage using cells from various sources, the question "why are we still looking for effective cartilage regeneration" should be more critically addressed by better understanding the data already available. Chondrocytes have long been the clinically used cell type since their first use in the autologous chondrocyte implantation procedures in 1994 [4]. While some success has been shown, the need for two operative procedures and a monolayer expansion phase that can be unpredictable and causes chondrocyte dedifferentiation into a collagen type I-producing cell [15] has led many to conclude improvements are still needed. When considering the use of mesenchymal stromal cells (MSCs), such as bone marrow derived MSCs, it is worth considering that the main issue to be overcome in vitro is often stated as being hypertrophy, whereby the cells progressed towards type X collagen expression, terminal differentiation, and bone formation [16, 17]. However, in clinical practice, treatments such as microfracture typically lead to fibrocartilage rich in collagen type I [18]. Despite the source of the cells being the same, in this case bone marrow, the outcomes are different. Yet the reasons for this are largely unknown [19]. It could be the monolayer expansion of the cells in vitro leads to a cell population vastly different to those seen in fresh marrow microfracture or that critical in vivo signals are missing in the in vitro studies, leading in vitro cells towards more hypertrophic fate. This area deserves more research as the answer is largely unresolved and the discrepancy between in vitro and clinical data needs more careful consideration.

When considering the microenvironment of the articulating joint, there have been a substantial number of studies investigating the development of biomaterials that can more accurately represent properties of the native tissue [20]. This leads to a number of considerations, such as accurate transmission of the load through the material to the cells contained within, adequate biodegradability rates to allow the scaffold to be slowly replaced by de novo matrix overtime, and the more general issue as to the requirement for a cell adhesion site. On the one hand, a cell adhesion site is beneficial in that it allows migration of cells into the scaffold material, and it enhances cell viability by reducing the possibility of anoikis, anchorage-dependent cell death [21]. However, attachment of chondrocytes or their progenitors to substrates has the potential to lead to cell spreading and actin cytoskeletal organization, which has been associated with the development of dedifferentiation and a loss of cartilage phenotype [22, 23]. Finding the correct balance between these two competing interests will be a major advance moving forward.

While in vitro studies have tended towards the utilization of a single cell type, there is increasing interest in coculture systems, which allow soluble factors or perhaps even cell to cell contact to lead to a more active response [24–26]. This has been apparent both in vitro and in therapies being more clinically applied such as the IMPACT trial which involves a combination of autologous chondrons with allogeneic stem cells [27]. The relative contribution of each cell type is still open for debate, but much data suggests the MSC population drives the differentiation of the implanted chondrons before undergoing cell death and being largely undetectable overtime [28]. The clinical data obtained suggests that studies comparing chondron biology to that of isolated chondrocytes, devoid of its interaction with its pericellular matrix, would yield valuable results. The need to develop enhanced coculture systems becomes more apparent when considering the whole joint as an organ [29]. While in vitro systems typically involve a single cell type and investigations into its chondrogenic potential, the reality in the in vivo joint is more complicated. Interactions with other tissues such as the subchondral bone, synovium, and fat pad lead to cross talk signals which will affect the result, leading to the need for more complex joint-on-a-chip platform technology [30]. Furthermore, damage to the joint is often associated with inflammation, and it is known the presence of inflammatory molecules can severely inhibit chondrogenic differentiation [31]. Future therapies for cartilage regeneration will likely need to take the inflammatory response into account. A key factor is to resolve the initial inflammation to allow healing to take place. Groups have been investigating the potential of soluble mediators, genetically modified cell types, or the potential of tissue engineered material itself to modulate the inflammatory response [32, 33]. The lack of inflammatory reactions within in vitro cell culture systems may be one of the major stumbling blocks when trying to translate new therapies into an in vivo animal model.

More recently the aim to boost the endogenous repair capability has led to a new field of regenerative medicine, with the idea to utilize the patient as a bioreactor and use rehabilitation as a means to regulate the repair response. This has led to an increase interest in cell secretome as a therapeutic [34]. The emerging field of regenerative rehabilitation combines surgical treatment and physical therapy [35, 36]. One issue, which becomes increasingly clear when considering cartilage regeneration, is the major role of mechanical stimulation in articular cartilage repair, which suggests the many in vitro culture models are overly simplistic, leading to issues of reproducibility when attempted in the kinematic environment of an articulating joint. Rehabilitation protocols after surgery offer an ideal opportunity to define the mechanics and direct the healing response. Both joint homeostasis and chondroprogenitor cell differentiation are highly regulated by mechanics and this should not be overlooked [37]. The TGF-β pathway is implicated in mechanical regulation of cartilage tissue and de novo constructs, both in vitro [38, 39] and in vivo [40].

As studies transition from in vitro to in vivo, there are additional aspects that need to be considered: not only the previously discussed cross talk between the multiple tissues of the joint [41] and the role of inflammation [42] but also the integration of the de novo tissue with surrounding bone and cartilage [43]. Delamination from the bone can occur, raising questions of how to sufficiently anchor the graft into the defect [44]. Additionally, the lateral integration to the surrounding cartilage is often an area where grafts fail. Also, considerations such as the simplicity of the standard culture media compared to the complexity of the synovial fluid composition should not be overlooked. Adaptation, such as inclusion of high molecular weight hyaluronic acid in the culture medium, has been shown to improve the chondrogenic phenotype of differentiating human MSCs [45, 46]. Another interesting observation is that after arthroscopy the synovial fluid is removed with the lavage solution being left, and that may damage chondrocytes or influence their metabolism [47].

Although in vitro models are evolving, they cannot replace animal models, which are still required for clinical transition as the full integration of multiple tissues, oxygen tension, angiogenesis, and immune regulation under complex load is not yet feasible within a laboratory setting. The choice and the duration of the study is a major challenge, especially when considering efficacy studies generally need larger animals [48, 49]. To truly obtain information on repair, long-term studies are needed, and in the case of cartilage, this could even be 1–2 years of follow-up [50]. This generates concerns regarding cost, and viable solutions are still yet to be found. In addition to all these issues, any potential therapy must have a financially viable route into the clinic if it is ever to be routinely used in patients [51].

Despite the large body of work studying focal cartilage injury and repair, it is clear that improvements can still be made. As we move to an era of single cell sequencing and combining cells to recreate a joint in the laboratory, there is hope the reasons for the current limitations will be clear.

References

1. Hunter W (1995) Of the structure and disease of articulating cartilages. Clin Orthop Relat Res (317):3–6
2. Gelber AC, Hochberg MC, Mead LA et al (2000) Joint injury in young adults and risk for subsequent knee and hip osteoarthritis. Ann Intern Med 133(5):321–328
3. Steadman JR, Rodkey WG, Rodrigo JJ (2001) Microfracture: surgical technique and rehabilitation to treat chondral defects. Clin Orthop Relat Res (391 Suppl):S362–S369
4. Brittberg M, Lindahl A, Nilsson A et al (1994) Treatment of deep cartilage defects in the knee with autologous chondrocyte transplantation. N Engl J Med 331(14):889–895
5. Pestka JM, Bode G, Salzmann G et al (2014) Clinical outcomes after cell-seeded autologous chondrocyte implantation of the knee: when can success or failure be predicted? Am J Sports Med 42(1):208–215. https://doi.org/10.1177/0363546513507768
6. Langer R, Vacanti JP (1993) Tissue engineering. Science 260(5110):920–926
7. Urlic I, Ivkovic A (2021) Cell sources for cartilage repair-biological and clinical perspective. Cell 10(9):2496. https://doi.org/10.3390/cells10092496
8. Vonk LA, de Windt TS, Slaper-Cortenbach IC et al (2015) Autologous, allogeneic, induced pluripotent stem cell or a combination stem cell therapy? Where are we headed in cartilage repair and why: a concise review. Stem Cell Res Ther 6:94. https://doi.org/10.1186/s13287-015-0086-1
9. Adkisson HD, Martin JA, Amendola RL et al (2010) The potential of human allogeneic juvenile chondrocytes for restoration of articular cartilage. Am J Sports Med 38(7): 1324–1333. https://doi.org/10.1177/0363546510361950
10. Rothweiler R, Basoli V, Duttenhoefer F et al (2020) Predicting and promoting human bone marrow MSC chondrogenesis by way of TGFbeta receptor profiles: toward personalized medicine. Front Bioeng Biotechnol 8: 618. https://doi.org/10.3389/fbioe.2020.00618
11. Dickinson SC, Sutton CA, Brady K et al (2017) The Wnt5a receptor, receptor tyrosine kinase-like orphan receptor 2, is a predictive cell surface marker of human mesenchymal stem cells with an enhanced capacity for chondrogenic differentiation. Stem Cells 35(11): 2280–2291. https://doi.org/10.1002/stem.2691
12. Craft AM, Rockel JS, Nartiss Y et al (2015) Generation of articular chondrocytes from human pluripotent stem cells. Nat Biotechnol 33(6):638–645. https://doi.org/10.1038/nbt.3210
13. Adkar SS, Wu CL, Willard VP et al (2019) Step-wise chondrogenesis of human induced pluripotent stem cells and purification via a reporter allele generated by CRISPR-Cas9

genome editing. Stem Cells 37(1):65–76. https://doi.org/10.1002/stem.2931

14. Wu CL, Dicks A, Steward N et al (2021) Single cell transcriptomic analysis of human pluripotent stem cell chondrogenesis. Nat Commun 12(1):362. https://doi.org/10.1038/s41467-020-20598-y
15. Benya PD, Shaffer JD (1982) Dedifferentiated chondrocytes reexpress the differentiated collagen phenotype when cultured in agarose gels. Cell 30:215–224
16. Pelttari K, Winter A, Steck E et al (2006) Premature induction of hypertrophy during in vitro chondrogenesis of human mesenchymal stem cells correlates with calcification and vascular invasion after ectopic transplantation in SCID mice. Arthritis Rheum 54(10): 3254–3266
17. Mueller MB, Tuan RS (2008) Functional characterization of hypertrophy in chondrogenesis of human mesenchymal stem cells. Arthritis Rheum 58(5):1377–1388
18. Knutsen G, Engebretsen L, Ludvigsen TC et al (2004) Autologous chondrocyte implantation compared with microfracture in the knee. A randomized trial. J Bone Joint Surg Am 86-A (3):455–464
19. Armiento AR, Alini M, Stoddart MJ (2018) Articular fibrocartilage – why does hyaline cartilage fail to repair? Adv Drug Deliv Rev 146: 289–305. https://doi.org/10.1016/j.addr.2018.12.015
20. Armiento AR, Stoddart MJ, Alini M et al (2018) Biomaterials for articular cartilage tissue engineering: learning from biology. Acta Biomater 65:1–20. https://doi.org/10.1016/j.actbio.2017.11.021
21. Karoubi G, Ormiston ML, Stewart DJ et al (2009) Single-cell hydrogel encapsulation for enhanced survival of human marrow stromal cells. Biomaterials 30(29):5445–5455. https://doi.org/10.1016/j.biomaterials.2009.06.035
22. Cao B, Peng R, Li Z et al (2014) Effects of spreading areas and aspect ratios of single cells on dedifferentiation of chondrocytes. Biomaterials 35(25):6871–6881. https://doi.org/10.1016/j.biomaterials.2014.04.107
23. Sliogeryte K, Botto L, Lee DA et al (2016) Chondrocyte dedifferentiation increases cell stiffness by strengthening membrane-actin adhesion. Osteoarthr Cartil 24(5):912–920. https://doi.org/10.1016/j.joca.2015.12.007
24. Kubosch EJ, Heidt E, Bernstein A et al (2016) The trans-well coculture of human synovial mesenchymal stem cells with chondrocytes leads to self-organization, chondrogenic differentiation, and secretion of TGFbeta. Stem Cell Res Ther 7(1):64. https://doi.org/10.1186/s13287-016-0322-3
25. Kim TW, Lee MC, Bae HC et al (2018) Direct coculture of human chondrocytes and synovium-derived stem cells enhances in vitro chondrogenesis. Cell J 20(1):53–60. https://doi.org/10.22074/cellj.2018.5025
26. Bian L, Zhai DY, Mauck RL et al (2011) Coculture of human mesenchymal stem cells and articular chondrocytes reduces hypertrophy and enhances functional properties of engineered cartilage. Tissue Eng Part A 17(7–8): 1137–1145. https://doi.org/10.1089/ten.TEA.2010.0531
27. de Windt TS, Vonk LA, Slaper-Cortenbach ICM et al (2017) Allogeneic MSCs and recycled autologous chondrons mixed in a one-stage cartilage cell transplantion: a first-in-man trial in 35 patients. Stem Cells 35(8): 1984–1993. https://doi.org/10.1002/stem.2657
28. de Windt TS, Vonk LA, Slaper-Cortenbach IC et al (2017) Allogeneic mesenchymal stem cells stimulate cartilage regeneration and are safe for single-stage cartilage repair in humans upon mixture with recycled autologous chondrons. Stem Cells 35(1):256–264. https://doi.org/10.1002/stem.2475
29. Brose TZ, Kubosch EJ, Schmal H et al (2021) Crosstalk between mesenchymal stromal cells and chondrocytes: the hidden therapeutic potential for cartilage regeneration. Stem Cell Rev Rep 17:1647–1665. https://doi.org/10.1007/s12015-021-10170-6
30. Paggi CA, Teixeira LM, Le Gac S et al (2022) Joint-on-chip platforms: entering a new era of in vitro models for arthritis. Nat Rev Rheumatol 18:217–231. https://doi.org/10.1038/s41584-021-00736-6
31. Wehling N, Palmer GD, Pilapil C et al (2009) Interleukin-1beta and tumor necrosis factor alpha inhibit chondrogenesis by human mesenchymal stem cells through NF-kappaB-dependent pathways. Arthritis Rheum 60(3): 801–812. https://doi.org/10.1002/art.24352
32. Li M, Yin H, Yan Z et al (2022) The immune microenvironment in cartilage injury and repair. Acta Biomater 140:23–42. https://doi.org/10.1016/j.actbio.2021.12.006
33. Fahy N, Farrell E, Ritter T et al (2015) Immune modulation to improve tissue engineering outcomes for cartilage repair in the osteoarthritic joint. Tissue Eng Part B Rev 21(1):55–66. https://doi.org/10.1089/ten.TEB.2014.0098

34. Stoddart MJ, Bara J, Alini M (2015) Cells and secretome--towards endogenous cell re-activation for cartilage repair. Adv Drug Deliv Rev 84:135–145. https://doi.org/10.1016/j.addr.2014.08.007
35. Ambrosio F, Russell A (2010) Regenerative rehabilitation: a call to action. J Rehabil Res Dev 47(3):xi–xv. https://doi.org/10.1682/jrrd.2010.03.0021
36. Willett NJ, Boninger ML, Miller LJ et al (2020) Taking the next steps in regenerative rehabilitation: establishment of a new interdisciplinary field. Arch Phys Med Rehabil 101(5): 917–923. https://doi.org/10.1016/j.apmr.2020.01.007
37. Hodgkinson T, Kelly DC, Curtin CM et al (2022) Mechanosignalling in cartilage: an emerging target for the treatment of osteoarthritis. Nat Rev Rheumatol 18(2):67–84. https://doi.org/10.1038/s41584-021-00724-w
38. Gardner OFW, Fahy N, Alini M et al (2017) Joint mimicking mechanical load activates TGFbeta1 in fibrin-poly(ester-urethane) scaffolds seeded with mesenchymal stem cells. J Tissue Eng Regen Med 11(9):2663–2666. https://doi.org/10.1002/term.2210
39. Li Z, Kupcsik L, Yao SJ et al (2010) Mechanical load modulates chondrogenesis of human mesenchymal stem cells through the TGF-beta pathway. J Cell Mol Med 14(6A):1338–1346
40. van der Kraan PM, Blaney Davidson EN, van den Berg WB (2010) A role for age-related changes in TGFbeta signaling in aberrant chondrocyte differentiation and osteoarthritis. Arthritis Res Ther 12(1):201. https://doi.org/10.1186/ar2896
41. Piluso S, Li Y, Abinzano F et al (2019) Mimicking the articular joint with in vitro models. Trends Biotechnol 37(10):1063–1077. https://doi.org/10.1016/j.tibtech.2019.03.003
42. Lee W, Nims RJ, Savadipour A et al (2021) Inflammatory signaling sensitizes Piezo1 mechanotransduction in articular chondrocytes as a pathogenic feed-forward mechanism in osteoarthritis. Proc Natl Acad Sci U S A 118(13):e2001611118. https://doi.org/10.1073/pnas.2001611118
43. Jelodari S, Ebrahimi Sadrabadi A, Zarei F et al (2022) New insights into cartilage tissue engineering: improvement of tissue-scaffold integration to enhance cartilage regeneration. Biomed Res Int 2022:7638245. https://doi.org/10.1155/2022/7638245
44. Friedman JM, Sennett ML, Bonadio MB et al (2018) Comparison of fixation techniques of 3D-woven poly(-caprolactone) scaffolds for cartilage repair in a weightbearing porcine large animal model. Cartilage 9(4):428–437. https://doi.org/10.1177/1947603517700953
45. Monaco G, El Haj AJ, Alini M et al (2021) Sodium hyaluronate supplemented culture medium combined with joint-simulating mechanical loading improves chondrogenic differentiation of human mesenchymal stem cells. Eur Cell Mater 41:616–632. https://doi.org/10.22203/eCM.v041a40
46. Monaco G, El Haj AJ, Alini M et al (2020) Sodium hyaluronate supplemented culture media as a new hMSC chondrogenic differentiation media-model for in vitro/ex vivo screening of potential cartilage repair therapies. Front Bioeng Biotechnol 8:243. https://doi.org/10.3389/fbioe.2020.00243
47. Mathies B (2006) Effects of Viscoseal, a synovial fluid substitute, on recovery after arthroscopic partial meniscectomy and joint lavage. Knee Surg Sports Traumatol Arthrosc 14(1): 32–39. https://doi.org/10.1007/s00167-005-0625-x
48. Pfeifer CG, Fisher MB, Saxena V et al (2017) Age-dependent subchondral bone remodeling and cartilage repair in a minipig defect model. Tissue Eng Part C Methods 23(11):745–753. https://doi.org/10.1089/ten.TEC.2017.0109
49. Pfeifer CG, Fisher MB, Carey JL et al (2015) Impact of guidance documents on translational large animal studies of cartilage repair. Sci Transl Med 7(310):310re9. https://doi.org/10.1126/scitranslmed.aac7019
50. Sennett ML, Friedman JM, Ashley BS et al (2021) Long term outcomes of biomaterial-mediated repair of focal cartilage defects in a large animal model. Eur Cell Mater 41:40–51. https://doi.org/10.22203/eCM.v041a04
51. Wang L, Guo X, Chen J et al (2022) Key considerations on the development of biodegradable biomaterials for clinical translation of medical devices: with cartilage repair products as an example. Bioact Mater 9:332–342. https://doi.org/10.1016/j.bioactmat.2021.07.031

Chapter 2

Chondrocyte Isolation and Expansion

Paul Harrison, Timothy Hopkins, Charlotte Hulme, Helen McCarthy, and Karina Wright

Abstract

Chondrocyte isolation requires a combination of enzymatic and mechanical digestion of cartilaginous tissues in order to release the chondrocytes. Extracted primary chondrocytes will then adhere to standard tissue culture plastics, typically in small clusters, over a period of a few days in monolayer culture. Chondrocyte populations are expanded in a basal medium containing serum, supplemented with ascorbic acid, antibiotics, and sometimes antifungal agents and growth factors. Here we describe the standard research grade and good manufacturing practice (GMP) protocols used for the isolation and expansion of chondrocytes by the Oswestry/Keele University Orthopaedic Research (OsKOR) group and John Charnley GMP and MHRA licensed laboratory, both based at the RJAH Orthopaedic Hospital, Oswestry, UK.

Key words Cartilage, Mechanical and enzymatic cartilage digestion, Chondrocytes, Tissue culture, Trypsinization, Autologous serum

1 Introduction

Chondrocytes are the only cell type that resides in healthy articular cartilage. Mechanical and enzymatic digestion has been used to release chondrocytes from healthy, injured, and diseased cartilage tissues for many years. Several established techniques exist that are used to grow isolated chondrocytes adhered to tissue culture plastics in monolayer. Trypsinization is used to passage and expand chondrocyte populations that can be used to study cartilaginous tissue pathologies and in order to manufacture chondrocytes for use as an advanced therapy medicinal product (ATMP) [1, 2].

Mechanical mincing of articular cartilage to increase the surface area exposure of the tissue to a collagenase type II enzyme is typically used to release chondrocytes from their extracellular matrix. Chondrocytes then gradually adhere to a standard tissue culture plastic over a period of several days, and during this time, in order to encourage adhesion, cultures are minimally manipulated.

Martin J. Stoddart et al. (eds.), *Cartilage Tissue Engineering*,
Methods in Molecular Biology, vol. 2598, https://doi.org/10.1007/978-1-0716-2839-3_2,

Chondrocyte populations are established within a week of isolation and are culture expanded in a standard basal medium supplemented with serum, ascorbic acid, and antibiotics/antifungal agents. Typically 5% CO_2 in air at 37 °C in a 90% humidified atmosphere is the condition used to expand chondrocyte populations. Although several studies have indicated that lower atmospheric oxygen tensions in the isolation and culture expansion phases help to maintain a more in vivo-like chondrocyte phenotype, slowing the growth of chondrocytes but also the dedifferentiation process typically observed following their monolayer expansion phase (*see* Chapter 5) [3]. In addition, supplementation of cultures with growth factors (e.g., fibroblast growth factor beta) during the expansion phase has been shown to enhance chondrogenesis and cartilaginous tissue formation in vitro [4].

Chondrocytes have been used as an autologous cell-based therapy for treatment of cartilage injuries in a series of research studies and clinical trials for more than 20 years [5–7]. In the UK autologous chondrocyte implantation (ACI) has been adopted as a first-line surgical approach for the treatment of certain cartilage injuries on the National Health Service, following the publication of a National Institute for Health and Care Excellence (NICE) technical appraisal of the technique [8, 9] in 2017. Chondrocytes used in ACI are classed as ATMPs and their manufacturing processes are evaluated and monitored by regulatory authorities such as the Medicines and Healthcare products Regulatory Agency (MHRA) in the UK to ensure they adhere to good manufacturing practice (GMP) guidance [10]. The GMP protocols used for the production of autologous chondrocyte ATMPs are usually very similar to the processes described above, with the exception of autologous serum often being used in place of animal derived serum in the expansion medium. In addition, all of the reagents and consumables used in the process must be GMP grade, and each of the processes is validated in a GMP facility and licensed, evaluated, and monitored by the relevant regulatory authority (e.g., the MHRA for the UK). The John Charnley GMP and MHRA-licensed laboratory, based at the RJAH Orthopaedic Hospital, Oswestry (whose manufacturing processes are described here), is one such facility that has been producing autologous chondrocytes for clinical use for more than 20 years and has treated over 500 patients.

2 Materials

2.1 Research Grade Mechanical and Enzymatic Extraction

All reagents should be warmed to 37 °C in a water bath prior to use.

1. Phosphate buffered saline (PBS; Gibco, Loughborough, UK).
2. Serum-free medium: Dulbecco's modified eagle medium/nutrient mixture F-12 (DMEM/F12; 1:1; Life Technologies, Paisley, UK) supplemented with 1% (v/v) penicillin-streptomycin (P/S; 5000 U/mL; Gibco).

3. Collagenase type II: Activity criteria are 150–350 IU/mg dry weight, caseinase activity 400–1900 IU/mg dry weight (Worthington, New Jersey, USA) in serum-free medium.
4. Chondrocyte culture medium: DMEM/F12 (1:1) (Life Technologies) supplemented with 10% (v/v) FBS (Life Technologies), 1% (v/v) P/S (5,000 U/mL; Gibco), and 62 μg/mL ascorbic acid (VWR, Pennsylvania, USA).
5. Tissue culture flasks (Falcon, Corning, NY, USA).
6. Centrifuge tubes (Type 128DBIRR, Sterilin, Newport, UK).
7. Reusable scalpel handle (Type 4L; Swann-Morton, Sheffield, UK) (*see* **Note 1**).
8. Disposable scalpel blades (Type 22A; Swann-Morton, Sheffield, UK).
9. Instrapac Sterile forceps (13 cm, Robinson Healthcare, Worksop, UK).
10. Chainex mail glove (Honeywell, Manchester, UK).
11. Benchtop tube roller (38 rpm with tilting; Benchmark Scientific, New Jersey, USA).
12. Cell strainers (40 μm; Thermo Fisher Scientific, Loughborough, UK).
13. Cell counting: trypan blue (Sigma-Aldrich) and hemocytometer (Hawksley, Lancing, UK).

2.2 Research Grade Chondrocyte Expansion

All reagents should be warmed to 37 °C in a water bath prior to use.

1. Chondrocyte culture medium (*see* Subheading 2.1).
2. PBS (*see* Subheading 2.1).
3. Trypsin: 0.05% (w/v) trypsin-ethylenediaminetetraacetic acid (trypsin-EDTA; Gibco).
4. Tissue culture flasks (*see* Subheading 2.1).
5. Centrifuge tubes (*see* Subheading 2.1).
6. Cell counting (*see* Subheading 2.1).

2.3 GMP Mechanical and Enzymatic Extraction

All reagents are brought up to room temperature prior to use.

1. GMP chondrocyte culture medium: Dulbecco's modified eagle medium/nutrient mixture F-12 (DMEM/F12; 1:1; Life Technologies) supplemented with 50 μg/mL gentamicin sulfate (Zentiva, Prague, Czech Rep), 2.5 μg/mL amphotericin (Bristol Myers Squibb, NY, US), 62 μg/mL ascorbic acid (VWR), and 10% or 20% (v/v) autologous patient serum, depending on the stage of culture expansion (*see* Subheading 3.3, **step 1** and **Note 2**).
2. Collagenase type II (*see* Subheading 2.1, using serum-free GMP chondrocyte medium).

3. Tissue culture plastics (*see* Subheading 2.1).
4. Disposable dissection equipment: sterile disposable forceps (Instrapac, Surbiton, UK) and scalpels (Type 22A; Swann-Morton).
5. Cell counting (*see* Subheading 2.1).

2.4 GMP Chondrocyte Expansion

All reagents are brought up to room temperature prior to use.

1. GMP chondrocyte culture medium (*see* Subheading 2.3).
2. PBS (*see* Subheading 2.1).
3. Trypsin/EDTA (*see* Subheading 2.2).
4. Tissue culture flasks (*see* Subheading 2.1).
5. Centrifuge tubes (*see* Subheading 2.1).
6. Blood tubes (coagulation tubes, Becton Dickinson, New Jersey, USA).
7. Syringes (Becton Dickinson).
8. Needles (Becton Dickinson).
9. 0.2 μm syringe filter (Sartorius, Epsom, UK).
10. Cell counting (*see* Subheading 2.1).

3 Methods

3.1 Research Grade Chondrocyte Isolation

1. Using sterile forceps transfer the cartilage into a sterile petri dish (*see* **Note 3**).
2. If necessary, rinse the tissue with PBS to remove blood (*see* **Note 4**).
3. Using a sterile scalpel handle and disposable blade, excise full thickness slices of cartilage into a new, sterile petri dish, taking care to not cut into the underlying bone (*see* **Notes 5** and **6**).
4. Using sterile forceps transfer the cartilage pieces into a sterile 50 mL tube and seal.
5. Weigh cartilage and record the weight (*see* **Note** 7).
6. Transfer the cartilage pieces into a new, sterile petri dish, and use a clean scalpel blade to dice the cartilage into small pieces (approx. 1 mm^3) and transfer into a clean, sterile 50 mL tube.
7. Prepare the required volume of collagenase type II working solution (1 mg/mL in serum-free media) by diluting the stock solution (10 mg/mL) in serum-free media. A 10 mL of working solution is required for every 100–300 mg of cartilage.
8. Add the required volume of collagenase type II working solution to the diced cartilage, and seal the tube (*see* **Note 8**).
9. Place the tube onto a tube roller inside the incubator, and incubate for 16 h at 37 °C with 5% CO_2 in air (*see* **Note 9**).

10. After 16 h, strain the cell suspension by passing through a 40 μm cell strainer into a sterile 50 mL tube (*see* **Notes 10** and **11**).
11. Add an equal volume of chondrocyte culture media to the cell suspension (*see* **Note 12**), and mix thoroughly.
12. Centrifuge the cell suspension at 179 g for 10 min to pellet the cells.
13. Discard the supernatant and resuspend the cell pellet in 20 mL fresh chondrocyte culture media (*see* **Note 13**).
14. Repeat **steps 12** and **13** twice more, the second time resuspending in 1 mL fresh chondrocyte culture media instead of 20 mL.
15. Perform a cell count and viability assessment by trypan blue exclusion.
16. Seed the chondrocytes at 5×10^3 chondrocytes/cm^2 in tissue culture flasks.
17. Allow 4–6 days for chondrocyte attachment before removing nonadherent cells by washing with PBS and replacing with fresh chondrocyte culture media (*see* **Note 14**).

3.2 Research Grade Chondrocyte Expansion

1. Replace the chondrocyte culture media every 2–3 days by aspirating the spent media and adding the required volume of fresh media (*see* **Note 15**).
2. Maintain chondrocytes in monolayer until 80% confluent.
3. Once 80% confluence is reached, carefully aspirate and discard the spent media.
4. Gently wash the chondrocyte monolayers with an appropriate volume of PBS (*see* **Note 16**).
5. Aspirate the PBS and discard.
6. Add an appropriate volume of 0.05% trypsin/EDTA to the culture flasks (*see* **Note 17**), and incubate at 37 °C with 5% CO_2 in air for 5 min.
7. After 5 min, check for chondrocyte detachment using the microscope. Gently tap the side of the tissue culture flask with an open palm to dislodge cells that are still attached (*see* **Note 18**).
8. Once all chondrocytes have become detached, add chondrocyte culture media to the flask, matching the volume of the trypsin/EDTA (*see* **Note 19**).
9. Aspirate the cell suspension and transfer to a clean, sterile 50 mL tube.
10. Centrifuge the tube(s) at 179 g for 10 min to pellet the cells.
11. Aspirate the media, taking care not to disturb the chondrocyte pellet, and discard.

12. Resuspend the pellet in 1 mL of chondrocyte culture media, and perform a cell count by trypan blue exclusion.
13. Re-seed the chondrocytes into new, sterile tissue culture flasks at a density of 5×10^3 chondrocytes/cm^2.
14. Repeat **steps 1–13** as required.

3.3 GMP Chondrocyte Isolation

1. Remove the cartilage biopsy from the transport medium (serum-free GMP chondrocyte culture medium) using sterile, disposable forceps, and place into a 100 × 20 mm sterile petri dish (*see* **Note 20**).
2. Wash the tissue twice with complete serum-free GMP chondrocyte medium and mince into small pieces (approx. 1 mm^3) using a sterile, disposable scalpel.
3. Divide the cartilage into two preweighed T12.5 vented tissue culture flasks containing 10 mL collagenase solution each.
4. Reweigh the flasks and record the weight of each biopsy half.
5. Place the flasks in a 90% humidified atmosphere at 37 °C with 5% CO_2 in air for 16–20 h.
6. Following the enzymatic digest, draw up the resulting cell suspension using a needle and syringe (*see* **Note 21**), dispense into a sterile centrifuge tube, and then centrifuge at 750 g for 10 min.
7. Following centrifugation, aspirate and discard the supernatant. Resuspend each resulting cell pellet in 2.4 mL GMP chondrocyte culture medium + 0.6 mL of the patients' own serum.
8. Perform a cell count and viability assessment by trypan blue exclusion.
9. Add each remaining cell suspension to a vented T12.5 flask, place in a 90% humidified atmosphere at 37 °C with 5% CO_2 in air to allow chondrocytes to adhere, and proliferate for 4–6 days (*see* **Note 14**).

3.4 GMP Chondrocyte Expansion in Autologous Serum Supplemented Medium

3.4.1 Autologous Serum Preparation

1. Fill 20 × 7 mL vacutainers with autologous blood via venepuncture, and allow to clot for 16 h, at room temperature.
2. Centrifuge clotted blood in the vacutainer collection tubes at 1000 g for 10 min.
3. After centrifugation, check the serum for evidence of hemolysis, i.e., a red coloration (*see* **Note 22**).
4. Aspirate nonhemolyzed serum from the uncapped vacutainers using a syringe and needle.
5. Filter the serum through a 0.2 μm syringe filter and store in prelabeled universal containers (*see* **Note 23**).
6. Store serum for immediate use at 5 °C; serum for future use is stored at −20 °C.

3.4.2 GMP Chondrocyte Expansion

1. After 4–6 days, replace the GMP chondrocyte medium by removing the culture medium using a sterile needle and syringe and replace with 2.4 mL of GMP chondrocyte medium + 0.6 mL autologous serum (*see* **Notes 24** and **25**).
2. Undertake a second medium replacement between days 7 and 9. If a flask at this stage contains only two colonies or less, discontinue the culture (*see* **Note 26**).
3. Assess the confluence of the cultures at approximately 10–14 days postdigest by using a confluence grid (*see* **Notes 27** and **28**).
4. At 70% ± 10% confluence, trypsinizc the flasks.
5. At the time of trypsinization, remove the culture media from the flasks (*see* **Note 25**), before washing the cells in 5 mL PBS.
6. Add 5 mL trypsin/EDTA to each of the flasks and incubate at 37 °C with 5% CO_2 in air for 5 min.
7. View under the microscope to observe cell detachment (*see* **Note 18**).
8. Once the cells have become detached, aspirate the trypsin/EDTA and cells using a syringe and needle and place into sterile universal containers.
9. Centrifuge the cell suspensions for 10 min at 750 g.
10. After centrifugation, remove and discard the supernatants and resuspend the cell pellets in in 9 mL GMP chondrocyte medium +1 mL autologous serum.
11. Add the cell suspensions into T80 tissue culture flasks.
12. Perform a cell count and viability assessment by trypan blue exclusion.
13. Record batch numbers, cell yields, and T80 seeding densities on the laboratory record sheet.
14. Examine and replace the medium in each flask every 3–4 days until either they require trypsinizing again, or the cells are harvested and returned to the referring surgeon.

3.4.3 Harvesting, Concentrating, and Releasing Cells for Autologous Transplantation

1. Inspect the culture flasks for evidence of bacterial/fungal contamination (macroscopic and microscopic visual inspection), remove the culture media, and wash the flask using 10 mL of PBS (*see* **Note 25**).
2. Add 10 mL of trypsin/EDTA and incubate at 37 °C with 5% CO_2 in air for 5 min.
3. View under the microscope to determine efficient cell detachment (*see* **Note 18**).
4. Aspirate the resulting cell suspension using a syringe and needle, and place into labeled sterile universal containers.
5. Centrifuge the cell suspension for 10 min at 750 g.

6. After centrifugation, aspirate and discard the supernatant.
7. Resuspend cell pellets by in 9 mL GMP chondrocyte medium +1 mL autologous serum.
8. Recentrifuge the resulting cell suspension for 10 min at 750 g.
9. Remove the supernatant and discard.
10. Resuspend the cell pellet in either 0.2 mL, 0.4 mL, or 0.6 mL of the patient's autologous serum, as defined by the prescribing clinician (*see* **Note 29**).
11. Perform a cell count and viability assessment by trypan blue exclusion.
12. Draw the cell suspension into a 2 mL syringe and cap with a sterile plug. Leave sufficient cell suspension to perform a cell count and endotoxin test (*see* **Note 30**).
13. Place the capped syringe into a labeled sterile centrifuge tube for dispatch to the operating theater (*see* **Note 31**).

4 Notes

1. Ensure that the reusable scalpel handle has been autoclaved in advance.
2. The DMEM/F12 medium undergoes quality control prior to its release and arrives prescreened for bacteria and fungi. Each new batch of media is supplied with a test certificate which is retained.
3. Maintain sterile conditions throughout. Examination and processing of the tissue should all take place inside a class II microbiological safety cabinet.
4. The amount of blood on tissue and therefore the requirement for an initial PBS wash depend on the surgical technique employed and will vary between surgeons and between patients.
5. Wear a chainmail glove whenever using a scalpel and primary tissue to prevent cuts and disease transmission.
6. At this stage, it is preferable to take large slices for ease of transfer into 50 mL tubes to be weighed.
7. Weigh the cartilage in the tube, using another tube of the same brand and size to zero the balance. The difference in weight between tubes is negligible at this stage.
8. Depending on the weight of cartilage and the corresponding volume of collagenase required, it may be necessary to split the mixture of tissue and enzyme between two or more tubes at this stage. We recommend a maximum of 40 mL collagenase type II (i.e., 1.2 g tissue) per 50 mL tube.

9. The tube roller should be cleaned thoroughly with 70% IPA before placing into the incubator.
10. After 16 h, the digestion mixture should appear turbid with only small pieces of undigested tissue remaining.
11. If the strainer becomes blocked, carefully pipette the cell suspension up and down. If the blockage persists, aspirate the cell suspension from the strainer, discard the used strainer, replace with a new strainer, and continue to filter.
12. It may be necessary to split the cell suspension/media mixture between tubes at this stage.
13. If the cell suspension is split between multiple tubes, these can be combined into a single tube at this stage.
14. It is important that flasks are minimally handled (ideally not at all) during the chondrocyte attachment phase.
15. Appropriate media volumes: T25 = 5 mL, T75 = 15 mL, T175 = 30 mL. Chondrocyte culture media is routinely made up fresh every 2 weeks.
16. Appropriate PBS volumes: T25 = 5 mL, T75 = 15 mL, T175 = 30 mL.
17. Appropriate trypsin/EDTA volumes: T25 = 3 mL, T75 = 8 mL, T175 = 12 mL.
18. If there still appears to be chondrocytes attached after the 5 min incubation and tapping of the flask, incubate at 37 °C for a further minute, and repeat the microscopic assessment and tapping steps.
19. At this stage, one can aspirate the cell suspension and use it to wash down the growth surface of the flask to ensure that all chondrocytes are in suspension.
20. GMP chondrocyte isolation and culture take place inside a class II microbiological safety cabinet. Cartilage biopsies will only be accepted if they are prearranged and if they arrive in the supplied, patient-labeled tube and are accompanied by a fully completed biopsy transit form and a knee map. Nonconforming harvests will not be processed but will be incinerated as clinical waste and will be reported through the nonconformance procedure. On receipt, the harvested cartilage is issued with a laboratory record sheet on which the patient identification details, sample number, date, time, and media/antibiotic/enzyme batch numbers currently in use are recorded. The transport medium is placed into sterile, labeled tubes and sent for microbial analysis, recording the results on a laboratory record sheet when received. The public health laboratory uses the closed Bactec System (Becton Dickinson) analyzed using a 14-day protocol for fastidious organisms. A positive test result will mean the culture being discarded and a nonconformance form issued (recorded in the Cell Harvesting and Culture Progress Log).

21. If undigested bone fragments are visible, these can be easily avoided by careful aspiration using a needle and syringe.
22. If hemolysis has occurred, the serum is discarded and a nonconformance form issued. A second collection of blood is requested and a note is made in the Cell Culture Progress Log.
23. A record is made in the Cell Harvesting and Culture Progress Log that serum has been successfully prepared.
24. Cultures are initiated and maintained until primary confluence in GMP chondrocyte medium containing 20% autologous serum. Passaged cells are maintained in GMP chondrocyte medium containing 10% autologous serum.
25. A sample of the spent culture fluid is placed into sterile, labeled tubes and sent for microbial analysis, recording the results on a laboratory record sheet when received. A positive test result will mean the culture being discarded and a nonconformance form issued (recorded in the Cell Harvesting and Culture Progress Log) (*see* **Note 20**). At the final harvesting stage, an aliquot of the culture medium is sent for microbial analysis at an independent and regulatory agency approved specialist laboratory. Again a 14-day protocol is used for fastidious organisms.
26. A colony is defined as a group of 20 or more cells in close proximity. If neither flask is continued, a nonconformance form is issued, and the nonconformance procedure followed.
27. The flasks will be approaching confluence approximately 10–14 days post culture initiation. It is important not to allow the culture to become over confluent as this will take the cells out of growth cycle, and it is therefore necessary to trypsinize the cells and move them into larger flasks. Time to confluence will vary from donor to donor.
28. To standardize the assessment of confluence and to estimate the coverage of growth area, a confluence grid is used. The confluence grid consists of ten vertical and ten horizontal lines contained within a flask outline drawn on a transparent sheet. When viewed under a flask on an inverted microscope, the number of cells at intersections of lines gives an estimation of confluence.
29. Typically either 0.4, 0.6, or 0.8 mL is prescribed, depending on the size and number of defects.
30. Endotoxin testing typically reports at <8.0 EU/mL; deviations would result in a QA/QP discussion prior to release.
31. If less than one million cells have been harvested or if the viability is less than 90%, a nonconformance report is issued; cells may still be released but only after consultation with the prescribing consultant. The laboratory record sheet,

nonconformance log, and electronic patient records are all reviewed for anomalies before the cells are released. Cell suspensions may be released by either the principal clinical scientist or technician after a full evaluation of in-process test results and culture profiles. The signed release is recorded on the laboratory record.

Acknowledgements

We would like to acknowledge the contributions made to this book chapter from all of the Oswestry/Keele University Orthopaedic Research (OSKOR) group and John Charnley GMP and MHRA licensed laboratory, both based at the RJAH Orthopaedic Hospital, Oswestry, UK.

References

1. Harrison PE, Ashton IK, Johnson WE et al (2000) The in vitro growth of human chondrocytes. Cell Tissue Bank 1(4):255–260. https://doi.org/10.1023/a:1010131729208
2. Garcia J, Mennan C, McCarthy HS et al (2016) Chondrogenic potency analyses of donor-matched chondrocytes and mesenchymal stem cells derived from bone marrow, infrapatellar fat pad, and subcutaneous fat. Stem Cells Int 2016:6969726. https://doi.org/10.1155/2016/6969726
3. Mennan C, Garcia J, McCarthy H et al (2019) Human articular chondrocytes retain their phenotype in sustained hypoxia while normoxia promotes their immunomodulatory potential. Cartilage 10(4):467–479. https://doi.org/10.1177/1947603518769714
4. Jakob M, Démarteau O, Schäfer D et al (2001) Specific growth factors during the expansion and redifferentiation of adult human articular chondrocytes enhance chondrogenesis and cartilaginous tissue formation in vitro. J Cell Biochem 81(2):368–377. https://doi.org/10.1002/1097-4644(20010501)81:2<368::aid-jcb1051>3.0.co;2-j
5. Brittberg M, Lindahl A, Nilsson A et al (1994) Treatment of deep cartilage defects in the knee with autologous chondrocyte transplantation. N Engl J Med 331(14):889–895. https://doi.org/10.1056/nejm199410063311401
6. Richardson JB, Caterson B, Evans EH et al (1999) Repair of human articular cartilage after implantation of autologous chondrocytes. J Bone Joint Surg Br 81(6):1064–1068. https://doi.org/10.1302/0301-620x.81b6.9343
7. Davies RL, Kuiper NJ (2019) Regenerative medicine: a review of the evolution of autologous chondrocyte implantation (ACI) therapy. Bioengineering (Basel) 6(1):22. https://doi.org/10.3390/bioengineering6010022
8. National Institute for Health and Care Excellence (NICE) (2017) Autologous chondrocyte implantation for treating symptomatic articular cartilage defects of the knee. Technology appraisal guidance [TA477]. https://www.nice.org.uk/guidance/ta477/resources/autologous-chondrocyte-implantation-for-treating-symptomatic-articular-cartilage-defects-of-the-knee-pdf-82604971061701
9. Mistry H, Connock M, Pink J et al (2017) Autologous chondrocyte implantation in the knee: systematic review and economic evaluation. Health Technol Assess 21(6):1–294. https://doi.org/10.3310/hta21060
10. EudraLex. Volume 4 – Good manufacturing practice (GMP) guidelines. (2011–2015). https://ec.europa.eu/health/documents/eudralex/vol-4_en

Chapter 3

Isolation of Chondrons from Hyaline Cartilage

Jasmijn V. Korpershoek, Margot Rikkers, and Lucienne A. Vonk

Abstract

In native healthy hyaline cartilage, the chondrocytes are surrounded by a pericellular matrix that has a distinct composition and function compared to the hyaline cartilage extracellular matrix. The chondrocyte together with its pericellular matrix is called a chondron. The type VI collagen, which is the main component of the pericellular matrix, is resistant to enzymatic digestion by pure collagenase and dispase that do digest the extracellular matrix. Therefore, this combination of enzymes can be used to enzymatically isolate chondrons from hyaline cartilage. Chondrons have a high potential for cartilage tissue engineering. This chapter describes in detail how chondrons can be isolated from hyaline cartilage for further use.

Key words Chondron, Hyaline cartilage, Pericellular matrix, Isolation, Type VI collagen, Chondrocyte

1 Introduction

A chondron can be described as the primary structural unit of cartilage. It consists of a chondrocyte and its surrounding pericellular matrix. The pericellular matrix of a chondrocyte consists of proteoglycans (e.g., aggrecan, biglycan, decorin), hyaluronan, glycoproteins (e.g., fibronectin, link protein, laminin), and various types of collagen, of which type VI collagen is the most specific to the pericellular matrix (Fig. 1) [1–4]. This pericellular matrix connects the chondrocyte to the territorial and interterritorial matrix and plays an important role in the joint homeostasis and mechanoregulation [3, 5, 6]. Initially, isolation of chondrons was performed mechanically using repeated low-speed homogenizations which led to damage and low viability of chondrons [1]. The later established enzymatic chondron isolation protocol increased the viability of chondrons while retaining the native pericellular matrix [7]. This protocol uses enzymatic digestion by a mild protease

Jasmijn V. Korpershoek and Margot Rikkers contributed equally with all other contributors.

Martin J. Stoddart et al. (eds.), *Cartilage Tissue Engineering*,
Methods in Molecular Biology, vol. 2598, https://doi.org/10.1007/978-1-0716-2839-3_3,

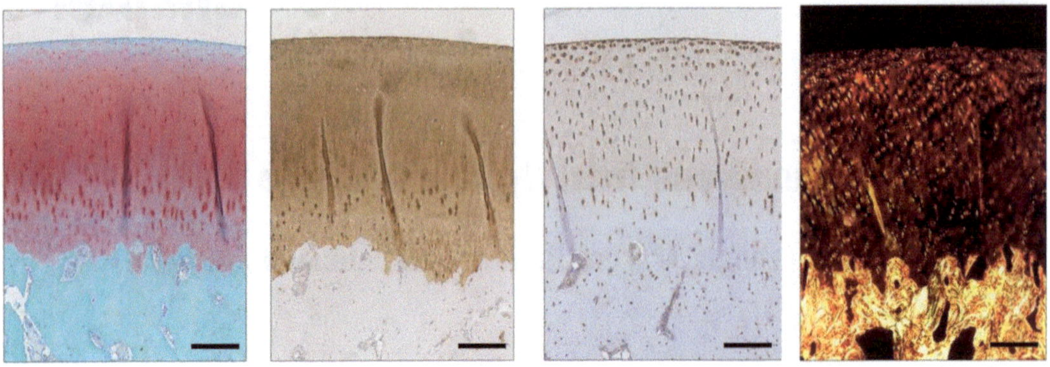

Fig. 1 Safranin-O/fast green, type II and VI collagen, and picrosirius red (with polarized light) (immuno)-stainings on healthy hyaline cartilage from horse, showing the pericellular localization of the type VI collagen. Scale bar represents 400 μm. (This figure is adapted from figure 2, Mouser et al. [21] as permitted by the Creative Commons Attribution 4.0 International license (http:/creativecommons.org/licenses/by/4.0/))

(dispase) and collagenase, which does not digest type VI collagen [8]. The current protocol was developed and successfully used to isolate chondrons from human, equine, porcine, bovine, and caprine hyaline cartilage [7, 9–13].

Preserving the pericellular matrix has advantages for cartilage tissue engineering as in vitro extracellular matrix production by chondrons is higher compared to chondrocytes [9, 10, 12, 14]. Both in vivo and in vitro, complementing chondrons with mesenchymal stromal cells (MSCs) increases cartilage formation even further [11, 13]. Isolation of chondrons can even be performed within 40 min for one-stage procedures in which harvesting, isolation, and cell therapy are performed within a single surgery [15–17].

2 Materials

Work in a sterile laminar flow cabinet and adhere to sterile working guidelines. Prepare and store all reagents at room temperature, unless stated otherwise. Follow all waste disposal regulations when disposing waste materials.

2.1 Chemicals and Materials

1. Freshly dissected hyaline cartilage.
2. Sterile phosphate-buffered saline (PBS; can be supplemented with antibiotics, *see* **Note 1**).
3. Dulbecco's Modified Eagle's Medium (DMEM), supplemented with 1% human serum albumin (*see* **Note 2**), 100 U/mL penicillin, and 100 μg/mL streptomycin (store all at 4 °C).
4. Dispase (>0.5 units/mg) (store at 4 °C).
5. Collagenase type II (>125 units/mg) (store at 4 °C).

6. 50 mL Falcon tubes.
7. Sterile petri dishes.
8. Sterile scalpel (*see* **Note 3**).
9. Sterile scalpel knives.
10. Sterile (anatomical) tweezers (*see* **Note 4**).
11. Tube roller/rocker/rotator, fitting 50 mL Falcon tubes.
12. Trypan blue solution.
13. 0.22 μm nylon filter.
14. Sterile syringe (50 mL).
15. Sterile cell strainer (70 μm pore).

2.2 Chondron Isolation Medium

1. Weigh 37.5 mg dispase and 25 mg collagenase type II in a 50 mL Falcon tube. Add 25 mL supplemented DMEM and allow to dissolve for 1 h under gentle agitation. Sterilize the solution by filtration through a 0.22 μm filter (*see* **Notes 4** and **5**).

3 Methods

Prepare a laminar flow cabinet for sterile working and set up all required materials. Make sure to have enough scalpel blades ready and that the enzymes have completely dissolved in the isolation medium before starting the procedure.

3.1 Processing Hyaline Cartilage

1. Transfer the pieces of tissue with sterile tweezers to a sterile petri dish inside a laminar flow cabinet, and rinse with PBS to remove any remaining blood.
2. In case the cartilage is attached to the underlying bone, trim the full thickness of the cartilage away using a sterile scalpel and store the pieces of cartilage in a new petri dish with PBS.
3. Use a new scalpel blade to mince the cartilage into 1- to 2 mm pieces. Adding a thin layer of PBS on the cartilage prevents it from drying and makes the mincing easier.

3.2 Chondron Isolation Procedure

1. Aspirate the PBS from the minced pieces of cartilage by tilting the petri dish and carefully removing the PBS.
2. Use the scalpel to bring together the pieces and transfer into the isolation medium. Using 25 mL of digestion medium is sufficient to digest at least 10 g cartilage tissue (*see* **Note 5**).
3. Close the tube and digest the cartilage overnight on a tube roller, rocker, or in a rotator at ~20 rpm at 37 °C.

3.3 Next Day

1. Place a 70 μm cell strainer on a 50 mL Falcon tube inside the laminar flow cabinet.
2. Pipette or pour the isolation medium containing the cartilage tissue digest over the strainer to remove undigested pieces of cartilage.
3. Centrifuge the cell suspension at 500 g for 5 min to pellet the chondrons.
4. Remove the supernatant, resuspend in 25 mL DMEM to wash the chondrons, and repeat the centrifugation step. Remove the supernatant and resuspend the chondrons in 10 mL of DMEM. Take a sample of 10 μL, mix with 10 μL trypan blue solution to count using a manual or automated cell counter (*see* **Note 6**).
5. Process the chondrons as desired for culture or cryopreservation. Samples of the primary isolate can be taken to determine the successful isolation of chondrons (*see* **Note 7**).

4 Notes

1. Antibiotics can be added to PBS. Concentrations of 100 U/mL penicillin, 100 μg/mL streptomycin, and 2.5 μg/mL amphotericin B have been used so far.
2. Instead of human serum albumin, other sources of albumin might be used such as bovine serum albumin or fetal bovine serum.
3. Scalpels and tweezers can be autoclaved sterile or submerged in 70% ethanol for a minimum of 30 min, followed by complete evaporation of the ethanol in a laminar flow hood.
4. Prepare the isolation medium fresh on the day of use. Alternatively, it can be prepared in advance and sterilized solutions can be stored at −20 °C for several months. However, storage stability and enzymatic activity after thawing have never been determined.
5. Digesting 10 g of cartilage tissue in 25 mL digestion medium for chondron isolation is a rough guideline translated from chondrocyte isolation protocols [18]. The isolation efficiency of chondrons is at least as good as the isolation efficiency of chondrocytes. However, the concentration of the enzymes and duration of isolation can be varied and optimized. Successful isolations have been performed with 1 [11] and 5 h [7] incubations with increased enzyme concentrations. However, it is important to monitor cell viability and isolation efficiency [11]. In our setup, an isolation time of 16–20 h was optimal for isolation of chondrons from human knee cartilage.

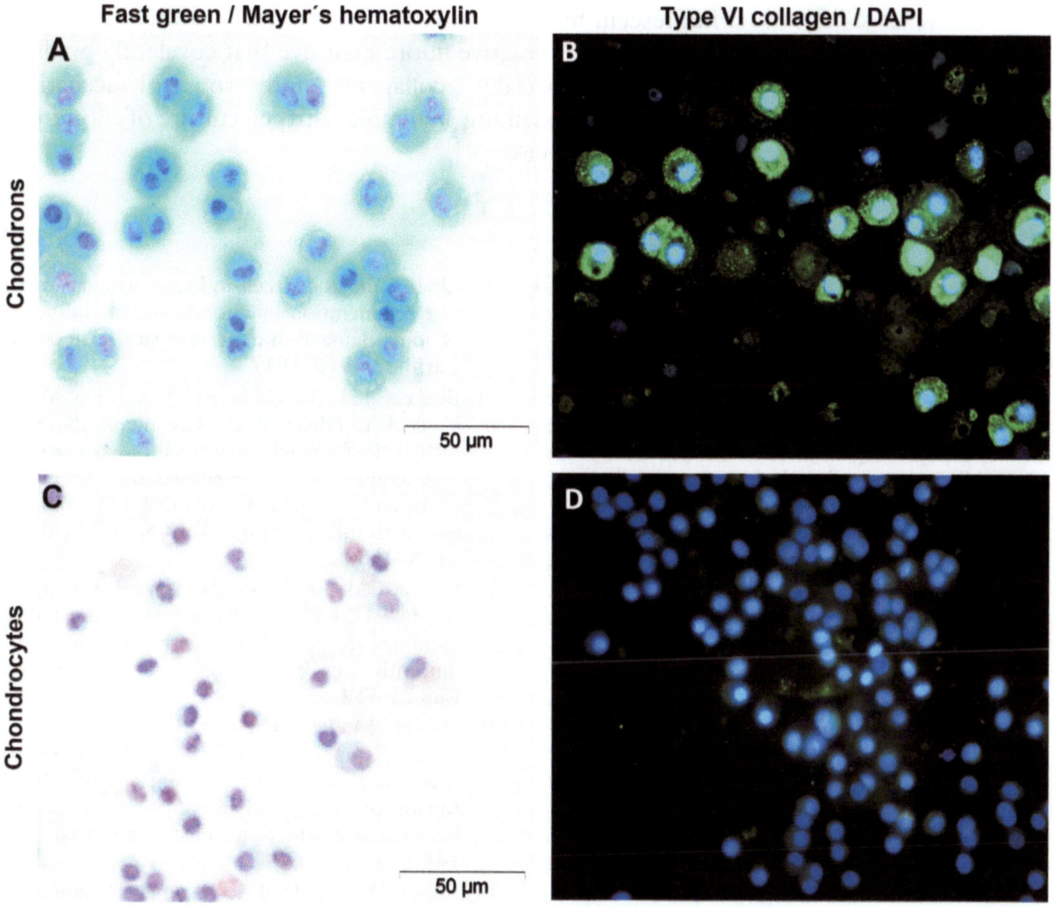

Fig. 2 Fast green/hematoxylin (**a**, **c**) and type VI collagen (Alexa-488 immuno, **b**, **d**) stainings on cytospins of freshly isolated chondrons (**a**, **b**) and chondrocytes (**b**, **d**). Scale bar represents 50 μm and is valid for all images. Panels (**a**) and (**c**) are a reproduction of figure 1, Vonk et al. [10] (with a license by Elsevier). Panels (**b**) and (**d**) are original microphotographs

6. When using an automated cell counter, set cell size to >6 μm to exclude remaining erythrocytes. Alternatively, if counting is hampered by erythrocytes, an erythrocyte lysis can be used prior to counting, or 3% acetic acid with methylene blue can be used to exclusively count nuclei from nucleated cells.

7. The pericellular matrix is not well visible using bright field microscopy, and additional stainings on the pericellular matrix can be performed to show the successful isolation of chondrons. For this, cytospins of approximately 20,000 chondrons can be used. The pericellular matrix can be stained with 0.4% aqueous fast green and counterstained with Mayer's hematoxylin, or a specific immunostaining for type VI collagen can be performed (Fig. 2). In addition, the pericellular matrix can be stained with 1 mg/mL of 5-([4,6-dichlorotriazin-2-yl]amino)

fluorescein hydrochloride (DTAF) in 0.2 N sodium bicarbonate. DTAF is a reactive fluorescent dye that covalently binds to proteins, especially collagens and to polysaccharides [19, 20]. This staining is suitable with cell culture of chondrons for at least 4 weeks [9].

References

1. Poole CA (1997) Articular cartilage chondrons: form, function and failure. J Anat 191: 1–13
2. Poole CA, Ayad S, Gilbert RT (1993) Chondrons from articular cartilage V. Immunohistochemical evaluation of type VI collagen organisation in isolated chondrons by light, confocal and electron microscopy. J Cell Sci 103:1101–1110
3. Guilak F, Hayes AJ, Melrose J (2021) Perlecan in pericellular mechanosensory cell-matrix communication, extracellular matrix stabilisation and mechanoregulation of load-bearing connective tissues. Int J Mol Sci 22:2716
4. Knudson W, Ishizuka S, Terabe K, Askew EB, Knudson CB (2019) The pericellular hyaluronan of articular chondrocytes. Matrix Biol 78: 32–46
5. Nguyen BV, Wang QG, Kuiper NJ, El Haj AJ, Thomas CR, Zhang Z (2010) Biomechanical properties of single chondrocytes and chondrons determined by micromanipulation and finite-element modelling. J R Soc Interface 7: 1723–1733
6. Guang Q, Nguyen B, Thomas CR, Zhang Z, El AJ, Kuiper NJ (2010) Biomaterials molecular profiling of single cells in response to mechanical force: comparison of chondrocytes, chondrons and encapsulated chondrocytes. Biomaterials 31:1619–1625
7. Lee GM, Poole CA, Kelley SS, Chang J, Caterson B (1997) Isolated chondrons: a viable alternative for studies of chondrocyte metabolism in vitro. Osteoarthr Cartil 5:261–274
8. Kielty CM, Lees M, Shuttleworth CA, Woolley D (1993) Catabolism of intact type VI collagen microfibrils: susceptibility to degradation by serine proteinases. Biochem Biophys Res Commun 191:1230–1236
9. Vonk LA, Doulabi BZ, Huang C, Helder MN, Everts V, Bank RA (2010) Preservation of the chondrocyte's pericellular matrix improves cell-induced cartilage formation. J Cell Biochem 110:260–271
10. Vonk LA, de Windt TS, Kragten AHM, Beekhuizen M, Mastbergen SC, Dhert WJA, Lafeber FPJG, Creemers LB, Saris DBF (2014) Enhanced cell-induced articular cartilage regeneration by chondrons; the influence of joint damage and harvest site. Osteoarthr Cartil 22:1910–1917
11. Bekkers JEJ, Tsuchida AI, Van Rijen MHP, Vonk LA, Dhert WJA, Creemers LB, Saris DBF (2013) Single-stage cell-based cartilage regeneration using a combination of chondrons and mesenchymal stromal cells: comparison with microfracture. Am J Sports Med 41: 2158–2166
12. Larson CM, Kelley SS, Blackwood AD, Banes AJ, Lee GM (2002) Retention of the native chondrocyte pericellular matrix results in significantly improved matrix production. Matrix Biol 21:349–359
13. Owida HA, De Las Heras Ruiz T, Dhillon A, Yang Y, Kuiper NJ (2017) Co-culture of chondrons and mesenchymal stromal cells reduces the loss of collagen VI and improves extracellular matrix production. Histochem Cell Biol 148:625–638
14. Fraser SA, Crawford A, Frazer A, Dickinson S, Hollander AP, Brook IM, Hatton PV (2006) Localization of type VI collagen in tissue-engineered. Tissue Eng 12:569–577
15. de Windt TS, Vonk LA, Slaper-Cortenbach ICM, van den Broek MPH, Nizak R, van Rijen MHP, de Weger RA, Dhert WJA, Saris DBF (2017) Allogeneic mesenchymal stem cells stimulate cartilage regeneration and are safe for single-stage cartilage repair in humans upon mixture with recycled autologous chondrons. Stem Cells 35:256–264
16. de Windt TS, Vonk LA, Slaper-Cortenbach ICM, Nizak R, van Rijen MHP, Saris DBF (2017) Allogeneic MSCs and recycled autologous chondrons mixed in a one-stage cartilage cell transplantion: a first-in-man trial in 35 patients. Stem Cells 35:1984–1993
17. Saris TFF, de Windt TS, Kester EC, Vonk LA, Custers RJH, Saris DBF (2021) Five-year outcome of 1-stage cell-based cartilage repair using recycled autologous chondrons and allogenic mesenchymal stromal cells: a first-in-human clinical trial. Am J Sports Med 49: 941–947

18. Otero M, Favero M, Dragomir C, El Hachem K, Hashimoto K, Plumb DA, Goldring MB (2005) Human chondrocyte cultures as models of cartilage-specific gene regulation. Methods Mol Med 107:69–95
19. Davison PF, Galbavy J (1985) Fluorescent dyes demonstrate the uniform expansion of the growing rabbit cornea. Invest Ophtalmol Vis Sci 26:1202–1209
20. Lee GM, Loeser RF (1998) Interactions of the chondrocyte with its pericellular matrix. Cells Mater 8:135–149
21. Mouser VHM, Dautzenberg NMM, Levato R, van Rijen MHP, Dhert WJA, Malda J, Gawlitta D (2018) Ex vivo model unravelling cell distribution effect in hydrogels for cartilage repair. ALTEX 35:65–76

Chapter 4

Articular Cartilage Chondroprogenitors: Isolation and Directed Differentiation

Ilyas M. Khan, Joshua McKenna, and Yadan Zhang

Abstract

Experimental data suggests that tissue-specific progenitors are present within hyaline articular cartilage with the potential to contribute to growth, maintenance, and repair. In this chapter, we show how colony-forming progenitor-like cells can be isolated from bovine articular cartilage using differential adhesion to fibronectin. Furthermore, we describe the optimal conditions and factors required to differentiate these progenitor cells to produce hyaline articular cartilage.

Key words Chondroprogenitors, Multipotent, Differentiation, Progenitors, Cell therapy

1 Introduction

For many years, cell-based biological repair of cartilage in damaged joints and tissues was viewed as an onerous task, especially for larger defects. The reasons for this pessimism were well-founded and based on the lack of a suitable cell type for cell culture expansion, an inability to direct appropriate chondrogenic differentiation of cells, and the lack of maturation of the resultant tissue [1]. In recent years, research into these areas of inquiry has brought us closer to the realization of the minimal key functional objectives required for cellular-based therapies and transplantation of tissue engineered cartilage (*see* **Note 1**). At the heart of any attempt at cell-directed repair of cartilage is the choice of cell type, and in this respect, the isolation of chondroprogenitor-like cells residing in adult cartilage tissues has been a significant discovery. In general, the longer they are passaged in culture, dedifferentiated chondrocytes isolated from adult cartilage progressively lose their potential to redifferentiate in three-dimensional culture settings in vitro or in vivo [2–4]. In contrast, tissue-specific chondroprogenitors isolated from the same tissues possess the characteristics of mesenchymal stem cells including the ability to be expanded in culture for many

Martin J. Stoddart et al. (eds.), *Cartilage Tissue Engineering*,
Methods in Molecular Biology, vol. 2598, https://doi.org/10.1007/978-1-0716-2839-3_4,

generations while retaining the ability to undergo chondrogenic differentiation [5, 6]. In this chapter we describe the isolation of chondroprogenitors from bovine articular cartilage using differential adhesion to fibronectin and their expansion and differentiation. Chondroprogenitors from other cartilaginous tissues (auricular, intervertebral disc) and from different species (rat, pig, rabbit, human, and equine) have also been isolated using the same technique [7–12].

Joint development involves the formation of an interzone, a strip of cells expressing growth and differentiation factor-5 (GDF-5) that marks the point of separation between a previously continuous stretch of cartilage. Genetic studies using knock-in Gdf-5(CreERT^2) mice show a continuous influx of progenitors from surrounding tissues into the interzone along with spatiotemporal dynamics of GDF-5 expression determine the order in which structural elements of the joint are formed, with articular chondrocytes specified relatively late in this developmental sequence [13]. Subsequently, knee joint progenitors generate small, nonmigratory progeny that produce distinct localized tissues [14]. The latter progeny form a permanent articular cartilage which lies above a transient epiphyseal cartilage [15]. During postnatal development epiphyseal cartilage is gradually resorbed and replaced with bone that is separated from the overlying permanent cartilage by a thin cement line of calcified cartilage. Using Gdf-5(CreERT^2) mated to R26-Confetti mice, Decker et al. showed perinatal and postnatal permanent cartilage growth occurs principally by a process of extracellular matrix deposition, cellular hypertrophy and rearrangement to form the characteristic articular cartilage morphology of superficial, middle, and deep zones, wherein the latter, chondrocyte columns are found [16]. Decker et al. conclude that cellular proliferation is a negligible or minor component of postnatal cartilage development; however, this does not negate the fact that cells within this compartment retain the ability to undergo cellular proliferation as described in experiments of neonatal and postnatal wounding of rat cartilage [17]. In the latter experiments, neonatal wounded cartilage underwent repair driven by cellular proliferation of surrounding chondrocytes whereas in more mature cartilage, where cell proliferation rates decline significantly, repair was poor in comparison. The same type of experiments in an ovine model of cartilage repair saw similar results with the additional finding that the lack of expression of inflammatory markers in wounded fetal compared to mature cartilages contributes to better healing [18]. Therefore, through studies of injured cartilage, the notion that postnatal articular cartilage consists solely of a population of terminally differentiated chondrocytes does not ring true [19]. There is, however, sufficient evidence to propose that progenitor populations exist in various postnatal tissue compartments within the joint, though their precise phenotype, location, frequency, and longevity are still a matter of debate.

Some further light was shed on the question of the existence of progenitors within articular cartilage via a set of experiments using the marsupial, *Monodelphis domestica* (short-tailed opossum). Marsupials are characterized by premature birth of their young who continue their development ex utero while attached to the nipples of their mother's lower belly. This feature allowed researchers to determine the proliferation kinetics of chondrocytes in a developing mammalian joint for the first time using bromodeoxyuridine pulse-chase labeling [20]. These experiments identified a subset of label-retaining cells with long cycling times, characteristic of a potential progenitor population, residing in the surface of articular cartilage. Thenceforth, attention was focused on the superficial zone of articular cartilage, and the work by Salter and colleagues showed that surface zone chondrocytes of the human fetal knee differentially expressed the fibronectin ED-A isoform—carrying a RGD integrin binding motif—compared to the underlying cartilage zones [21]. Unsurprisingly, Dowthwaite et al. showed that surface zone chondrocytes were also associated with higher expression of integrin receptor α5β1 binding fibronectin [5]. These findings formed the basis for the rationale to enrich for surface zone chondrocytes containing subpopulations of chondroprogenitors by differential adhesion to fibronectin, a technique that had previously been used to successfully enrich for clonal populations of keratinocyte stem cells from skin epithelia [22].

Dowthwaite et al. then demonstrated that cartilage-derived surface zone chondrocytes differentially enriched for cells with greater fibronectin binding capacity had higher clone-forming efficiency than cells from either the mid or deep zones [5]. Clonal isolates from adhesion assays were able to undergo more than 35 population doublings while retaining the ability to undergo chondrogenic differentiation using TGFβ1 [23]. Furthermore, clonal isolates fulfilled the minimal criteria to be identified as mesenchymal stromal cells including developmental plasticity by differentiating into osteogenic and adipogenic lineages [24, 25]. When compared to bone-derived mesenchymal stem cells, fibronectin-isolated chondroprogenitors did not express epiphyseal cartilage markers such as Runx2 and collagen type X proteins following TGFβ1-induced chondrogenic differentiation [8], conforming to their tissue-specific origin that precludes terminal chondrocyte differentiation, at least in an early postnatal in vitro developmental context. The obvious utility of chondroprogenitors led to their isolation from human articular cartilage from adult donors where experiments showed they exhibit similar properties to their bovine counterparts, including the extensive proliferative and redifferentiation capacity of cells [11]. One important feature of human chondroprogenitors compared to "classical" dedifferentiated populations of expanded chondrocytes is the maintenance of average telomere length following extensive expansion whereas telomere erosion is evident in unselected dedifferentiated cell populations correlating with a reduced or lack of telomerase activity [11].

The ability to expand chondroprogenitors for more than 30 population doublings and maintain their ability to redifferentiate is an important property for cell therapies and tissue engineering of hyaline cartilages. Allogeneic adult chondroprogenitors offer a step change in the ability of surgeons to treat larger defects using an off-the-shelf solution. The challenges of generating billions of cells for cell therapies are process-related and surmountable. Expanded chondroprogenitors can also be used for tissue engineering replacement cartilage where the challenges are more complex, but, given the pace of discoveries in our field also surmountable [26, 27].

2 Materials

The cattle feet containing the metacarpophalangeal (MCP) joint used in our studies are freshly retrieved from local abattoirs from animals that are part of the human food chain (*see* **Note 2**) and are scrutinized by veterinary surgeons. We require institutional ethical permission to conduct research on retrieved animal parts and the requisite regulatory permissions from National Authorities, which in our case were obtained upon application to The Department for Environment, Food & Rural Affairs (DEFRA), UK. Similar permissions may or may not be required in different countries.

1. Surgical blade handle size No.4 and No.23 blades (Swann-Morton, Sheffield, England). A smaller blade handle (No.3) and blades (No.10) are available, but the larger combination work best for both dissection and tissue extraction work.
2. Pronase digestion buffer: 0.1% w/v pronase from *Streptomyces griseus* (Roche, Cat. No. 11459643001) in DMEM high glucose (Corning, Cat. No. 10-017-CV contains L-glutamine and 4.5 g/L glucose) with 50 μg/mL gentamicin sulfate (Merck: Cat. No. 345815) and 10 mM HEPES pH 7.5. This pronase solution is prepared fresh and filter sterilized (*see* **Note 3**).
3. Collagenase digestion buffer: 0.06% w/v collagenase from *Clostridium histolyticum* (Sigma-Aldrich, Cat. No. C0130) in DMEM high glucose (Corning, 10-017-CVR) with 50 μg/mL gentamicin sulfate (Merck, Cat. No. 345815), 10 mM HEPES pH 7.5, and 5% fetal bovine serum (FBS). This digestion solution is prepared fresh and filter sterilized (*see* **Note 4**).
4. Aluminum foil.
5. Parafilm.
6. Tube roller or rocker plate.

7. Plasticware: 15 and 50 mL Falcon conical centrifuge tubes. 90 mm diameter cell culture dishes. 6-well culture dishes. Polystyrene cloning cylinders, 8 mm diameter (Merck, Cat. No. C7983). Corning cell strainer 40 μm (Corning, Cat. No. 4317500). T75 and T175 tissue culture flasks (various suppliers). 5, 10, and 25 mL glass or plastic Stripettes.
8. Hemocytometer or automatic cell counting slides (various suppliers).
9. Fibronectin solution from bovine plasma, sterile-filtered (~1 mg/mL; Merck, Cat. No. F1411). Make the fibronectin-coated 6-well culture dishes as follows: in 9 mL PBS containing 1 mM $MgCl_2$ and 1 mM $CaCl_2$ (sterilized by filtration), add 90 μL of fibronectin solution, mix, and then aliquot 1.5 mL of this into each well. Incubate overnight at 4 °C. On the following day, coated plates are blocked with a solution of PBS containing 1% w/v bovine serum albumin for 1 h after which they are washed once with prewarmed DMEM wash buffer. Leave the plates in the 37 °C tissue culture incubator until needed (*see* **Note 5**).
10. Cell culture expansion medium: DMEM/F-12 (1:1) (Invitrogen, Cat. No. 31330-038) supplemented with 10% FCS, 100 μg/mL 2-phospho-L-ascorbic acid trisodium salt, 50 μg/mL gentamicin, 10 mM HEPES pH 7.5.
11. Sterile polyester cloning cylinders (Merck, Cat. No. C7983).
12. Vaseline.
13. Differentiation medium: DMEM high glucose (Corning, 10-017-CV) with 50 μg/mL gentamicin sulfate (Merck, Cat. No. 345815), 10 mM HEPES pH 7.5, 100 mg/mL 2-phospho-L-ascorbic acid trisodium salt, 1 × ITS-A (contains sodium pyruvate; Fisher Scientific, UK Cat. No. 12067589) (*see* **Note 6**).
14. Transforming growth factor beta 1 (Peprotech, 100-21C).
15. Bone morphogenetic protein 9 (also known as GDF2; Peprotech, 120-07).
16. Cryo-SFM (PromoCell, Cat. No. C20012).

3 Methods

3.1 ChondroprogenitorChondroprogenitors Isolation

1. The cattle feet are usually covered in mud and farm detritus, and they must first undergo a general rinse in warm water to clean the hoof. A strong bristled brush is a useful tool to help remove loosely attached material, but we find a surgical blade handle (without blade) is also useful to help dislodge mud and well-engrained dirt particularly in older animals (*see* **Note 7**).

2. A dilute solution of household detergent is then sprayed onto the foot and a stiff bristled brush used under an open cold water tap to thoroughly clean the hoof of dirt.
3. The cattle feet are disinfected by liberally spraying them with 70% ethanol and laid in a supine position, i.e., the anterior aspect of the joint facing upwards.
4. Removal of skin: using a No.23 surgical blade, make a longitudinal incision down the ventral surface, starting from top of the foot down to the hoof, just deep enough to break the surface of the skin, and create a visible score line—the ventral surface is the side of the joint that bends towards you (Fig. 1a).
5. Now make a deeper incision circumferentially just above the hoof line.
6. At the superior end of the foot, carefully cut deeper into the score line made in **step 4**, as you do so pull the skin back to expose the tissue underneath. You will see the skin is attached to underlying structures through a superficial facia whose strands of connective tissue can be easily cut with the blade—the blade can point away from the foot as you are doing this. Pulling back the skin while cutting ensures that joint integrity, i.e., the synovial enclosure around the joint maintaining its sterility, is preserved. Keep removing the skin until you reach just above the hoof line, where the circumferential incision line will help to guide the removal of skin around the back of the joint. Once one side of the score incision line has had the skin drawn back by dissection, you can start on the opposing side repeating the process. When both sides of skin have been dissected back, you can flip the foot onto its side and begin to methodically to remove skin from the dorsal side, again taking care to pull the skin and stretch the connective tissue and using the blade to sever the strands and sheets of this tissue without bringing the blade anywhere near the joint.
7. Once the skin has been removed, the foot is covered in household detergent and rinsed under the tap using the brush to remove hairs that have become entangled with the exposed flesh (Fig. 1b).
8. The cattle foot should now be liberally sprayed with 70% ethanol. To reduce the chance of bacterial infection, we cover the exposed parts of the foot with laboratory gloves; covering the cut part of the foot also prevents blood from escaping.
9. Joint dissection: this procedure takes place in a tissue culture class II hood, and as normal all materials entering the hood are sterilized with 70% ethanol.

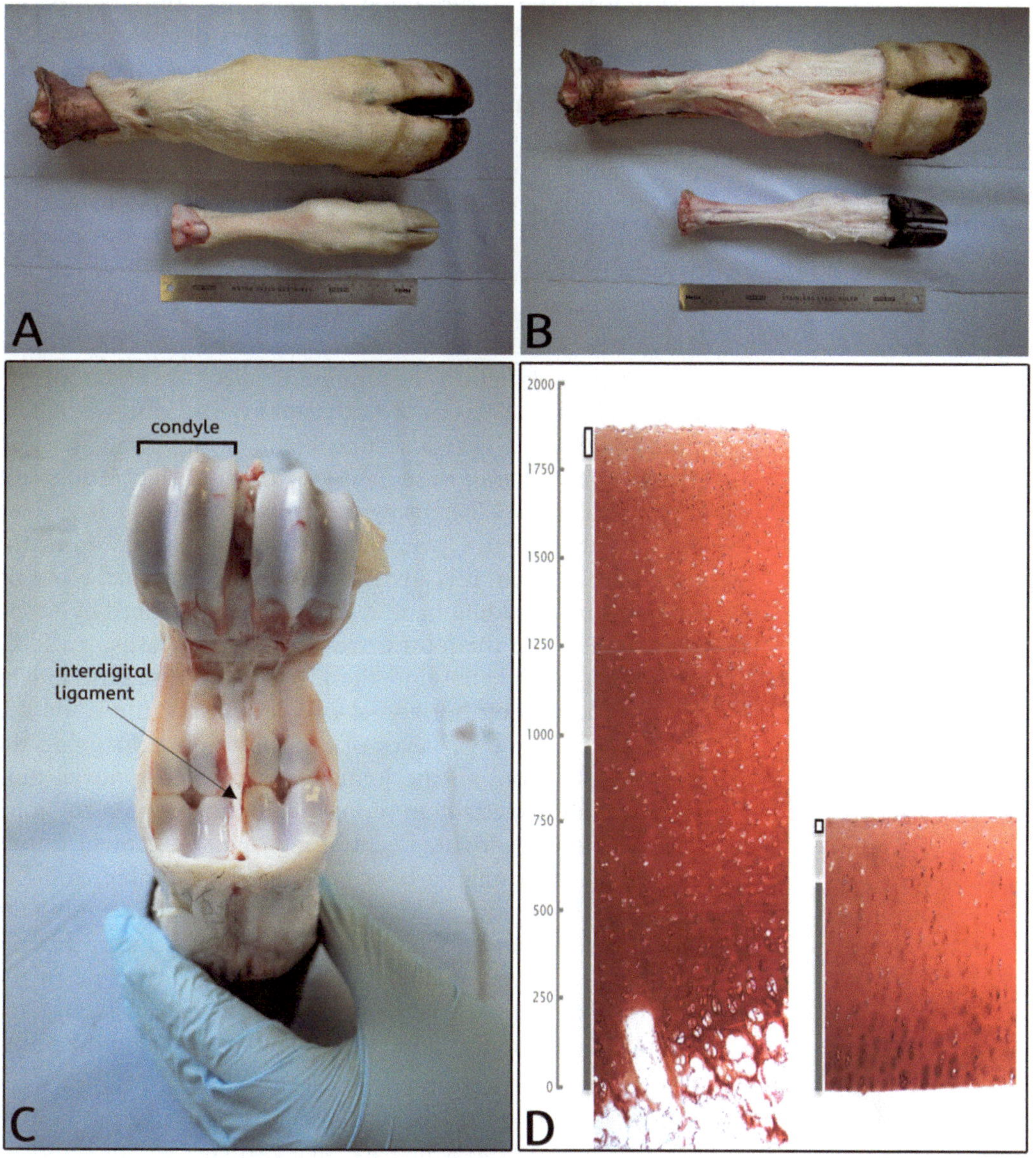

Fig. 1 Dissection of bovine steer feet for the isolation of full depth chondrocytes. (**a**) The bovine feet, 18–24 month mature (above), 7-day-old immature (below), after a rinse and clean. (**b**) Image shows the feet after the skin has been dissected, the dorsal side of the joint is facing upwards. (**c**) The joint is opened with a scalpel blade following the joint line from the dorsal side. To open out the joint fully, the interdigital ligament (*arrowed*) must be cut. Shavings of articular cartilage can be cut from the condylar ridges and trochlear grooves. (**d**) Full-depth coronal sections from mature and immature bovine MCP joints showing the superficial (open box), mid (light grey box) and deep (dark grey box) zones. Immature joint cartilage is mostly transient epiphyseal cartilage (which is soft) and therefore only thin shavings should be taken, whereas cartilage from the full-depth of tissue can be extracted from mature joints because the growth plate has closed and the blade will not penetrate the subchondral plate

10. First prepare the culture hood for joint dissection; the following items should be laid out: a few squares of absorbent paper (blue roll) above which an aluminum foil is placed, two 15 mL Falcon tubes, one filled with 70% ethanol and the other with DMEM wash medium—these tubes will be used to disinfect blade holders and blades, and then rinse them before use—blade and blade holders, one 90 mm dish per joint dissected containing ~15 mL of DMEM wash medium, the pronase digestion buffer prealiquoted into either 15 mL or 50 mL Falcon tubes.
11. The cattle foot is again liberally sprayed with 70% ethanol before being placed on the aluminum sheet in the hood with the dorsal side facing upward - the apex of the joint when bent should face you. The dissecting blade is dipped in the 70% ethanol then ethanol rinsed away by dipping the blade in the DMEM solution. When using the scalpel, always cut away from you and always place your fingers and hands away from the plane of dissection. It is advisable to wear a chainmail glove on the non-blade holding hand if you are inexperienced in dissection. By bending the joint backwards and forwards, it will be possible to see the joint line; the aim is to cut along this line by slowly making long horizontal cuts; eventually the blade will slip between the two halves of the joint and help guide the blade slowly exposing the joint. Straw-colored synovial fluid will seep out, especially in more mature animals. As the joint becomes more exposed, a central interdigital ligament tightly connecting opposing parts of bone becomes visible, and this should be slowly cut; by doing so the whole joint, including the condylar ridges and trochlear grooves of the metacarpal bone, will be partially exposed (Fig. 1c). Following the line of the bone on the upper part of the joint with the blade will cause the top half of the joint to flip backwards fully exposing the joint.
12. With the whole joint exposed, hold the bone of the superior part of the joint; this part contains the metacarpal convex condylar ridges from which articular cartilage will be removed for cell isolation. The inferior part of the joint contains the phalangeal bones, whose joint faces are concave and therefore difficult to dissect (*see* **Note 8**).
13. Using the same blade, which has been freshly disinfected in the 70% ethanol-containing tube and rinsed in DMEM, chips of articular cartilage are removed from the joint surface. In immature animals the blade can penetrate the calcified zone quite easily as the subchondral bone plate has not formed; blood may ooze out as a consequence (Fig. 1d). Make less deep incisions as you remove the articular cartilage as you gauge the depth and softness of the joint tissues. The excised cartilage chips or explants are placed in a 90 mm diameter cell culture dish filled

with DMEM wash solution. When all the cartilage from the surface has been removed, cover the 90 mm tissue culture dish, and then remove the cows foot from the hood.

14. Next the cartilage chips need to be minced to increase their surface area in order to speed the process of digestion. Using the same, disinfected blade, the cartilage chips are cut into approximately 1 mm^3 pieces. The dish is then held at an angle and the minced cartilage moved to the lower edge; then by placing the blade on the lower edge of the dish, the DMEM is slowly poured into the empty lid of the dish, taking care that the cartilage chips remain in the dish. Use the blade to gather the minced cartilage together, and again using the blade carefully tip the minced cartilage into a 50 mL Falcon containing pronase digestion buffer (*see* **Note 9**).
15. Cover the lid of the Falcon tube with a short strip of parafilm; this is to prevent spillage should a leak occur. Leaks occur because a small chip of cartilage lodge in the inner part of the cap; by removing them when you switch over to the collagenase solution, you reduce the chance of leakage.
16. Place the Falcon tube on a tube roller (ideally a tube roller and mixer) in a 37 °C incubator for 2 h at 60–80 rpm. If you don't have a tube roller, a rocker platform or tube rotator will work just as well.
17. Following pronase digest, remove the parafilm strip from the tubes, and very carefully decant the digestion buffer leaving the minced and partially digested cartilage in the tube, and then add 20 mL collagenase solution. Reseal the closed lid with a strip of parafilm and replace on the roller-mixer for 12–16 h at 37 °C.
18. Following overnight incubation, the collagenase digestion should leave the solution appearing a lighter shade of pink and cloudy. If the solution is bright yellow, an infection has occurred, and the digest should be immediately disinfected with virkon and disposed of.
19. The parafilm strip should be removed in the tissue culture hood, and the contents of the digestion decanted through a 40 μm cell strainer to obtain a single cell suspension. The cells should be counted with trypan blue to obtain their concentration and quantify their viability (*see* **Note 10**).
20. Differential fibronectin adhesion: place ~1000 cells/mL in DMEM high-glucose medium containing 100 μg/mL 2-phospho-L-ascorbic acid trisodium salt, 50 μg/mL gentamicin, 10 mM HEPES pH 7.5, and 1 × ITS-A. 9 mLs of cells in total are required to fill a 6-well culture dish to a volume of 1.5 mL per well (~1500 cells). Add 1.5 mL of these cells to each well of the fibronectin-coated culture dish, and incubate for 20 min in a tissue culture incubator at 37 °C.

21. Using a Stripette remove the unattached cells and medium from the 6-well culture dish, and immediately flood the wells with 3 mL of culture expansion medium. After 3–4 days you will note the profusion of groups of 4–8 cells on the dishes. In the subsequent days, the cells will double in number every 24 h, and therefore you should monitor colony growth daily. When the colonies have grown larger >50 cells, they can be isolated using cloning rings (*see* **Note 11**).
22. Colony cloning: the location of isolated colonies can be marked on the underside of the 6-well culture dish using a thick marker pen; 3–4 colonies per well can be isolated in this way. The culture medium is removed, and the dish is washed in DMEM wash medium. Sterile polyester cloning rings (dipping the cleaned rings in ethanol is sufficient to sterilize them) are dipped in silicone grease (or Vaseline) and then placed over the marked location on the top of the culture dish. When all colonies have been covered with the cloning ring, 50 μL of trypsin solution is added and the dish placed in a tissue culture incubator for 5 min. Following incubation, a pipette is used to transfer the trypsin solution; it is useful to pipette up and down a few times to dislodge cells from the colony, directly to another 6 well-dish containing 3 mL cell culture expansion medium. Isolated colonies are incubated for further a week until the well is confluent.
23. Up until 20 population doublings (PD), the chondroprogenitors grow rapidly, one division per day, and therefore they will need to be transferred relatively quickly to T75 and T175 dishes for expansion (*see* **Note 12**).
24. Long-term cryogenic storage of chondroprogenitors: we have found that using CryoSFM, which is a protein and animal product-free cryopreservation freezing medium, allows good revivification rates of cells that are frozen gradually ~1 °C/min using a Mr. Frosty™ freezing container (ThermoFisher Scientific: Cat. No. 5100-0001) at a concentration of 1×10^6 cells/mL.

3.2 Cellular Differentiation of Chondroprogenitors

1. Culture expanded chondroprogenitors up to 35 population doublings can be efficiently differentiated to form cartilage.
2. In addition to the differentiation medium, growth factor addition is critical for efficient induction of chondrogenic differentiation, and two members of the bone morphogenetic family, transforming growth factor-β1 (TGFβ1) and bone morphogenetic protein-9 (BMP-9); also known as growth and differentiation factor-2 (GDF-2), can be used. By using TGFβ1, the differentiated construct will be more fibrous, stiff, and compact. In contrast when using BMP-9, the construct will be relative more voluminous, less fibrous, and less stiff.

3. Pellet cultures: chondrogenic differentiation can be induced in many cellular configurations, on flattened plastic culture dishes, but in general, efficient induction of differentiation occurs in high density three-dimensional culture settings such as pellet or micromass culture. In pellet culture, culture-expanded chondroprogenitors are trypsinized, washed in serum containing medium, and counted. The cells are then spun down at 250 × *g*, and medium removed and resuspended at 0.5×10^6 cells/mL in chondrogenic medium including the chosen growth factor, TGFβ1 10 ng/mL, or, BMP-9 100 ng/mL (*see* **Note 13** and ref. 27). 1 mL of the latter cell suspension is aliquoted into 1.5 mL Eppendorf tubes that are then microcentrifuged at 500 rpm (20 × *g*) for 5 min at room temperature.
4. The pelleted chondroprogenitors are then incubated at 37 °C in a 5% CO_2 incubator for 72 h before the medium is exchanged. Thereafter the medium is exchanged every 72 h for 21–28 days.
5. After the culture period, the pellets, with chondrogenic medium removed, can be flash frozen and stored (−70 °C) for RNA extraction or biochemical composition (−20 °C) analyses of sulfated glycosaminoglycan, DNA, and hydroxyproline content [27]. Pellets can also be fixed in 10% neutral-buffered formyl saline (NBFS) at 4 °C for 12 h for wax embedding, sectioning, and histological staining (Fig. 2).
6. Micromass cultures: culture-expanded chondroprogenitors are trypsinized, washed, and counted, then resuspended in cell culture expansion medium at a concentration of 1×10^7 cells/mL. 10–20 μL of this cell suspension is placed as a droplet in the center of a 12-well tissue culture dish. The periphery of the well can be filled a little (~250 μL) chondrogenic medium to stop the droplet of cells in the center of the dish from drying out, or you can add a 1 mL of medium to empty wells to maintain humidity. The plate is then incubated in a tissue culture incubator, 37 °C and 5% CO_2, for 3–4 h. In this time, the droplet of cells will produce sufficient extracellular matrix to allow it to maintain its shape and integrity when 2 mL of chondrogenic medium including growth factor (*see* **Note 14**) is carefully aliquoted into each well. The medium is changed every 72 h thereafter for the required culture period, 7–14 days.
7. Micromass cultures can be removed from the plate using forceps tweezers and flash frozen in liquid nitrogen for later RNA extraction (storage at −70 °C) or biochemical compositional analyses (storage at −20 °C). RNA extraction in our lab uses dismembranation of intact micromass cultures placed in metal chambers with a metal ball and 0.5 mL TRIzol reagent, all

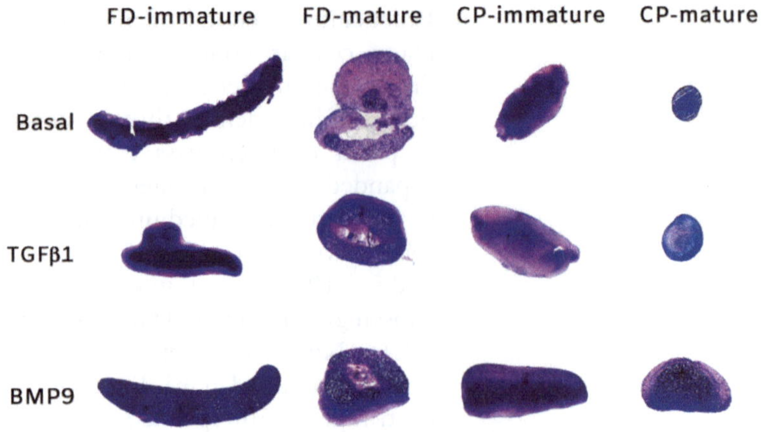

Fig. 2 High-density pellet cultures of chondrocytes and chondroprogenitors. Pellets cultures using 500,000 cells from freshly isolated full-depth (FD) and culture expanded chondroprogenitor (CP) clonal isolates of immature and mature bovine MCP joints grown in basal chondrogenic medium alone, or, supplemented with TGFβ1 (10 ng/mL) or BMP-9 (100 ng/mL) for 21 days and sectioned and stained with toluidine blue to visualize proteoglycan deposition. The most obvious difference is seen in mature CP pellets where the use of BMP-9 appears to drive chondrogenic differentiation to the same extent seen in FD-mature pellets. This observation indicates developmental or culture-induced changes in the response to growth factors in mature CPs not evident in their immature counterparts that can be circumvented through the use of BMP-9

components flash frozen in liquid nitrogen (Merck, Cat. No. T9424) which are together placed in a vibrating grinding mill (Mikro-dismembrator, B. Braun Biotech, GmbH). The ground material is removed from the metal chamber and processed for total RNA isolation by addition of chloroform followed by centrifugation and retention of the aqueous phase. The aqueous phase is then processed through the RNA Easy Plus kit to rapidly isolate total RNA using spin columns (Qiagen, Cat. No. 73404). Chondrogenic differentiation is visualized and quantified by staining micromasses with Alcian Blue pH 1.0–2.5. Micromasses are fixed for 5 min in 10% NBFS at room temperature, washed in water, then stained with Alcian Blue, enough to cover the micromass, for 12 h. The unbound dye is removed with multiple washes with water. The remaining dye can be extracted using 6 M guanidine hydrochloride overnight at room temperature and quantified by measuring the solution's optical density at 595 nm.

4 Notes

1. The term "key functional objectives" (KFO) for a biologically engineered construct was first used by Professor Robert Brown to reduce the growing complexity of tissue engineering to simpler ideals, i.e., the minimal requirements for successful function of implanted tissues [28]. For example, for cartilage repair in a weight-bearing joint, the KFOs could be, in order, (a) a production of chondrogenic tissue that (b) resists compression and (c) tensile forces and (d) displays vertical and (e) horizontal integration. Once these objectives have been achieved, then the next set of more challenging KFOs can be iteratively tackled, such as the structural organization of collagen, cellular organisation and long-term durability.
2. Metacarpophalangeal joints are found between the metacarpal bones of the hand and the proximal phalanges of the fingers. These joints are condyloid and are formed of the condylar heads of the distal ends of the metacarpal bone and the shallow concave ends of the proximal phalanges bone. In humans this joint is referred to as the "knuckle" bone, and in quadrupeds, e.g., horses and cows, it is known as the fetlock.
3. Penicillin/streptomycin can also be used in place of gentamicin.
4. Insulin-transferrin-selenium (ITS) is used to substitute for FBS in good manufacturing practice (GMP) protocols in order provide a nonserum, zoonotic-free medium. The FBS is used in the collagenase buffer to maintain cell viability as digestion proceeds. If digestion is not complete following overnight incubation, then try lowering the FBS concentration to 2.5%.
5. The fibronectin solution cannot be reused as cloudy precipitates form upon storage.
6. Sodium pyruvate is an essential intermediate in many biosynthetic pathways, but also, importantly, acts as an antioxidant [29].
7. It is advisable to first hold the joint with the exposed fleshy part facing down towards the sink and manipulate the joint to force any remaining noncongealed blood in the foot to be ejected. Failure to do this often leads to an unwelcome squirt of blood whilst cleaning or dissecting the feet. The latter is more of a problem with cattle feet from mature animals (18- to 23-month-old steers) than immature (7-day-old steers).
8. If there is blood within the joint, then discard it immediately. The sterility of the joint has been affected, and, in most cases, using cartilage from a joint like this will result in bacterial infection during the digestion phase.

9. We use 20 mL of pronase solution per 7.5 g minced cartilage; in general, the amount of cartilage retrieved from a single joint does not exceed this weight, so it's not something we actively monitor. However, for larger joints and those with cartilage of greater depth, e.g., human knee cartilage, the total tissue mass should be weighed.
10. It is possible to isolate over 50×10^6 chondrocytes from a single immature steer MCP joint, whereas the figure is much lower, ~$1–2 \times 10^6$ chondrocytes from mature steer MCP joints.
11. Polyclonal isolates are probably more useful for studies of chondrogenesis and tissue engineering, as they require less time to expand, and any possible inherent variation of cells which will be amplified in monoclonal lines will be averaged out in polyclonal populations.
12. In this chapter we describe the isolation and expansion of bovine-derived articular chondroprogenitors, and we have found that supplementing medium with serum alone is sufficient to drive their proliferation. There are however species-specific differences in culture media requirements for chondroprogenitor proliferation; human chondroprogenitors require, in addition to serum, fibroblast growth factor 2 (FGF2: 5 ng/mL), and transforming growth factor β1 (TGFβ1: 1 ng/mL) for optimal growth [11], whereas equine articular chondroprogenitors only require FGF2 (10 ng/mL) in addition to serum in medium for optimal growth and proliferation—for over 50 population doublings [8]. Absence of exogenously added growth factor(s) can lead to either significantly slower growth kinetics, early senescence, or apoptosis of cells.
13. BMP-9 has a profound effect on the chondrogenic differentiation of articular chondroprogenitors and differs significantly in the mode of chondrogenesis when compared to TGFβ1 and other defined chondrogenic factors [27]. First, it has a pronounced positive effect in inducing chondrogenesis in mature CPs compared to TGFβ1 (*see* Fig. 2), and secondly it induces morphogenic changes that include structural changes in collagen orientation that mimic tissue maturation [27].
14. We recommend using TGFβ1 when culturing micromasses, as they produce copious amount of proteoglycan and collagen. BMP-9 induces the production of significantly larger constructs that are replete with collagen but have much less proteoglycan. The morphology, however, of cells is very different with chondrocytes in TGFβ1 micromasses appearing small and flattened, whereas cells in BMP-9 micromasses are rounder and hypertrophic (not terminally differentiated—just larger), thereby contributing to the larger average size constructs.

References

1. Hunziker EB (2009) The elusive path to cartilage regeneration. Adv Mater 21(32–33): 3419–3424
2. Benya PD, Shaffer JD (1982) Dedifferentiated chondrocytes reexpress the differentiated collagen phenotype when cultured in agarose gels. Cell 30(1):215–224
3. Dell'Accio F, De Bari C, Luyten FP (2001) Molecular markers predictive of the capacity of expanded human articular chondrocytes to form stable cartilage in vivo. Arthritis Rheum 44(7):1608–1619
4. Passaretti D et al (2001) Cultured chondrocytes produce injectable tissue-engineered cartilage in hydrogel polymer. Tissue Eng 7(6): 805–815
5. Dowthwaite GP et al (2004) The surface of articular cartilage contains a progenitor cell population. J Cell Sci 117(Pt 6):889–897
6. Alsalameh S et al (2004) Identification of mesenchymal progenitor cells in normal and osteoarthritic human articular cartilage. Arthritis Rheum 50(5):1522–1532
7. Li Y et al (2016) Intermittent hydrostatic pressure maintains and enhances the chondrogenic differentiation of cartilage progenitor cells cultivated in alginate beads. Develop Growth Differ 58(2):180–193
8. McCarthy HE et al (2012) The comparison of equine articular cartilage progenitor cells and bone marrow-derived stromal cells as potential cell sources for cartilage repair in the horse. Vet J 192(3):345–351
9. Neumann AJ et al (2015) Human articular cartilage progenitor cells are responsive to mechanical stimulation and adenoviral-mediated overexpression of bone-morphogenetic protein 2. PLoS One 10(8): e0136229
10. Tong W et al (2015) In vivo identification and induction of articular cartilage stem cells by inhibiting NF-kappaB signaling in osteoarthritis. Stem Cells 33(10):3125–3137
11. Williams R et al (2010) Identification and clonal characterisation of a progenitor cell sub-population in normal human articular cartilage. PLoS One 5(10):e13246
12. Xue K et al (2015) Isolation and identification of stem cells in different subtype of cartilage tissue. Expert Opin Biol Ther 15(5):623–632
13. Shwartz Y et al (2016) Joint development involves a continuous influx of Gdf5-positive cells. Cell Rep 15(12):2577–2587
14. Rux D et al (2019) Joints in the appendicular skeleton: developmental mechanisms and evolutionary influences. Curr Top Dev Biol 133: 119–151
15. Hunziker EB, Kapfinger E, Geiss J (2007) The structural architecture of adult mammalian articular cartilage evolves by a synchronized process of tissue resorption and neoformation during postnatal development. Osteoarthr Cartil 15(4):403–413
16. Decker RS et al (2017) Cell origin, volume and arrangement are drivers of articular cartilage formation, morphogenesis and response to injury in mouse limbs. Dev Biol 426(1):56–68
17. Wagner W et al (2001) Neonatal rat cartilage has the capacity for tissue regeneration. Wound Repair Regen 9(6):531–536
18. Ribitsch I et al (2018) Fetal articular cartilage regeneration versus adult fibrocartilaginous repair: secretome proteomics unravels molecular mechanisms in an ovine model. Dis Model Mech 11(7):dmm033092
19. Tew SR et al (2000) The reactions of articular cartilage to experimental wounding: role of apoptosis. Arthritis Rheum 43(1):215–225
20. Hayes AJ et al (2001) The development of articular cartilage: evidence for an appositional growth mechanism. Anat Embryol (Berl) 203(6):469–479
21. Salter DM, Godolphin JL, Gourlay MS (1995) Chondrocyte heterogeneity: immunohistologically defined variation of integrin expression at different sites in human fetal knees. J Histochem Cytochem 43(4):447–457
22. Jones PH, Watt FM (1993) Separation of human epidermal stem cells from transit amplifying cells on the basis of differences in integrin function and expression. Cell 73(4):713–724
23. Khan IM et al (2009) Clonal chondroprogenitors maintain telomerase activity and Sox9 expression during extended monolayer culture and retain chondrogenic potential. Osteoarthr Cartil 17(4):518–528
24. Dominici M et al (2006) Minimal criteria for defining multipotent mesenchymal stromal cells. The International Society for Cellular Therapy position statement. Cytotherapy 8(4):315–317
25. Pittenger MF et al (1999) Multilineage potential of adult human mesenchymal stem cells. Science 284(5411):143–147
26. Khan IM et al (2011) Fibroblast growth factor 2 and transforming growth factor beta1 induce

precocious maturation of articular cartilage. Arthritis Rheum 63(11):3417–3427

27. Morgan BJ et al (2020) Bone morphogenetic protein-9 is a potent chondrogenic and morphogenic factor for articular cartilage chondroprogenitors. Stem Cells Dev 29(14):882–894
28. Brown RA, ProQuest (2013) Extreme tissue engineering: concepts and strategies for tissue fabrication. Wiley, Hoboken. 1 online resource (270p)
29. Ramos-Ibeas P et al (2017) Pyruvate antioxidant roles in human fibroblasts and embryonic stem cells. Mol Cell Biochem 429(1–2): 137–150

Chapter 5

Physioxic Culture of Chondrogenic Cells

Girish Pattappa, Brandon D. Markway, Denitsa Docheva, and Brian Johnstone

Abstract

Cartilage resides under a low oxygen tension within articulating joints. The oxygen tension within cartilage of the knee joint has been measured to be between 2% and 5% oxygen. Although the literature has historically termed this level of oxygen as hypoxia, particularly when doing experiments in vitro in this range, this is actually the physiological oxygen tension experienced in vivo and is more accurately termed physioxia. In general, culture of chondrogenic cells under physioxia has demonstrated a donor-dependent beneficial effect on chondrogenesis, with an upregulation in cartilage genes (SOX9, COL2A1, ACAN) and matrix deposition (sulfated glycosaminoglycans (sGAGs), collagen II). Physioxia also reduces the expression of hypertrophic markers (COL10A1, MMP13). This chapter will outline the methods for the expansion and differentiation of chondrogenic cells under physioxia using oxygen-controlled incubators and glove box environments, with the typical assays used for qualitative and quantitative assessment of chondrogenesis.

Key words Physioxia, Hypoxia, Tissue culture, Chondrocytes, Mesenchymal stem cells, Pellet

1 Introduction

In vivo, physiological oxygen tension (physioxia) within human articular cartilage ranges from 2% to 5%, while bone marrow physioxia has been measured at 7% oxygen [1–4]. Thus, the atmospheric oxygen level (20% oxygen) at which most typical tissue culture incubators operate is actually nonphysiological or hyperoxic for chondrogenic cells. To produce stable cartilage from chondrogenic cells, including differentiated chondrocytes, scientists now routinely use environmental stimuli relevant to articular cartilage, including culture at physioxia, for both proliferation and differentiation with cartilage matrix production. The transcription factors (hypoxia-inducible factors or HIFs) that control cellular responses to changes in oxygen level respond very rapidly to increases in oxygen level, meaning that consistent control of oxygen is important for maintaining the cells and for analysis of gene and protein

Martin J. Stoddart et al. (eds.), *Cartilage Tissue Engineering*,
Methods in Molecular Biology, vol. 2598, https://doi.org/10.1007/978-1-0716-2839-3_5,

expression. The development of oxygen-controlled incubators and glove box chambers that enable media changes and subculture to be performed under constant physiological oxygen tension has made such techniques practical with limited perturbation to oxygen levels.

Chondrocytes redifferentiated in 3D (as aggregates, pellets, or in biomaterials) after monolayer expansion have significantly higher gene expression of chondrogenic genes (e.g., ACAN, COL2A1), which leads to greater extracellular matrix deposition. If this is done under physioxia, chondrocytes from both healthy and osteoarthritic donors produce even more matrix [5–7]. Cells with chondrogenic potential, such as MSCs, expanded under physioxia have increased colony-forming units (CFU-Fs), and this subsequently leads to shorter population doubling time during expansion [8–12]. Moreover, chondrogenic differentiation of MSCs under physioxia has been demonstrated to increase cartilage matrix deposition (e.g., sGAG and collagen II) and upregulate cartilage matrix genes compared to hyperoxic cultures, even in cytokine-supplemented media used to simulate a degenerative joint environment [8, 13–25]. However, there is donor variation in this chondrogenic response under physioxia [14, 26].

In addition to the more rapid expansion and greater matrix production by chondrogenic cells in physioxic culture, in vitro models of chondrogenesis indicate that physioxia induces a more stable hyaline cartilage phenotype, downregulating hypertrophy-related genes. MSCs respond to physioxia with lowered hypertrophy-related gene expression and matrix production, but those genes remain higher compared to chondrocytes by comparison [14, 19, 20, 25, 27]. Furthermore, the use of physioxia cultured MSCs or chondrogenic implants has been shown to improve osteochondral repair in vivo [12, 18, 21, 28].

In summary, physioxia has been demonstrated to induce an increased anabolic response in both chondrocytes and chondrogenically differentiated MSCs, while suppressing the expression of hypertrophic genes. The following chapter outlines the protocols for physioxic culture of chondrogenic cells and qualitative and quantitative assessment of chondrogenesis comparing physioxic and hyperoxic culture. A summary of the workflow is described in Fig. 1.

2 Materials

2.1 Hypoxia/ Physioxia-Controlled Work Environment

1. Oxygen-controlled incubator.
2. Oxygen-controlled glove box workstation/chamber (BioSpherix, Lacona, USA).

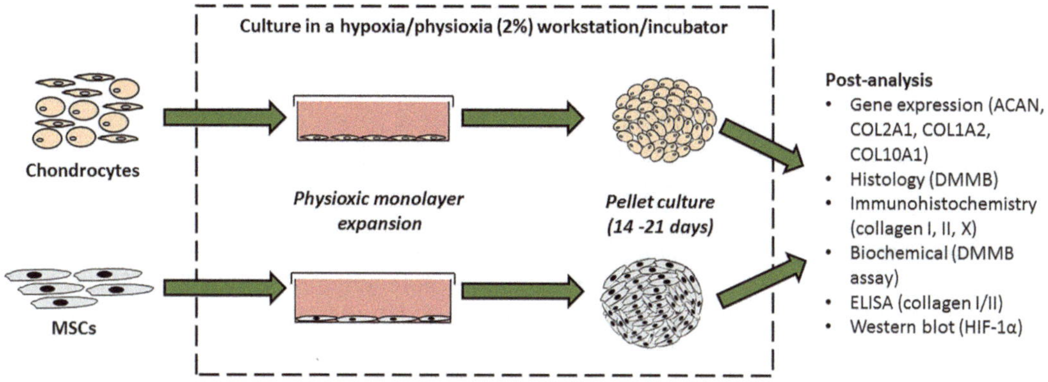

Fig. 1 Summary workflow for the physioxic culture of chondrocytes and bone marrow MSCs with subsequent postanalysis outputs for evaluating chondrogenic cultures

2.2 Chondrocyte Isolation and Culture

1. Cartilage samples—cartilage chips or pieces from operating theater. Transported in serum-free Dulbecco's Modified Eagle's Medium high glucose (4.5 g/L) (DMEM-HG).
2. 15 and 50 mL centrifuge tubes.
3. Pipettes and tissue culture dishes and flasks.
4. Scalpel blades and forceps.
5. Chondrocyte expansion medium: supplement DMEM low glucose (1 g/L) (DMEM-LG) with 10% FBS. Store at 4 °C for 8 weeks.
6. 0.05% trypsin/0.02% EDTA in PBS.
7. Pronase solution: add DMEM-LG + 5% FBS + 1% pen/strep and make to a concentration of 10 mg/mL or 70 U/mL (e.g., 900 mg pronase powder in 90 mL of DMEM-LG). Add appropriate media volume and sterile filter the solution under the biological safety cabinet. Make 10 mL aliquots and freeze at −20 °C.
8. Penicillin/streptomycin (Sigma-Aldrich).
9. Collagenase II solution: add DMEM-LG + 5% FBS + 1% pen/-strep and make to concentration of 1 mg/mL or 200 U/mL (e.g., 100 mg collagenase powder in 75 mL DMEM-LG). Add appropriate media volume and sterile filter the solution under the biological safety cabinet. Make 10 mL aliquots and freeze at −20 °C.
10. Cell strainer (70 μm pores).

2.3 MSC Isolation and Culture

1. Complete culture medium: DMEM-LG with 10% FBS and 1% penicillin/streptomycin. Store at 4 °C for 8 weeks.
2. Percoll gradient: create a 70% Percoll (Sigma, St. Louis, MO) gradient in a 50 mL centrifuge tube by mixing 22.05 mL of Percoll with 2.45 mL of 1.5 M NaCl and 10.5 mL of Tyrode's

solution. The gradient is formed by centrifuging this solution at 20,000 g for 15 min. The formed gradient can be stored at 4 °C for up to 4 weeks.

3. Fibroblast growth factor 2 (basic fibroblast growth factor) (FGF2/bFGF): prepare a 25 μg/mL stock solution (5000×) of FGF2 in sterile H_2O. Store at −80 °C.

2.4 Chondrogenic Differentiation

1. Dexamethasone stock solution (100×): dissolve dexamethasone at 1 mM in absolute ethanol and dilute 1:100 with DMEM-HG to obtain a 10 μM stock solution. Store at −20 °C.
2. Ascorbate-2-phosphate stock solution (100×): prepare a 5 mg/mL of ascorbate-2-phosphate in H_2O. Store at −20 °C.
3. Transforming growth factor beta 1 (TGF-β1): prepare a 2 μg/mL stock solution (200×) of TGF-β1 (R and D systems) in 4 mM hydrochloride acid (HCl) and 1% bovine serum albumin. Store at −80 °C.
4. Chondrogenic differentiation medium: supplement DMEM-HG with 10% ITS+ premix tissue culture supplement, 100 nM dexamethasone, 50 μg/mL of ascorbate-2-phosphate, 1 mM sodium pyruvate, and 10 ng/mL of transforming growth factor beta 1 (TGF-β1).
5. Nunc polypropylene V-bottom 96-well plate.
6. Electronic repeater pipette.
7. Repeater pipette tips.

2.5 Histological and Immunohistological Analysis

1. 4% paraformaldehyde.
2. 0.2 M phosphate buffer: dissolve 0.2 M sodium phosphate dibasic dihydrate ($Na_2HPO_4.2H_2O$) and sodium phosphate monobasic dihydrate ($NaH_2PO_4.\ H_2O$) in distilled water; adjust to pH 7.4.
3. Sucrose solutions (10%, 20%, and 30%)—sucrose solutions are formulated in 0.1 M phosphate buffer.
4. Tissue-Tek solution.
5. Plastic molds for embedding.
6. Cryostat.
7. DAKO pen.
8. Blocking buffer: 0.08 M Tris–HCl, 0.8% NaCl; pH 7.
9. Anti-collagen type II, mouse monoclonal antibody (Sigma, CP18) used diluted 1:100 in blocking buffer.
10. Anti-collagen type I, mouse monoclonal antibody (Sigma, C2456) used diluted 1:50 in blocking buffer.
11. Anti-collagen type X, mouse monoclonal antibody (Thermo-Scientific, X53, 14-9771-82) used diluted 1:300 in blocking buffer.

12. Wash buffer: 0.2 M Tris–HCl, 0.1 M NaCl; pH 7.
13. Fluorescein isothiocyanate (FITC)-conjugated goat anti-mouse immunoglobulin secondary antibody used diluted 1: 200 in wash buffer.
14. McIlvaine or citrate-phosphate buffer (pH 3.6): mix 67.8 mL of 0.1 M acetic acid with 32.2 mL of 0.2 M disodium hydrogen phosphate solution, and adjust to pH 3.6.
15. Pepsin solution: prepare 1 mg/mL solution in McIlvaine buffer.
16. Hyaluronidase solution: prepare 1 mg/mL in PBS.
17. 0.1% DMMB (1,9-dimethylmethylene blue) dye: dissolve 1 g DMMB powder in 1 L of distilled water.
18. 4′,6-Diamidino-2-phenylindole (DAPI): prepare from fresh stock solution (20 mg/mL in distilled water) by diluting 1: 10,000 in PBS.
19. Mowiol 4-88 mounting medium.
20. Ethanol (70%, 96%, 100%).
21. Xylene.
22. Glass coverslips.
23. DEPEX mounting medium.

2.6 DNA and sGAG Assay

1. Papain digestion buffer: dissolve 6 mM L-cysteine and 6 mM EDTA in 50 mL of PBS. Adjust to pH 6. Add 10 U/mL of papain enzyme to the buffered solution.
2. DMMB dye: Dissolve 16 μg/mL of DMMB dye in 5 mL of ethanol, and add 950 mL of distilled water and 30 mM sodium formate. Add 2 mL of formic acid and adjust to pH 3.0.
3. Chondroitin sulfate solution: prepare a 40 mg/mL of chondroitin sulfate A (bovine, Sigma) stock solution in papain digestion buffer.
4. Quant-iT dsDNA assay kit (Invitrogen).
5. Spectrophotometer.
6. 96-well plate.
7. Pipette tips.
8. Micropipette.

2.7 Gene Expression Analysis

1. Liquid nitrogen for snap freezing.
2. Micropestle or mechanical homogenizer (Precelly's homogenizer system, Bertin Instruments).
3. QIAzol reagent.
4. RNeasy Mini Kit (QIAGEN).
5. RNA/DNA nanodrop.

6. Transcriptor first-strand cDNA synthesis kit (Roche).
7. PCR primers for genes of interest (e.g., SOX9, ACAN, COL2A1) and housekeeping genes (e.g., PSMB4, RPL13a).
8. Real-time PCR system.

3 Methods

3.1 Isolation and Seeding of Human Chondrocytes

Traditionally, chondrocytes have been cultured in a hyperoxic incubator (37 °C/5% CO_2) following isolation from cartilage samples [6]. However, isolated chondrocytes can also be cultured in a physioxia incubator set at 37 °C/5% CO_2/2% O_2 as described here.

1. Prepare biological safety cabinet for aseptic work.
2. Measure the weight of a sterile petri dish.
3. Bring tube containing cartilage sample into the biological safety cabinet.
4. Carefully remove cartilage samples from tube and place onto petri dish using forceps. Measure the weight of the petri dish to calculate wet weight of cartilage sample.
5. Using a pair of scalpel blades, mince the cartilage tissue into fine pieces. Keep tissue moist using serum-free DMEM-LG during this process.
6. Once the cartilage has been chopped into fine pieces, divide into ~0.5 g portions, and add each to a separate sterile 15 mL centrifuge tube.
7. Add 10 mL of pronase solution to each tube, and then place tube(s) on a roller in a 37 °C incubator or dry oven. Incubate for 45 min.
8. While the sample is being incubated, prepare.
9. Remove cartilage sample from oven following pronase incubation time, and centrifuge tube at 500 × *g* for 5 min.
10. Carefully aspirate the supernatant leaving the partially digested cartilage tissue.
11. Add 10 mL of collagenase solution to the cartilage tissue, and then place tube on a roller or shaker in a 37 °C incubator or dry oven. Incubate overnight at normal oxygen tension (20% oxygen).
12. The following day, prepare the aseptic hood, and place a 70 μm filter on a 50 mL centrifuge tube containing 10 mL chondrocyte expansion medium.
13. Carefully remove collagenase supernatant including cartilage pieces, and place onto 70 μm filter to enable filtration of chondrocytes, and remove undigested tissue. Centrifuge tube at 500 × *g* for 5 min.

14. Aspirate supernatant from the tube, leaving only the chondrocyte pellet at the bottom of the tube. Resuspend the chondrocytes in 5–10 mL chondrocyte expansion medium.
15. Determine the number of chondrocytes with a hemacytometer.
16. Plate the cells in tissue culture flasks at a density of 5×10^3 cells/cm^2 in chondrocyte expansion medium.
17. Place the flasks in the physioxia incubator. Media changes should be conducted twice per week using an oxygen-controlled workstation and oxygen pre-equilibrated chondrocyte expansion media (*see* Subheading 3.3).

3.2 Isolation and Seeding of Human Mesenchymal Stromal Cells (Bone Marrow MSCs)

MSCs are isolated from many other cell types found in bone marrow due to their adherence to tissue culture plastic. The following protocol uses Percoll gradient to isolate the mononucleated cell population within the bone marrow prior to culture, but some groups directly plate bone marrow into tissue culture flasks. In both cases, subsequent media changes and subculturing removes unattached cells (e.g., erythrocytes, leukocytes) [29, 30]. The bone marrow MSC isolation and expansion is performed at physioxia in an incubator set at 37 °C/5% CO_2/2% O_2 (*see* **Notes 1** and **2**).

1. Prepare biological safety cabinet for aseptic work.
2. Collect heparin containing bone marrow samples, and place under the biological safety cabinet.
3. In 50 mL centrifuge tubes, add 10 mL of complete medium to each tube.
4. Eject bone marrow (5–10 mL) into medium containing tube, and then resuspend bone marrow sample using a 25 mL pipette.
5. Centrifuge bone marrow sample at $500 \times g$ for 5 min.
6. Carefully remove supernatant above bone marrow pellet leaving approximately 5 mL above the pellet.
7. Carefully pipet the cell suspension into Percoll tube to form a layer on top of the Percoll gradient. (Load no more than 2×10^8 cells/tube [29]).
8. Transfer tube to a centrifuge and spin at $400 \times g$ for 15 min without brake.
9. Remove the top cell layer (≈15 mL), transfer to another sterile 50 mL tube, and resuspend in 50 mL of complete culture media.
10. Centrifuge at $500 \times g$ for 5 min.
11. Aspirate the supernatant and resuspend in 5 mL of complete media. Transfer 200 μL to a 1.5 mL microcentrifuge tube for counting mononuclear cell population.

12. Transfer 50 μL of cell suspension and add 50 μL of complete medium.
13. Add 100 μL of 4% acetic acid to mix for lysis of erythrocytes.
14. Count the number of nucleated cells with a hemacytometer, and calculate the total number of cells in the volume of resuspended bone marrow.
15. Plate cells at a seeding density of 1.6×10^5 cells/cm^2 in complete culture media, and place into physioxia incubator (*see* **Note 1**).
16. Perform first media changes at 5 days postseeding. Supplement complete media with 5 ng/mL of FGF2, and visualize colonies until 70–80% confluence.

3.3 Physioxic Culture of Chondrogenic Cells

Media for cultures expanded or differentiated under physioxia requires appropriate media to be equilibrated to the same oxygen level and medium changes performed in a controlled oxygen glove box/workstation to ensure that the cells experience the same oxygen tension throughout the culture period. Prior to each media change and under a sterile hood, take a high volume cell culture flask (e.g., 175 cm^2 flask) or centrifuge tube, and add an appropriate medium volume to the flask. Place the flasks or tubes into the oxygen-controlled incubator, and leave overnight for oxygen equilibration (*see* **Note 1**). Medium changes for physioxia cultured cells should be performed in a glove box environment set at appropriate oxygen tension (e.g., 2% oxygen) if available (*see* **Notes 1** and **2**). The following protocol is used for general subculture of all chondrogenic cells under physioxia, using a glove box or closed chamber environment if available (*see* **Notes 1** and **3**).

1. During cell expansion phase and for physioxic culture, place a large volume cell culture flask (e.g., 175 cm^2) containing cell culture media (calculated based on the number of flasks), and put in an oxygen-controlled incubator set at the appropriate oxygen tension.
2. For cell subculture, wash cells with an appropriate volume (5 mL for a 175 cm^2 flask) of PBS.
3. Aspirate PBS and then add an appropriate volume of trypsin-EDTA (8 mL for a 175 cm^2 flask).
4. Place culture in the physioxia incubator, and monitor cell detachment over a period of 7 min.
5. Once the majority of cells have detached from the surface, return to glove box environment, and then add an equal volume of culture media.
6. Draw up cell suspension and carefully place into an appropriate centrifuge tube. Using culture medium, wash the flask to remove remaining cells, and add to the cell suspension.

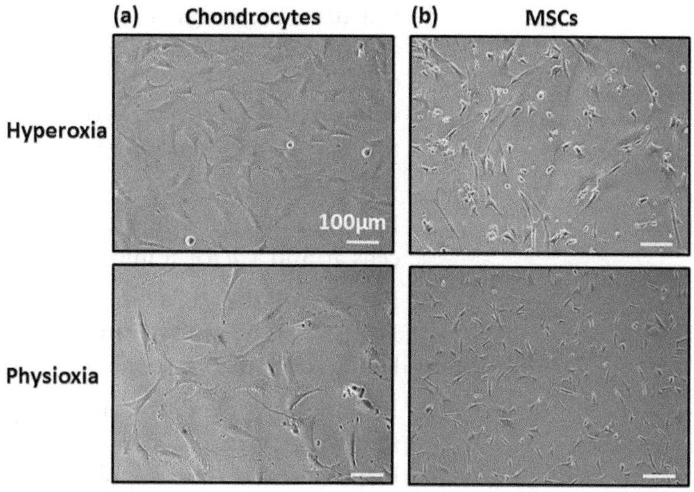

Fig. 2 Representative images of monolayer expanded human (**a**) chondrocytes and (**b**) bone marrow-derived MSCs cultured under hyperoxia and physioxia

7. Spin the tubes on a benchtop centrifuge at 500 × *g* for 5 min.
8. Determine cell number using a hemacytometer.
9. Cells are then seeded into flasks with oxygen-equilibrated cell culture media (*see* above) and returned to the physioxia incubator. For MSCs, cell should be seeded at a density of 2×10^3 cells/cm^2. Chondrocytes should be seeded at a density of 5×10^3 cells/cm^2.

Chondrocytes change from a rounded form in the tissue to a flattened fibroblastic morphology upon culture in monolayer (Fig. 2a). This shift in morphology is accompanied by a phenotypic modulation (dedifferentiation). While the morphological change occurs regardless of oxygen tension, the culture at physioxia slows the phenotypic modulation [31]. In contrast, MSCs form colonies after a week in culture that increase in size over a period of 7–10 days with a fibroblastic morphology shown at the start of culture. Upon expansion, culturing MSCs under physioxia result in smaller cells with shorter doubling time compared with those cultured at hyperoxia (i.e., regular incubator condition of 20% oxygen; Fig. 2b) [9–13]. Under physioxia, MSCs are typically confluent at 8–10 days and chondrocytes at approximately 14 days postseeding using the provided protocols (*see* **Note 4** and Fig. 2).

3.4 Physioxia-Mediated Induction of Chondrogenesis

Induction of chondrogenesis is most effective with cell–cell contacts and a high cell density. A classical model for chondrogenic induction or chondrocyte redifferentiation is the pellet culture model [32, 33]. Studies have also described chondrogenic induction within cell-seeded scaffolds or hydrogels [24, 34, 35] (*see* **Note 5**). The method described here uses pellet cultures in V-bottomed 96-well plates to enable a high-throughput approach [6, 36].

1. Calculate the number of pellets to be created, and formulate appropriate volume of chondrogenic differentiation medium (i.e., 250 μL per 2×10^5 cells/pellet). Place the appropriate volume of serum-free DMEM-HG into the physioxia incubator for oxygen equilibration (24 h prior to starting the experiment).
2. Harvest cells (according to Subheading 3.3) and centrifuge the cells at $500 \times g$ for 5 min.
3. Resuspend the cells in 10 mL serum-free pre-equilibrated DMEM-HG and count the number of cells using a hemacytometer.
4. Prepare physioxia preconditioned chondrogenic media, by removing physioxia conditioned basal media from physioxia incubator and then supplementing with ITS+ premix, dexamethasone, ascorbate, and TGF-β1 as described in Subheading 2.4 (*see* **Note 6**).
5. Centrifuge the cells at $500 \times g$ and then resuspend them at a density of 1×10^6 cells/mL in the physioxia pre-equilibrated chondrogenic differentiation medium.
6. Using a repeater pipette, dispense 250 μL aliquots of cell suspension (2×10^5 cells/well) into the 96-well plate.
7. Spin plates at $500 \times g$ for 5 min.
8. Place plates in the appropriate incubators for 14–21 days.
9. Change chondrogenic differentiation medium every other day in a hypoxia workstation set at incubator oxygen levels with media pre-equilibrated prior to media change.
10. Carefully aspirate media using either a sterile 1000 μL pipette tip or using either a pipette or a glass pipette attached to a vacuum system. Aliquot 250 μL of fresh chondrogenic media to each well.

3.5 Qualitative Evaluation of Chondrogenic Differentiation: Collagen I and II Staining

Using the described protocols, pellets are collected on day 14 or day 21 during culture. This section describes a typical analysis of chondrogenic pellets following culture under physioxia. The following protocol is for collagen I and collagen II immunostaining of the pellets (Fig. 3). All steps are at room temperature unless stated [26].

1. Collect chondrogenic pellets and fix in 4% buffered paraformaldehyde for 1 h in glove box/workstation.
2. Aspirate paraformaldehyde solution and replace with phosphate buffer to wash pellets. Incubate for 30 min.
3. For cryosections, apply increasing sucrose concentrations for 1 h each or overnight at 4 °C.

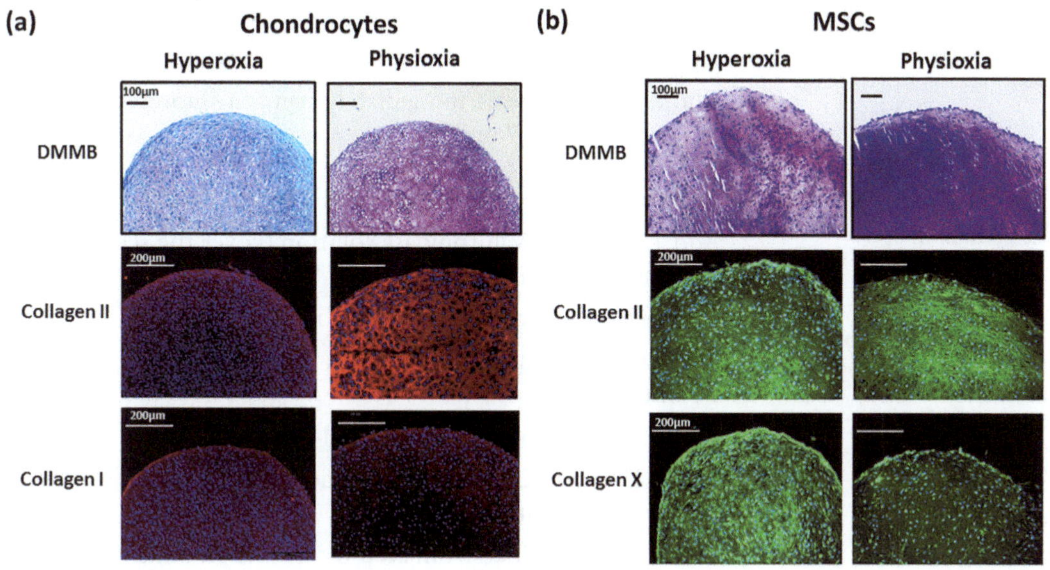

Fig. 3 Representative histological staining for hyperoxic and physioxic (**a**) chondrocyte pellets stained for DMMB, collagen II and I, and (**b**) chondrogenic MSC pellets stained for DMMB, collagen II and X. (**b** image is reproduced from Pattappa et al. [26])

4. Upon reaching 30% sucrose, place pellets in deep plastic molds or equivalent, and then carefully add Tissue-Tek solution. Place either within cryostat or on a copper plate placed on a dry ice to freeze the blocks.
5. Using a cryostat, create 10 μm sections and place in a −20 °C freezer until use.
6. Place sections in a cuvette and add wash buffer. Incubate at room temperature for 10 min.
7. Once washed and partially dried, draw a ring using DAKO pen around the section to be stained.
8. Add 1 mg/mL of pepsin-McIlvaine solution (cover the section), and incubate for 15 min.
9. Wash three times in wash buffer for 5 min each.
10. Add blocking buffer to each section and leave for an hour.
11. Place primary antibody solution (collagen I or collagen II) on sections, and incubate overnight (16 h) at room temperature. Negative control sections should be incubated with blocking buffer solution only.
12. Wash sections in wash buffer three times for 5 min each.
13. Add secondary antibody (FITC conjugated anti-mouse diluted 1:200 in wash buffer) and incubate for an hour.
14. Wash three times in wash buffer for 5 min.
15. Counterstain sections with DAPI for 5 min.

16. Wash three times with PBS for 5 min each.
17. Apply a coverslip with a small drop of Mowiol for mounting.
18. Photograph sections immediately using a fluorescent microscope (*see* **Notes 7** and **8**).

3.6 Qualitative Evaluation of Chondrogenic Differentiation: Collagen X Staining

1. Repeat as described in **steps 1** and **2** in Subheading 3.5.
2. Add 1 mg/mL hyaluronidase solution (cover the section), and incubate for 60 min at 37 °C.
3. Wash section three times in wash buffer.
4. Add pepsin solution (cover the section) and incubate for 15 min.
5. Repeat as described in **steps 4** and **5** in Subheading 3.5.
6. Apply collagen X antibody to sections and incubate overnight at room temperature.
7. Proceed as described from **step 8** onwards in Subheading 3.5.

3.7 Qualitative Evaluation of Chondrogenic Differentiation: DMMB Staining

To visualize the proteoglycan content of the pellets, sections can be stained using DMMB solution for assessing extracellular matrix of pellets (*see* **Note 9**). A cartilaginous matrix that has a high proteoglycan content stains purple (the dye binding to glycosaminoglycans causes a metachromatic shift from blue to purple). In contrast, fibrous tissue or undifferentiated areas stain blue. The following protocol is used for DMMB staining of sections (Fig. 3).

1. Place sections in a cuvette and place in distilled water to wash sections.
2. Incubate at room temperature for 10 min.
3. Place sections in DMMB for 2 min.
4. Remove sections and rinse in tap water for 2 min.
5. Dehydrate sections in ascending series of ethanol (70%, 96%, and 100%, 1 min each), and then clear sections in xylene for 1 min.
6. Mount the sections using DEPEX.

3.8 Quantitative Evaluation of Chondrogenic Differentiation

To quantify the amount of sGAGs within the matrix, the DMMB dye (*see* Subheading 2.6) can also be used in a microplate assay. The methodology described below can be used to simultaneously measure both sGAG and DNA content within the pellets (*see* **Notes 10–14**).

1. Transfer individual pellets at predetermined timepoints using either a thin spatula or using a micropipette to preweighed 1.5 mL microcentrifuge tubes.
2. Carefully aspirate excess medium, and then measure the wet weight of the individual pellets. Calculate the pellet wet weight.

3. Pool three pellets together into a tube. If using at a later point, store pellets dry at −20 °C.
4. Add 150 μL of papain digestion buffer to each tube containing pellets.
5. Place tubes either on a hot plate or dry oven set at 60 °C and leave overnight.
6. Briefly vortex samples to ensure the pellets are completely digested and no longer macroscopically visible. Digested samples can be stored at −20 °C prior to analysis.
7. Serially dilute chondroitin sulfate A stock solution in digest buffer without papain to have a standard curve ranging from 80 to 1 μg/mL (*see* **Note 12**).
8. Pipette 25 μL of samples and standards into a 96 well plate (*see* **Note 13**).
9. Add 200 μL of DMMB reagent to all wells using the multipipette.
10. Transfer the plate to a spectrophotometer set at 575 nm wavelength. Read the plate, and check to see if samples are within standard curve.
11. Measure DNA content using an appropriate assay kit (*see* **Note 14**).

In addition to the stated glycosaminoglycan assay, specific analyses for collagen can also be performed (*see* **Notes 10**, **15**, and **16**).

3.9 Gene Expression Analysis

Gene analysis of chondrogenic genes *SOX9*, *COL2A1*, and *ACAN* is routinely measured for chondrogenic preparations, while *COL10A1*, *RUNX2*, and *MMP13* are used to test hypertrophy in these cultures. *COL1A1* is an indicator of chondrogenic progression, since undifferentiated cells produce collagen I and chondrogenically differentiated cells should express collagen II. The methodology described below is for RNA extraction, cDNA synthesis, and gene expression analysis of cultures.

1. Snap freeze 3–6 pellets (depending upon starting cell number) for RNA extraction. For physioxic culture, keep pellets under physioxia until snap frozen.
2. Add 1 mL QIAzol, and then carefully crush the pellets using either a micropestle or mechanical homogenizer.
3. Isolate RNA using QIAzol method according to the manufacturer's instructions [26] (*see* **Note 10**). Purify and remove genomic DNA from samples using RNeasy Mini Kit according to the manufacturer's instructions.
4. Measure RNA purity and quantity using RNA/DNA spectrophotometer, and store at −80 °C for long-term storage.

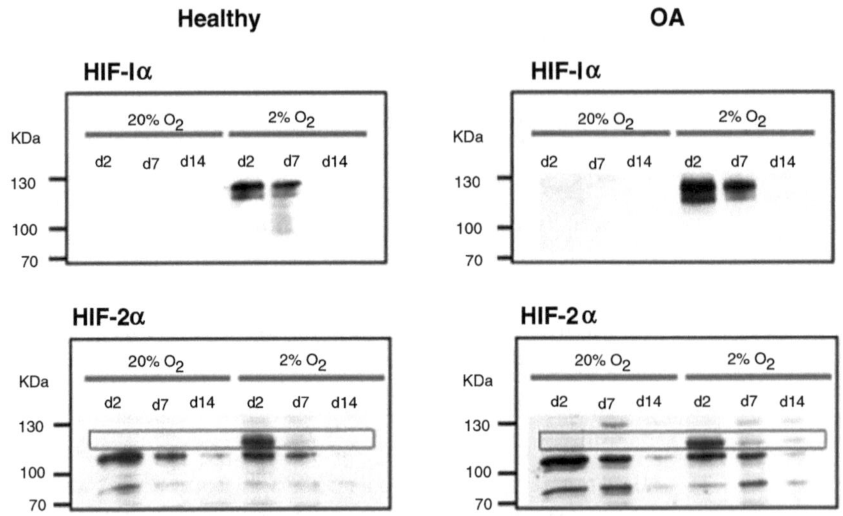

Fig. 4 Western blot image for HIF1α and HIF-2α protein expression from healthy and osteoarthritic human chondrocytes cultured under hyperoxia (20% oxygen) and physioxia (2% oxygen). (Image reproduced from Markway et al. [6])

5. Reverse transcribe 200–500 ng total RNA using an appropriate cDNA synthesis kit according to manufacturer's instructions [14, 26].
6. Evaluate genes using a real-time PCR analyzer (*see* **Note 17**).
7. Calculate the fold change in gene expression using the $\Delta\Delta C_t$ method (*see* **Note 17**).

An additional molecular validation that can be performed to the demonstrate that chondrogenic cells respond to physioxia is through examination of hypoxia inducible factors (HIFs) [6, 25, 37] (*see* **Note 18**). Specifically HIF-1α, HIF-2α, and HIF-3α have all variously been shown to be important to gene regulation for chondrogenic cells and can be studied using either Western blot or gene expression analysis (*see* **Notes 19–22** and Fig. 4) [6, 37].

4 Notes

1. Independent oxygen sensors that measure both media and gaseous oxygen tension can be used to check that sensors on incubator and glove box environment are correctly set at the oxygen concentration measured. Oxygen sensors are available from various companies including BioSpherix (New York, USA) and PreSens (Regensburg, Germany).

2. If there is no access to a hypoxia workstation or glove box chamber, pre-equilibrate medium in incubator and close cap upon removal from incubator. Keep the duration for media changes for physioxic cultures to a minimum to shorten the exposure time to hyperoxia. Additionally, individual components for chondrogenic differentiation should be added to this pre-equilibrated media.
3. Do not allow cells to reach 100% confluence during expansion phase. Especially in the case of MSCs, previous studies have demonstrated that differentiation potential is inhibited.
4. Keep chondrocytes and MSCs to passages below 5, unless the experimental question requires higher passages. Chondrocytes dedifferentiate and senescence is induced with extended passage [7]. Similarly, chondrogenesis is one of the first lineages to be lost in MSCs with extensive monolayer passaging [38].
5. Chondrogenesis within biomaterials (e.g., scaffolds or hydrogels) is more efficient in high cell seeding density. For both chondrocytes and mesenchymal stem cells, 30–60 × 10^6 cells/mL are optimal for chondrogenic processes [24, 34, 35].
6. L-Proline (Sigma) can also be added to the chondrogenic media at a final concentration of 40 μg/mL [14].
7. Fluorescently labeled sections fade with time. Preferably, sections should be imaged on the same day as the staining protocol is finished or sections should be stored at 4 °C in the dark.
8. Chromogenic staining (e.g., DAB staining) can be used instead of using fluorescent-tagged antibodies. Protocol should be adjusted for each staining according to the manufacturer's instructions for the specific detection system. An advantage of this staining method is that slides can be imaged without loss of information for a longer time period compared to fluorescently labeled sections.
9. Alternative histological stains for matrix produced during chondrogenic differentiation are safranin-O/fast green and toluidine blue. Safranin-O stains orange red for proteoglycans in the matrix. Toluidine blue staining demonstrates a similar reaction to DMMB staining, whereby cartilaginous matrix develops metachromasia upon interaction with sulfated proteoglycans.
10. The protocols described are specifically for pellet culture models. If using cell-seeded biomaterials, extraction protocols for sGAG, collagen, RNA, and protein need to be adjusted for the chosen biomaterial, prior to analysis.
11. DMMB stock solution for sGAG assay is light sensitive. Wrap bottle containing stock solution with foil and store in a dark place until use.

12. Stock solution and standards should be prepared fresh each time.
13. sGAG and DNA assay requires sample dilution to be within the standard curve generated from the samples. For the sGAG assay, recommended sample dilution include 1:10, 1:100, and 1:1000.
14. For DNA measurement, PicoGreen assay (Quant-iT dsDNA kit, Invitrogen, Carlsbad, CA) is performed according to the manufacturer's instructions or alternative assays (e.g., Hoechst 33258) can be used [14, 26].
15. Total collagen can be estimated with a hydroxyproline assay. For more specific analysis, commercial ELISAs for collagens I and II (Chondrex) are available. Digestion of the pellets needs to be performed (*see* **Notes 10** and **16**, e.g., digestion protocol) and then ELISA conducted according to the manufacturer's instructions.
16. Collagen I and II ELISA requires extraction using pepsin and elastase enzymes to solubilize the collagens produced in the matrix. Our protocol utilizes 10 mg/mL pepsin in CH3COOH/NaCl (pH 2.9) on shaker for 48 h at 4 °C. Following this step, 1 mg/mL elastase in 0.1 M Tris–HCl + 0.2 M NaCl (pH 7.8–8) is added to pepsin solution and incubated at 4 °C for 24 h; samples are centrifuged and supernatant removed for ELISA analysis. A mechanical homogenization device is used to accelerate the solubilization process [25].
17. There are several housekeeping/reference genes that can be used for evaluation of the stated genes under physioxia. Based on our previous work, *TBP*, *RPL13a*, *HPRT*, and *PSMB4* are housekeeping genes that have demonstrated stable expression under both hyperoxia and physioxia. The geometric mean for two stably expressed genes across conditions and/or time points is used for subsequent ΔC_t calculations.
18. Analysis of the hypoxia inducible factors (HIF-1α and -2α) expression can be evaluated to demonstrate that the cells experienced lowered oxygen if they are being switched from hyperoxic culture.
19. HIFs gene expression can be evaluated by PCR, although due to rapid degradation upon increase in oxygen tension, western blot is the best tool to evaluate HIF expression.
20. Pre-equilibrating wash buffers and rapidly processing lysis buffer under a controlled oxygen environment are recommended for protein extraction. Two efficient methods for protein extraction of cells or pellets are to use an enhanced RIPA buffer to enable to isolation of the different cell fractions, especially the nuclear fraction [6, 39].

21. Commercially available kits for separating cellular components and isolating nuclear proteins can be used for the processing of pellets or cell-seeded scaffolds. An example is the Qproteome nuclear protein kit (37582, QIAGEN) that also contains relevant controls for analysis.
22. Stable HIF expression requires translocation within the nucleus and thus, the nuclear fraction Is required for assessment [29]. For analysis of nuclear proteins in chondrogenic pellets, histone H3 can be used as a control for protein loading [22].

Acknowledgements

This work was supported by the DFG Research network (FOR2407/1): ExCarBon, the European Union Horizon 2020 research and innovation program (EU No.: 814444 (MEFISTO), and a Shriners Hospitals for Children research grant.

References

1. Lafont JE (2010) Lack of oxygen in articular cartilage: consequences for chondrocyte biology. Int J Exp Pathol 91(2):99–106. https://doi.org/10.1111/j.1365-2613.2010.00707.x
2. Brighton CT, Heppenstall RB (1971) Oxygen tension in zones of the epiphyseal plate, the metaphysis and diaphysis. An in vitro and in vivo study in rats and rabbits. J Bone Joint Surg Am 53(4):719–728
3. Lund-Olesen K (1970) Oxygen tension in synovial fluids. Arthritis Rheum 13(6):769–776
4. Grant JL, Smith B (1963) Bone marrow gas tensions, bone marrow blood flow, and erythropoiesis in man. Ann Intern Med 58:801–809
5. Strobel S, Loparic M, Wendt D et al (2010) Anabolic and catabolic responses of human articular chondrocytes to varying oxygen percentages. Arthritis Res Ther 12(2):R34. https://doi.org/10.1186/ar2942
6. Markway BD, Cho H, Johnstone B (2013) Hypoxia promotes redifferentiation and suppresses markers of hypertrophy and degeneration in both healthy and osteoarthritic chondrocytes. Arthritis Res Ther 15(4):R92. https://doi.org/10.1186/ar4272
7. Egli RJ, Bastian JD, Ganz R et al (2008) Hypoxic expansion promotes the chondrogenic potential of articular chondrocytes. J Orthop Res 26(7):977–985. https://doi.org/10.1002/jor.20603
8. Acharya C, Adesida A, Zajac P et al (2012) Enhanced chondrocyte proliferation and mesenchymal stromal cells chondrogenesis in coculture pellets mediate improved cartilage formation. J Cell Physiol 227(1):88–97. https://doi.org/10.1002/jcp.22706
9. Bornes TD, Jomha NM, Mulet-Sierra A et al (2015) Hypoxic culture of bone marrow-derived mesenchymal stromal stem cells differentially enhances in vitro chondrogenesis within cell-seeded collagen and hyaluronic acid porous scaffolds. Stem Cell Res Ther 6:84. https://doi.org/10.1186/s13287-015-0075-4
10. Sheehy EJ, Buckley CT, Kelly DJ (2012) Oxygen tension regulates the osteogenic, chondrogenic and endochondral phenotype of bone marrow derived mesenchymal stem cells. Biochem Biophys Res Commun 417(1):305–310. https://doi.org/10.1016/j.bbrc.2011.11.105
11. Weijers EM, Van Den Broek LJ, Waaijman T et al (2011) The influence of hypoxia and fibrinogen variants on the expansion and differentiation of adipose tissue-derived mesenchymal stem cells. Tissue Eng Part A 17(21–22):2675–2685. https://doi.org/10.1089/ten.tea.2010.0661
12. Pattappa G, Krueckel J, Schewior R et al (2020) Physioxia expanded bone marrow derived mesenchymal stem cells have improved cartilage repair in an early osteoarthritic focal

defect model. Biology (Basel) 9(8):230. https://doi.org/10.3390/biology9080230

13. Pattappa G, Johnstone B, Zellner J et al (2019) The importance of physioxia in mesenchymal stem cell chondrogenesis and the mechanisms controlling its response. Int J Mol Sci 20(3): 484. https://doi.org/10.3390/ijms20030484
14. Anderson DE, Markway BD, Bond D et al (2016) Responses to altered oxygen tension are distinct between human stem cells of high and low chondrogenic capacity. Stem Cell Res Ther 7(1):154. https://doi.org/10.1186/s13287-016-0419-8
15. Bae HC, Park HJ, Wang SY et al (2018) Hypoxic condition enhances chondrogenesis in synovium-derived mesenchymal stem cells. Biomater Res 22:28. https://doi.org/10.1186/s40824-018-0134-x
16. Meyer EG, Buckley CT, Thorpe SD et al (2010) Low oxygen tension is a more potent promoter of chondrogenic differentiation than dynamic compression. J Biomech 43(13): 2516–2523. https://doi.org/10.1016/j.jbiomech.2010.05.020
17. O'HEireamhoin S, Buckley CT, Jones E et al (2013) Recapitulating aspects of the oxygen and substrate environment of the damaged joint milieu for stem cell-based cartilage tissue engineering. Tissue Eng Part C Methods 19(2):117–127. https://doi.org/10.1089/ten.TEC.2012.0142
18. Bornes TD, Adesida AB, Jomha NM (2018) Articular cartilage repair with mesenchymal stem cells after chondrogenic priming: a pilot study. Tissue Eng Part A 24(9–10):761–774. https://doi.org/10.1089/ten.TEA.2017.0235
19. Muller J, Benz K, Ahlers M et al (2011) Hypoxic conditions during expansion culture prime human mesenchymal stromal precursor cells for chondrogenic differentiation in three-dimensional cultures. Cell Transplant 20(10): 1589–1602. https://doi.org/10.3727/096368910X564094
20. Khan WS, Adesida AB, Hardingham TE (2007) Hypoxic conditions increase hypoxia-inducible transcription factor 2alpha and enhance chondrogenesis in stem cells from the infrapatellar fat pad of osteoarthritis patients. Arthritis Res Ther 9(3):R55. https://doi.org/10.1186/ar2211
21. Leijten J, Georgi N, Moreira Teixeira L et al (2014) Metabolic programming of mesenchymal stromal cells by oxygen tension directs chondrogenic cell fate. Proc Natl Acad Sci U S A 111(38):13954–13959. https://doi.org/10.1073/pnas.1410977111
22. Markway BD, Tan GK, Brooke G et al (2010) Enhanced chondrogenic differentiation of human bone marrow-derived mesenchymal stem cells in low oxygen environment micropellet cultures. Cell Transplant 19(1):29–42. https://doi.org/10.3727/096368909X478560
23. Lee HH, Chang CC, Shieh MJ et al (2013) Hypoxia enhances chondrogenesis and prevents terminal differentiation through PI3K/Akt/FoxO dependent anti-apoptotic effect. Sci Rep 3:2683. https://doi.org/10.1038/srep02683
24. Zhu M, Feng Q, Bian L (2014) Differential effect of hypoxia on human mesenchymal stem cell chondrogenesis and hypertrophy in hyaluronic acid hydrogels. Acta Biomater 10(3):1333–1340. https://doi.org/10.1016/j.actbio.2013.12.015
25. Adesida AB, Mulet-Sierra A, Jomha NM (2012) Hypoxia mediated isolation and expansion enhances the chondrogenic capacity of bone marrow mesenchymal stromal cells. Stem Cell Res Ther 3(2):9. https://doi.org/10.1186/scrt100
26. Pattappa G, Schewior R, Hofmeister I et al (2019) Physioxia has a beneficial effect on cartilage matrix production in interleukin-1 beta-inhibited mesenchymal stem cell chondrogenesis. Cell 8(8):936. https://doi.org/10.3390/cells8080936
27. Meretoja VV, Dahlin RL, Wright S et al (2013) The effect of hypoxia on the chondrogenic differentiation of co-cultured articular chondrocytes and mesenchymal stem cells in scaffolds. Biomaterials 34(17):4266–4273. https://doi.org/10.1016/j.biomaterials.2013.02.064
28. Portron S, Merceron C, Gauthier O et al (2013) Effects of in vitro low oxygen tension preconditioning of adipose stromal cells on their in vivo chondrogenic potential: application in cartilage tissue repair. PLoS One 8(4): e62368. https://doi.org/10.1371/journal.pone.0062368
29. Solchaga LA, Penick KJ, Welter JF (2011) Chondrogenic differentiation of bone marrow-derived mesenchymal stem cells: tips and tricks. Methods Mol Biol 698:253–278. https://doi.org/10.1007/978-1-60761-999-4_20
30. Johnstone B, Yoo J (2000) Bone marrow-derived mesenchymal progenitor cells. Methods Mol Biol 137:313–315. https://doi.org/10.1385/1-59259-066-7:313
31. Duval E, Leclercq S, Elissalde JM et al (2009) Hypoxia-inducible factor 1alpha inhibits the fibroblast-like markers type I and type III

collagen during hypoxia-induced chondrocyte redifferentiation: hypoxia not only induces type II collagen and aggrecan, but it also inhibits type I and type III collagen in the hypoxia-inducible factor 1alpha-dependent redifferentiation of chondrocytes. Arthritis Rheum 60(10):3038–3048. https://doi.org/10.1002/art.24851

32. Johnstone B, Hering TM, Caplan AI et al (1998) In vitro chondrogenesis of bone marrow-derived mesenchymal progenitor cells. Exp Cell Res 238(1):265–272. https://doi.org/10.1006/excr.1997.3858
33. Yoo JU, Barthel TS, Nishimura K et al (1998) The chondrogenic potential of human bone-marrow-derived mesenchymal progenitor cells. J Bone Joint Surg Am 80(12):1745–1757
34. Erickson IE, Huang AH, Chung C et al (2009) Differential maturation and structure-function relationships in mesenchymal stem cell- and chondrocyte-seeded hydrogels. Tissue Eng Part A 15(5):1041–1052. https://doi.org/10.1089/ten.tea.2008.0099
35. Huang AH, Stein A, Tuan RS et al (2009) Transient exposure to transforming growth factor beta 3 improves the mechanical properties of mesenchymal stem cell-laden cartilage constructs in a density-dependent manner. Tissue Eng Part A 15(11):3461–3472. https://doi.org/10.1089/ten.TEA.2009.0198
36. Penick KJ, Solchaga LA, Welter JF (2005) High-throughput aggregate culture system to assess the chondrogenic potential of mesenchymal stem cells. BioTechniques 39(5):687–691. https://doi.org/10.2144/000112009
37. Markway BD, Cho H, Zilberman-Rudenko J et al (2015) Hypoxia-inducible factor 3-alpha expression is associated with the stable chondrocyte phenotype. J Orthop Res 33(11):1561–1570. https://doi.org/10.1002/jor.22930
38. Banfi A, Muraglia A, Dozin B et al (2000) Proliferation kinetics and differentiation potential of ex vivo expanded human bone marrow stromal cells: implications for their use in cell therapy. Exp Hematol 28(6):707–715. https://doi.org/10.1016/s0301-472x(00)00160-0
39. Holden P, Horton WA (2009) Crude subcellular fractionation of cultured mammalian cell lines. BMC Res Notes 2:243. https://doi.org/10.1186/1756-0500-2-243

Chapter 6

Isolation and In Vitro Chondrogenic Differentiation of Human Bone Marrow-Derived Mesenchymal Stromal Cells

Angela R. Armiento, Yann D. Ladner, Elena Della Bella, and Martin J. Stoddart

Abstract

Bone marrow-derived mesenchymal stromal cells (BM-MSC) are widely studied in the field of cartilage regeneration due to their capacity to differentiate into chondrocytes under specific in vitro culture conditions. This chapter describes the isolation of MSC from bone marrow aspirate, their expansion in monolayer, and the chondrogenic differentiation in pellet culture.

Key words Bone marrow, MSC, TGF-β1, Chondrogenesis, Pellet culture

1 Introduction

Multipotent mesenchymal stromal cells (MSC) have been defined by the International Society for Cellular Therapy as cells adherent to tissue culture plastic, expressing CD73, CD90, and CD105 surface markers, lacking hematopoietic surface markers and able to differentiate into osteoblasts, chondrocytes, and adipocytes [1].

The isolation, expansion, and trilineage potential of human MSCs have been reported for the first time by Pittenger and colleagues in 1999 [2]. MSCs can be isolated from bone marrow (BM) aspirate as part of the mononuclear cell fraction either using a density separation or a red blood cell lysis buffer. The mononuclear fraction will contain a number of cell types but only MSCs will adhere to plastic and form colonies from which they proliferate in monolayer. To maintain cell proliferation rate and multilineage differentiation potential during in vitro expansion basic fibroblast growth factor (FGF2) is nowadays widely used [3–5]. It is accepted that use of FGF2 induces changes in certain surface makers [6] although the biological significance of these surface markers

Martin J. Stoddart et al. (eds.), *Cartilage Tissue Engineering*,
Methods in Molecular Biology, vol. 2598, https://doi.org/10.1007/978-1-0716-2839-3_6,

remains unclear [7]. Furthermore, it is recognized that MSCs are a heterogenous population of cells with respect to marker distribution and differentiation potential [8, 9].

The cellular steps from undifferentiated MSC to fully differentiated chondrocyte can be studied using an in vitro model of 3D pellet culture in polypropylene centrifuge tubes [10, 11]. The cells are cultured in a chemically defined (serum free) medium containing TGF-β1, dexamethasone, and ascorbic acid for up to 4 weeks. A more time and cost efficient approach is the use of a 96-well plate format allowing for high-throughput screening assays [12].

To date, MSCs can only be defined by in vitro culture conditions and by a minimal panel of cell surface markers, although the research on their elusive nature is constantly evolving [13]. Surface markers offer limited information regarding cell function, although TGFβ receptor profiles have been proposed as a predictive functional marker [14]. Together with their intrinsic cellular heterogeneity, the lack of a universal isolation and in vitro expansion protocol, and challenges in stable differentiation, the clinical translation of MSCs has yet to evolve into an established reality. Here we describe the protocol routinely used in our laboratory to isolate BM-MSCs by gradient centrifugation and plastic adherence and the use of BM-MSCs for chondrogenic differentiation in pellet culture.

2 Materials

2.1 Cell Isolation and Expansion

1. Fully equipped biosafety level 2 (BSL2) laboratory.
2. Human bone marrow aspirate obtained with ethics approval and informed patient consent.
3. 70 μm cell strainers.
4. Histopaque-1077.
5. Sterile phosphate buffered saline (PBS).
6. Penicillin/streptomycin solution (Pen/Strep): 10,000 U/mL of penicillin and 10,000 μg/mL of streptomycin. Aliquot and store at −20 °C.
7. Isolation medium: dissolve 10.08 g of minimal essential medium alpha (α-MEM) and 2.2 g of sodium hydrogen bicarbonate in 1 L of Milli-Q water. Mix by stirring, sterilize using a 0.2 μm filter paper, and store at 4 °C. Just before use, supplement the α-MEM with 10% fetal bovine serum (FBS) and 1% Pen/Strep (*see* **Note 1**).
8. Scepter™ Handheld Cell Counter™ or another automated cell counter.

9. Tissue culture flasks. A growth surface of 150 cm^2 or more is recommended.
10. Basic fibroblast growth factor (bFGF): add 20 mL of Milli-Q water to 1 mg of lyophilized bFGF to obtain a stock of 50 μg/mL. Aliquot and store at −20 °C.
11. Growth medium for MSCs: add 5 ng/mL of bFGF to the isolation medium. Mix by pipetting and use it soon after preparation.
12. Trypsin-EDTA solution: use PBS to dilute to a working solution of 0.05% w/v trypsin and 0.02% w/v EDTA.

2.2 Chondrogenic Differentiation

1. PBS.
2. 0.2 μm syringe filters and filter papers.
3. Pen/Strep: *see* **step 6** in Subheading 2.1.
4. ITS+ premix universal culture supplement: 100× ready-to-use solution containing 625 μg/mL of Insulin, 625 μg/mL of transferrin, 625 ng/mL selenous acid, 125 mg/mL bovine serum albumin, and 535 μg/mL linoleic acid. Aliquot and store at 4 °C.
5. MEM nonessential amino acids (NEAA): 100× ready-to-use solution (*see* **Note 2**). Aliquot and store at 4 °C.
6. L-Ascorbic acid 2-phosphate sesquimagnesium salt hydrate stock (AA2P): dissolve 5 mg/mL of AA2P in Milli-Q water, mix by stirring, and sterilize using a 0.2 μm syringe filter. Aliquot and store at −20 °C (*see* **Note 3**).
7. Dexamethasone–cyclodextrin complex, water-soluble formulation (DEXA): 0.1 mM in Milli-Q water, sterilize it using a 0.2 μm syringe filter. Aliquot and store at −20 °C (*see* **Note 4**).
8. Citric acid: prepare a 10 mM, pH 3 solution in Milli-Q water, and sterilize using a 0.2 μm syringe filter. Store at 4 °C.
9. Bovine serum albumin: prepare a 0.2% solution in 1× PBS and sterilize using a 0.2 μm syringe filter. Store at 4 °C.
10. TGF-β1: add 100 μL of 10 mM citric acid and 900 μL of 0.2% BSA to 10 μg of lyophilized TGF-β1 to obtain a 10 μg/mL stock solution. Aliquot and store at −20 °C.
11. DMEM base medium: dissolve 13.38 g of Dulbecco's modified Eagle's medium (DMEM) containing 4.5 g/L of glucose and L-glutamine and 3.7 g of sodium hydrogen carbonate in 1 L of Milli-Q water. Mix by stirring, sterilize using a 0.2 μm filter, and store at 4 °C.
12. Chondro-permissive medium: to DMEM base medium supplement 1% Pen/Strep solution, 1% ITS, 1% NEAA, 50 μg/mL of AA2P, and 100 nM DEXA (Table 1). Mix by pipetting and use soon after preparation.

Table 1
Composition of chondro-permissive medium

	Stock concentration	Working concentration
DMEM	N/A	N/A
Pen/Strep	10,000 U/mL penicillin 10,000 μg/mL streptomycin	100 U/mL penicillin 100 μg/mL streptomycin
ITS+	100×	1×
NEAA	100×	1×
AA2P	5 mg/mL	50 μg/mL
DEXA	0.1 mM	100 nM

Table 2
Composition of chondrogenic medium

	Stock concentration	Working concentration
DMEM	N/A	N/A
Pen/Strep	10,000 U/mL penicillin 10,000 μg/mL streptomycin	100 U/mL penicillin 100 μg/mL streptomycin
ITS+	100×	1×
NEAA	100×	1×
AA2P	5 mg/mL	50 μg/mL
DEXA	0.1 mM	100 nM
TGF-β1	10 μg/mL	10 ng/mL

13. Chondrogenic medium: to the chondro-permissive medium supplement 10 ng/mL of TGF-β1 (Table 2). Mix by pipetting and use soon after preparation (*see* **Note 5**).
14. V-bottom 96 well plates (*see* **Note 6**).
15. Centrifuge equipped with a rotor for multi well plate.
16. Micro-spoon spatulas.

3 Methods

3.1 Isolation of MSC from Bone Marrow Aspirate

1. Estimate the volume of bone marrow aspirate using a pipette, and transfer it to a 50 mL tube (*see* **Note 7**).
2. Add two volumes of PBS to the bone marrow aspirate.
3. Use a 70 μm cell strainer to filter the solution into a new 50 mL tube.
4. Rinse the cell strainer using 1 volume of PBS.

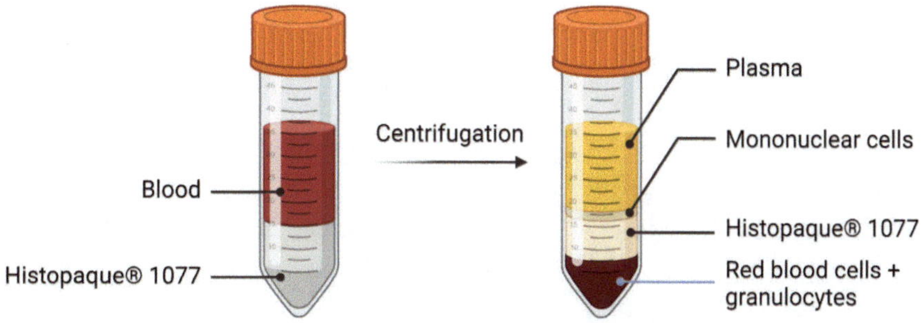

Fig. 1 Illustration of a gradient separation before and after centrifugation. (Created with Biorender.com)

5. Carefully pipette the Histopaque-1077 on the bottom of a 50 mL tube avoiding that the liquid touches the side of the tube. Pipette 2.6 mL per each mL of undiluted bone marrow aspirate, with a maximum volume of 7 mL of undiluted aspirate per each 50 mL tube (*see* **Note 8**).
6. Very carefully layer the diluted bone marrow on top of the Histopaque-1077 (*see* Fig. 1).
7. Centrifuge at 800 × *g* for exactly 20 min at room temperature with acceleration, and brake set at zero or at the minimum allowed from the centrifuge.
8. The mononucleated cells now form an interphase that can be collected using a pipette. Collect 1–2 mm above and below the interphase (*see* Fig. 1).
9. Add 5 mL isolation medium to each mL of collected interphase, and mix gently by pipetting.
10. Centrifuge at 400 × *g* for 15 min at room temperature (standard acceleration/brake settings).
11. Remove the liquid phase and repeat **steps 9** and **10**.
12. Remove the liquid phase and resuspend the pellet in 10–20 mL of isolation medium.
13. Assess the cell number using a Scepter™ equipped with a 40 μm sensor (*see* **Note 9**).
14. Seed the cells at a density of ≈53,000/cm^2 in growth medium (e.g., 16 × 10^6 cells in 300 cm^2 flasks) and place them in incubator (37 °C, 5% CO_2, 95% relative humidity) for 4 days. Avoid moving the flasks during this time.
15. After 4 days, change the medium and repeat the medium change three times per week.
16. When most of the colonies have reached ≈80% confluence, passage the cells (this is usually after 10–14 days in culture; *see* Fig. 2 for images of colonies).

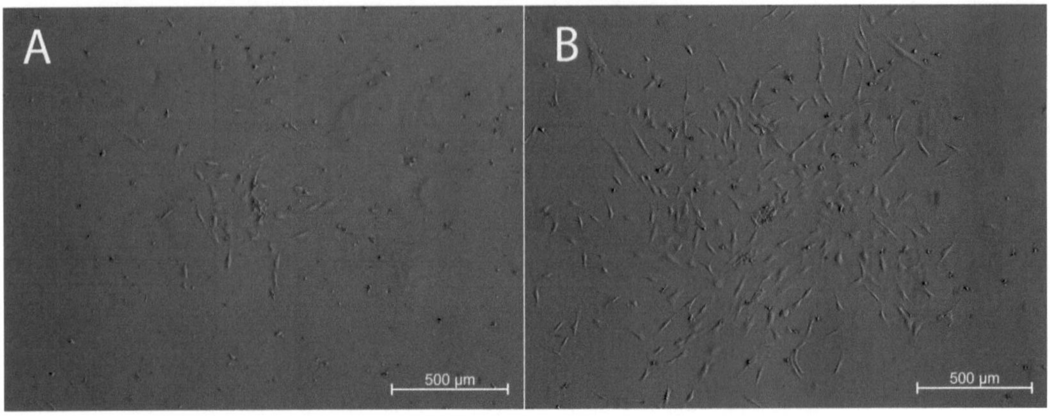

Fig. 2 Brightfield image of a colony formation at the same spot (**a**) 4 days and (**b**) 6 days after MSC isolation from bone marrow aspirate

17. Count the cells with a Scepter™ cell counter equipped with a 60 μm sensor.
18. Seed the cells (now passage 1) at 3,000/cm^2 in growth medium.
19. Monitor the cells regularly and passage them, or freeze them for storage in liquid nitrogen when ≈80% confluence is reached (*see* **Note 10**).

3.2 Chondrogenesis in Pellet Culture

1. Culture human BM-MSCs in tissue culture flasks until they reach 80% confluence.
2. Detach and count the cells.
3. Calculate the number of cells required (2 × 10^5/pellet), and suspend the cells in chondro-permissive or chondrogenic medium at a density of 1 × 10^6 cells/mL.
4. Add 200 μL of cell suspension to each well of a 96-well plate. To avoid the edge effect [15], add PBS to the outer wells of the 96-well plate (red wells in Fig. 3), and seed the cells only in the inner wells (green wells in Fig. 3).
5. Centrifuge for 5 min at 500 × *g*.
6. Place the plate in incubator (37 °C, 5% CO_2), and let a pellet to form overnight.
7. Aspirate the medium, and replace with fresh media (chondro-permissive or chondrogenic) three times per week (*see* **Note 11**).
8. At the desired time point, use a micro-spoon spatula to collect the pellets for qPCR analysis, histology, and sulfated glycosaminoglycans (sGAG) quantification (*see* Fig. 4 for an example of sGAG quantification for TGF-β1-treated pellets and a histological section after 28 days).

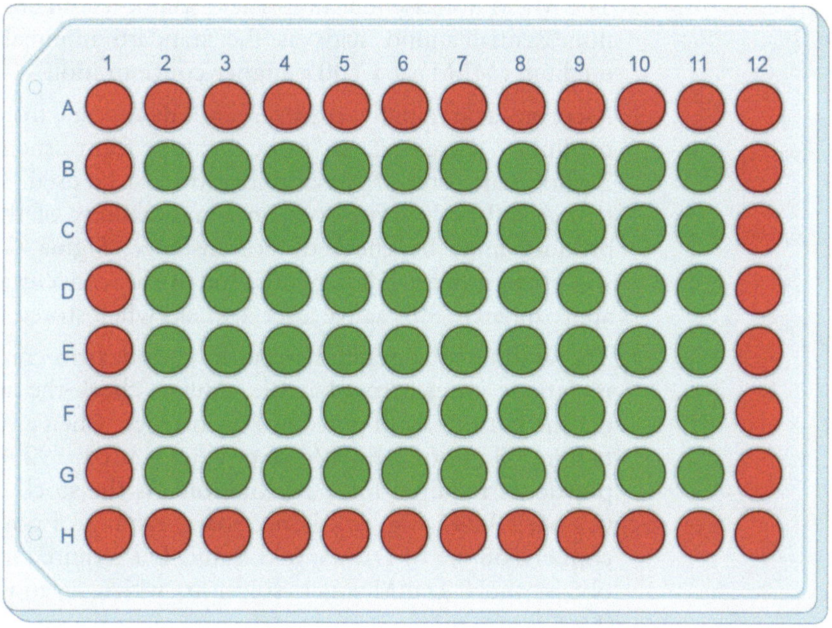

Fig. 3 Seeding layout in a 96-well plate. In green wells containing cells and in red wells containing PBS. (Created with Biorender.com)

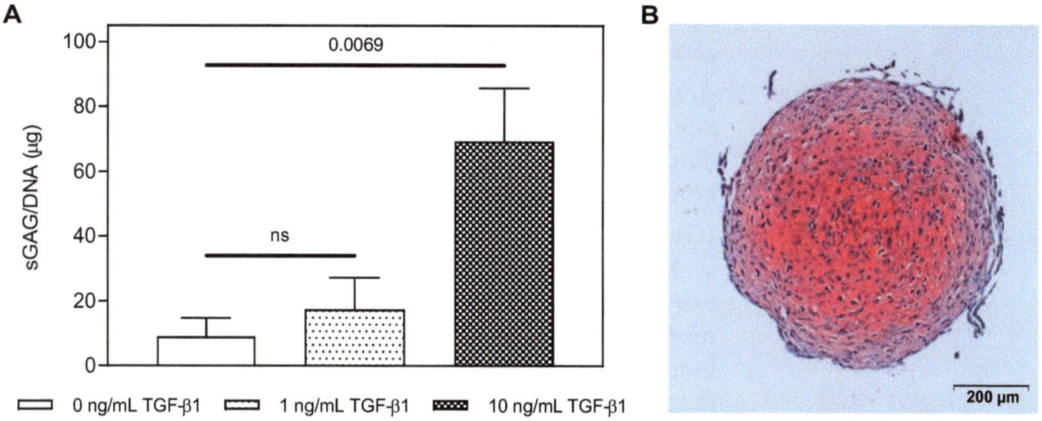

Fig. 4 (**a**) Expected sGAG content in pellets after 28 days of differentiation (figure reproduced with changes from Behrendt et al. [16] © 2019 The Authors) and (**b**) Safranin O/Fast Green staining after 21 days of differentiation

4 Notes

1. Serum represents a great source of variability in MSC expansion and subsequent differentiation. Therefore, it is highly recommended to perform a batch characterization starting from the isolation step to demonstrate capability of the cells to differentiate towards chondrogenic, osteogenic, and adipogenic lineage.

2. The MEM nonessential amino acid solution contains the same nonessential amino acids as the standard minimal essential medium (MEM) at a 100× higher concentration.
3. Ascorbic acid rapidly oxidizes and disappears from culture medium; therefore, a more stable form such as the 2-phosphate ester of L-ascorbic acid must be used. It remains that AA2P hydrolyzes over time and stability of the specific product must be checked. For instance, Sigma CA8960 is soluble in water at 50 mg/mL, and at this concentration, it is approximately 80% active after 4 weeks when stored at 5 °C.
4. The amount of dexamethasone in cyclodextrin-complex formulations varies across batches. Always check the amount of dexamethasone/g of powder. For instance, when a batch containing 65 mg of DEXA/g of powder is used, 392.47 g/L of powder is required for 1 M solution. As the stock should be 0.1 M, 39.247 mg/mL is necessary. As 1 g of powder only contains 65 mg of DEXA, 603.8 mg/L is required in order to obtain the 0.1 mM stock solution, which is equivalent to 0.6038 mg/mL.
5. TGF-β1 is a very sticky protein. Do not filter media containing 10 ng/mL. TGF-β1 as the protein will adsorb to the filter.
6. Use non-tissue-culture-treated plates (e.g., V-bottom 96 well plate—Thermo Fisher plate #612V96/lid #642000 or Ratio-Lab plate #6018313/lid #6018401).
7. A maximum volume of 12 mL of undiluted bone marrow aspirate can be added to a 50 mL tube as this volume will be diluted in PBS. Use multiple tubes in case larger volumes of bone marrow aspirates are to be processed.
8. If the bone marrow aspirate is less than 2 mL, use a 15 mL tube as this will increase the depth of the interphase making the cell extraction easier. If the aspirate is more than 2 mL, use one or more 50 mL tubes, depending on the volume of the original bone marrow aspirate (*see* **Note 6**).
9. Generally, a 1:20 or 1:50 dilution in PBS gives an appropriate concentration for cell counting. Gate the read out to exclude anything smaller than 8 μm to obtain the correct cell number for seeding.
10. It is imperative cells do not reach confluency. This leads to a loss of multipotency.
11. Pellets are generally maintained in culture up to 4 weeks.

References

1. Dominici M, Le Blanc K, Mueller I et al (2006) Minimal criteria for defining multipotent mesenchymal stromal cells. The International Society for Cellular Therapy position statement. Cytotherapy 8(4):315–317. https://doi.org/10.1080/14653240600855905
2. Pittenger MF, Mackay AM, Beck SC et al (1999) Multilineage potential of adult human mesenchymal stem cells. Science 284(5411): 143–147
3. Martin I, Muraglia A, Campanile G et al (1997) Fibroblast growth factor-2 supports ex vivo expansion and maintenance of osteogenic precursors from human bone marrow. Endocrinology 138(10):4456–4462. https://doi.org/10.1210/endo.138.10.5425
4. Bianchi G, Banfi A, Mastrogiacomo M et al (2003) Ex vivo enrichment of mesenchymal cell progenitors by fibroblast growth factor 2. Exp Cell Res 287(1):98–105. https://doi.org/10.1016/s0014-4827(03)00138-1
5. Tsutsumi S, Shimazu A, Miyazaki K et al (2001) Retention of multilineage differentiation potential of mesenchymal cells during proliferation in response to FGF. Biochem Biophys Res Commun 288(2):413–419. https://doi.org/10.1006/bbrc.2001.5777
6. Hagmann S, Moradi B, Frank S et al (2013) FGF-2 addition during expansion of human bone marrow-derived stromal cells alters MSC surface marker distribution and chondrogenic differentiation potential. Cell Prolif 46(4): 396–407. https://doi.org/10.1111/cpr.12046
7. Barry FP, Murphy JM, O'Brien T et al (2005) Mesenchymal stem cell transplantation for tissue repair. Semin Plast Surg 19(3):229–239. https://doi.org/10.1055/s-2005-919718
8. Buhring HJ, Treml S, Cerabona F et al (2009) Phenotypic characterization of distinct human bone marrow-derived MSC subsets. Ann N Y Acad Sci 1176:124–134. https://doi.org/10.1111/j.1749-6632.2009.04564.x
9. Torensma R, Prins HJ, Schrama E et al (2013) The impact of cell source, culture methodology, culture location, and individual donors on gene expression profiles of bone marrow-derived and adipose-derived stromal cells. Stem Cells Dev 22(7):1086–1096. https://doi.org/10.1089/scd.2012.0384
10. Johnstone B, Hering TM, Caplan AI et al (1998) In vitro chondrogenesis of bone marrow-derived mesenchymal progenitor cells. Exp Cell Res 238(1):265–272. https://doi.org/10.1006/excr.1997.3858
11. Yoo JU, Barthel TS, Nishimura K et al (1998) The chondrogenic potential of human bone-marrow-derived mesenchymal progenitor cells. J Bone Joint Surg Am 80(12):1745–1757. https://doi.org/10.2106/00004623-199812000-00004
12. Penick KJ, Solchaga LA, Welter JF (2005) High-throughput aggregate culture system to assess the chondrogenic potential of mesenchymal stem cells. BioTechniques 39(5):687–691. https://doi.org/10.2144/000112009
13. Nombela-Arrieta C, Ritz J, Silberstein LE (2011) The elusive nature and function of mesenchymal stem cells. Nat Rev Mol Cell Biol 12(2):126–131. https://doi.org/10.1038/nrm3049
14. Rothweiler R, Basoli V, Duttenhoefer F et al (2020) Predicting and promoting human bone marrow MSC chondrogenesis by way of TGFbeta receptor profiles: toward personalized medicine. Front Bioeng Biotechnol 8: 618. https://doi.org/10.3389/fbioe.2020.00618
15. Mansoury M, Hamed M, Karmustaji R et al (2021) The edge effect: a global problem. The trouble with culturing cells in 96-well plates. Biochem Biophys Rep 26:100987. https://doi.org/10.1016/j.bbrep.2021.100987
16. Behrendt P, Ladner Y, Stoddart MJ et al (2020) Articular joint-simulating mechanical load activates endogenous TGF-beta in a highly cellularized bioadhesive hydrogel for cartilage repair. Am J Sports Med 48(1): 210–221. https://doi.org/10.1177/0363546519887909

Chapter 7

Adipose-Derived Stromal Cells: Isolation, Expansion, and Differentiation

Justyna M. Buchert, Benedict Lotz, Solvig Diederichs, and Wiltrud Richter

Abstract

Adipose-derived stromal cells (ASC) are a promising alternative cell source to chondrocytes as well as to bone marrow-derived mesenchymal stromal cells (BMSC) in cartilage tissue engineering and repair. Here we describe ASC isolation from liposuction by-products by collagenase-based tissue digestion combined with cell filtration and followed by monolayer attachment and expansion culture. Quality control requires confirmation of correct surface marker expression and multilineage differentiation potential by a trilineage differentiation assay.

Key words Adipose-derived stromal cells, Adipose tissue, Chondrogenesis, MSC, Cartilage

1 Introduction

As cartilage lacks intrinsic regenerative capacity, autologous chondrocyte transplantation is a current gold standard for cell-based tissue repair. Due to the low number of chondrocytes in articular cartilage as well as the necessity for an additional invasive procedure for cell harvest, alternatives are intensively explored.

Mesenchymal stromal cells (MSC) are a promising cell source for the repair of cartilage tissue [1] which are suitable for autologous transplantation and are known to possess the potential to differentiate into chondrogenic progenitor cells [2–5]. Bone marrow and adipose tissue are two major sites in the human body from which stromal cells can be isolated [4, 6].

Adipose-derived stromal cells (ASC) have first been described in 1976 [7]. In 2004 the International Fat Applied Technology Society (IFATS) reached a consensus to predominantly use the term ASC [8]. According to the criteria of the IFATS, plastic adherent cells within the stromal vascular fraction (SVF) are termed ASC [9]. In contrast to bone marrow, adipose tissue is much more abundant and can be harvested less invasively, and ASC have been

Martin J. Stoddart et al. (eds.), *Cartilage Tissue Engineering*,
Methods in Molecular Biology, vol. 2598, https://doi.org/10.1007/978-1-0716-2839-3_7,

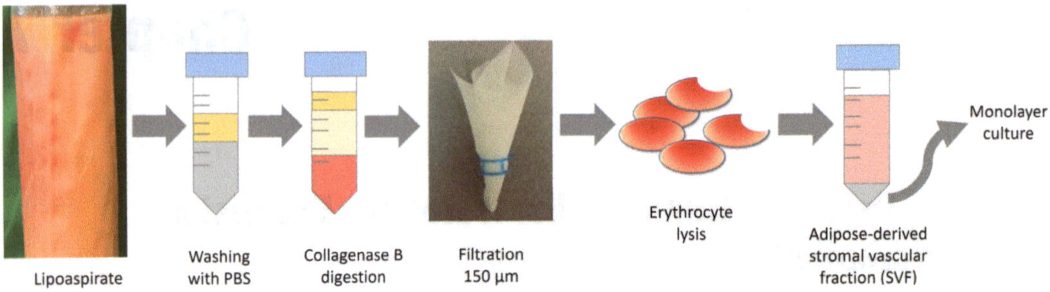

Fig. 1 Scheme of ASC isolation procedure. Lipoaspirates are obtained as by-products of liposuction operations and can be readily used for ASC isolation. After washing, tissue is digested with collagenase and filtered and erythrocytes are lysed. Adherent cells from adipose-derived stromal vascular fraction (SVF) are expanded in monolayer culture (for details *see* Subheading 3.1)

also shown to possess high proliferative capacity and delayed senescence compared to bone marrow-derived MSC (BMSC) [4, 6, 10]. Liposuction by-products can be used to obtain large amounts of ASC that can be further expanded in culture (Fig. 1). About 300-fold more stromal cells can be obtained from a lipoaspirate compared to an equal volume of bone marrow aspirate, which reduces the length of the cell expansion phase [11, 12].

While the isolation principle for both, BMSC and ASC, is different, they are both based on the plastic adherent properties of the cells. For lipoaspirates, collagenase-based digestion is followed by cell filtration through a polypropylene mesh. Next, ASC are enriched by attachment culture and expanded in a culture medium, which is known to enhance progenitor cell proliferation [2, 13]. ASC of passages two to three are known to be a morphologically homogenous population with expression of characteristic MSC markers and to be able to differentiate along the chondrogenic, osteogenic, and adipogenic lineage [5, 6]. Multilineage differentiation capacity is routinely used to confirm the functional progenitor status of cells as quality control. In addition, surface marker analysis is necessary to demonstrate that ASC are CD73, CD90, and CD105 positive and negative for CD34, a typical endothelial cell marker. The expression levels of these surface markers should be comparable with BMSC [2, 14] (Fig. 2). Differences in expression have been observed only in few markers, including CD146, a multipotency marker, as well as CD106, a vascular cell adhesion molecule (VCAM-1), both of which were reported to be higher expressed in BMSC compared to ASC [2, 6, 10, 15].

The current protocol consists of two parts: ASC isolation and enrichment (*see* Subheading 3.1) as well as differentiation of ASC into the chondrogenic (*see* Subheading 3.2), osteogenic, or adipogenic (*see* Subheading 3.3) lineages.

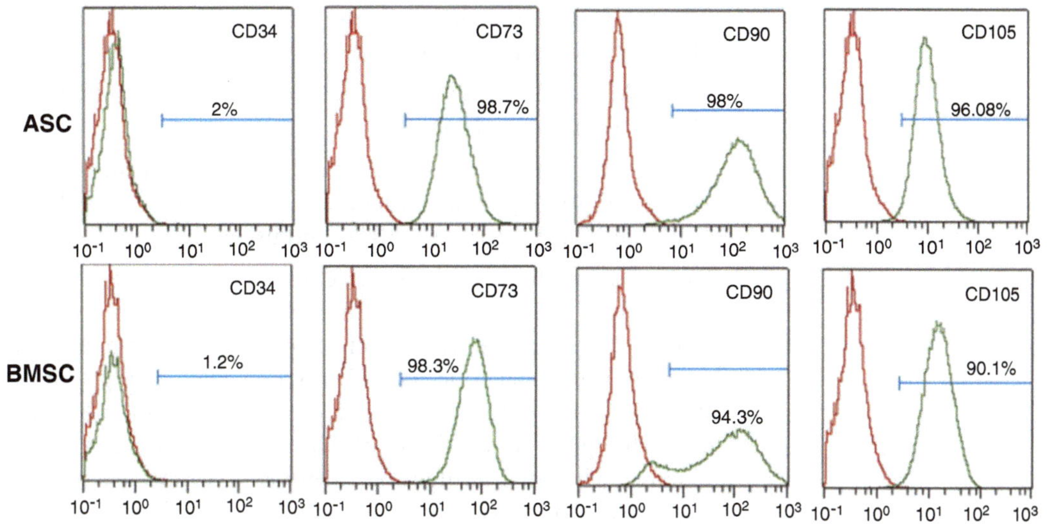

Fig. 2 Flow cytometry analysis of surface markers CD34, CD73, CD90, and CD105 expression in ASC and BMSC. Cells were stained with appropriate antibodies (red) or isotype control antibody (green). Histograms depict representative results

2 Materials

General remark: prepare cell culture medium, perform sterile filtration of buffers, and open autoclaved solutions only under a sterile laminar flow bench. Use only autoclaved ultrapure water to prepare buffers and solutions.

2.1 Tissue

1. Adipose tissue obtained from liposuction: 200–2000 mL lipoaspirate can be stored up to one night at +4 °C.
2. Adipose tissue obtained from subcutaneous tissue: upon harvest, fragments of about 2–15 g are transferred into PBS and stored.

2.2 Media

1. 100 μg/mL sodium selenite stock: dissolve 100 mg sodium selenite in 10 mL of ultrapure water, and then dilute 10 μL in 990 μL ultrapure water. Prepare aliquots of 15, 30, 100, and 200 μL, store at −20 °C, and use within a few months.
2. 10^{-5} M dexamethasone stock: dissolve 100 mg dexamethasone in 25.5 mL of ultrapure water, then dilute 50 μL with 49.95 mL ultrapure water, and prepare 2 mL aliquots.
3. 173 mM ascorbic acid stock: dissolve 2.5 g in 10 mL of ultrapure while mixing (milky suspension), and then add water up to 50 mL in total. Mix until the solution is clear. Prepare aliquots of 300 μL, store at −20 °C, and use within a few months. Use ascorbic acid (L-ascorbic acid-2-phosphate) preparation of at least 99.9% purity.

4. 20 mM indomethacin stock: dissolve 0.032 g in 4.5 mL of 100% ethanol (*see* **Note 1**).
5. 500 mM isobutyl-methylxanthine stock: dissolve 0.066 g in 600 μL DMSO (*see* **Note 2**).
6. 10 mg/mL transferrin stock: dissolve 100 mg transferrin in 10 mL of ultrapure water. Prepare aliquots of 150 μL, 300 μL, 1 mL, and 2 mL, store at −20 °C and use within a few months.
7. Expansion medium: add 75 mL fetal calf serum (FCS; *see* **Note 3**), 6 mL of NEAA (0.01 mM final conc. for each amino acid), 6 mL of L-glutamine (final conc. 2 mM), 6 mL of penicillin/streptomycin (final conc. 100 U/mL penicillin, 100 μg/mL of streptomycin), 600 μL of 2-mercaptoethanol 50 mM (final conc. 50 μM) to 500 mL of DMEM containing 4.5 g/L glucose (without L-glutamine); store at 4 °C and use within a few weeks. Add recombinant human FGF-2 freshly prior to use (final conc. 4 ng/mL; Miltenyi Biotec GmbH, Bergisch Gladbach, Germany; *see* **Note 4**).
8. Basal chondrogenic medium: add 150 μL of 10 mg/mL transferrin (final conc. 5 μg/mL), 15 μL of 100 μg/mL sodium selenite (final conc. 5 ng/mL), 3 mL of 10^{-5} M dexamethasone (final conc. 0.1 μM), 300 μL of 50 mg/mL ascorbic acid (final conc. 0.17 mM), 3 mL of 100 mM sodium pyruvate (final conc. 1 mM), 300 μL of 0.35 M proline (final conc. 0.35 mM), 3.75 mL of 100 mg/mL BSA (final conc. 1.25 mg/mL), and 3 mL penicillin/streptomycin (final conc. 100 U/mL penicillin and 100 μg/mL of streptomycin) to 286 mL of the DMEM containing 4.5 g/L glucose and 3.97 mM L-glutamine. Sterilize using 0.22 μm pore filter, aliquot, store at −20 °C, and use within a few weeks (*see* **Note 5**).
9. Complete chondrogenic medium: reconstitute growth factors according to manufacturer's instructions. Add 55 μL of 3.64 mg/mL insulin (final con. 5 μg/mL; Sanofi-Aventis, Frankfurt am Main, Germany), 40 μL of 10 μg/mL TGFβ-1 (final conc. 10 ng/mL; BD Biosciences GmbH, Heidelberg, Germany), and 40 μL of 10 μg/mL BMP6 (final conc. 10 ng/mL; R&D Systems, Wiesbaden, Germany) to 40 mL of basal chondrogenic medium. Store medium at 4 °C and use within a few days.
10. Osteogenic medium: add 30 mL FCS, 3 mL of 10^{-5} M dexamethasone (final conc. 0.1 μM), 300 μL of 50 mg/mL ascorbic acid (final conc. 0.17 mM), 3 mL of 1 M β-glycerophosphate (final conc. 10 mM) to 261 mL of the DMEM containing 4.5 g/L glucose, and 3.97 mM L-glutamine. Sterilize using 0.22 μm pore filter, aliquot and store at −20 °C, and use within a few weeks.

11. Adipogenic medium: of the DMEM containing 4.5 g/L glucose and 3.97 mM L-glutamine, add 3 mL penicillin/streptomycin (final conc. 100 U/mL penicillin and 100 μg/mL of streptomycin). 3 mL 10^{-4} M dexamethasone (final conc. 1 μM), 3 mL of 20 mM indomethacin (final conc. 0.2 mM), 300 μL of 500 mM isobutyl-methyl-xanthine (final conc. 0.5 mM) to 260 mL. Sterilize using 0.22 μm pore filter, aliquot, store at −20 °C, and use within a few weeks. Add 82.5 μL of 3.64 mg/mL (100 I.E./mL) insulin (final conc. 0.01 mg/mL) for each 30 mL of medium prior to use.

2.3 Buffers

1. Krebs-Ringer buffer: dissolve 0.1 g $MgCl_2 \times 6\, H_2O$ (final conc. 0.5 mM), 0.34 g KCl (final conc. 4.5 mM), 7 g NaCl (final conc. 120 mM), 0.12 g $Na_2HPO_4 \times 2\, H_2O$ (final conc. 0.7 mM), 0.204 g $NaH_2PO_4 \times H_2O$ (final conc. 1.7 mM), 1.8 g D-glucose (final conc. 1 mM), 1.26 g $NaHCO_3$ (final conc. 15 mM), 5.95 g of HEPES (final conc. 25 mM), and 20 g BSA (final conc. 20 mg/mL) in 1 L of ultrapure water (*see* **Note 6**). Adjust pH to 7.4, use 0.22 μm pore filter for sterilizing, store at 4 °C, and use within a few weeks.
2. Erythrocyte lysis buffer: dissolve 8.24 g NH_4Cl (final conc. 154 mM), 1 g $KHCO_3$ (final conc. 10 mM), and 200 μL of 0.5 M EDTA pH 8.0 (final conc. 0.1 mM) in 1 L of ultrapure water, adjust pH to 7.3, and use 0.22 μm pore filter for sterilizing. Store at 4 °C and use within a few weeks.
3. 10× PBS stock: dissolve 80 g NaCl (final conc. 1.37 M), 2 g KCl (final conc. 27 nM), 11.5 g Na_2HPO_4 (final conc. 81 mM), and 2 g KH_2PO_4 (final conc. 15 mM) in 1 L of ultrapure water, adjust pH to 7.4, and autoclave. For ready-to-use PBS, dilute stock PBS in a ratio 1:10 with ultrapure water, adjust pH to 7.4, and autoclave.
4. PBS pH 7.4 supplemented with 20 mg/L BSA: dissolve 20 g of BSA in 1 L of 1× PBS. Sterilize using 0.22 μm pore filter and store at 4 °C, and use within a few weeks.
5. 4% formaldehyde: dissolve 40 g of paraformaldehyde powder in 800 mL of 1× PBS preheated to 60 °C. Solve by constant stirring and dropwise addition of 1 N NaOH until the solution is clear. Next, fill up with PBS to the total volume of 1 L, and adjust pH to 6.9 using small amounts of HCl. Freeze small aliquots at −20 °C, and use within a few months.

2.4 Other Reagents

1. Collagenase type B (15 mg/mL in PBS; from *Clostridium histolyticum*) (Roche Diagnostics GmbH, Mannheim, Germany).
2. Türk's solution (Merck KGaA, Darmstadt, Germany).

3. Gelatin (Merck KGaA, Darmstadt, Germany).
4. Trypan blue solution.
5. Trypsin/EDTA.

2.5 Utensils

1. Disposable scalpel.
2. Sterile pipettes 10, 25, and 50 mL.
3. Sterile pipette tips 20, 100, 200, 1000 μL.
4. 1.5 mL microcentrifuge tubes.
5. Glass bottles.
6. 50 mL polypropylene tubes.
7. Polypropylene membrane (pore size 150 μm; neoLab GmbH, Heidelberg, Germany).
8. T75 and T175 cell culture flasks.
9. 96-well round bottom plates.
10. 25G needle.

2.6 Equipment

1. Centrifuge.
2. 37 °C incubator with platform shaker.
3. Hemocytometer.

3 Methods

If not stated otherwise, perform all the steps at room temperature.

3.1 ASC Isolation

1. Transfer the liposuction aspirate to a glass reagent bottle. Discard the lower blood phase using a 25 or 50 mL serological pipet. Wash adipose tissue twice using an equal volume (v/v) of sterile PBS supplemented with 20 mg/mL BSA and mix well. Let the suspension settle for 3 min on the bench top to separate the phases, then transfer the buffer (bottom phase) to a separate polypropylene tube for centrifugation upon **step 4**, and continue with **step 3** using only the lipoaspirate.
2. If applicable, wash tissue fragments twice using 40 mL PBS with 20 mg/mL BSA. Remove fibrotic tissue and blood vessels with a scalpel. Chop tissue pieces into small fragments of about 10–20 mg using a scalpel.
3. Dilute collagenase type B in Krebs-Ringer buffer in a ratio 1:25 (the final concentration of collagenase should be 0.6 mg/mL), and sterilize using 0.22 μm pore filter. Add equal volume of collagenase solution to the lipoaspirate or adipose tissue (the final concentration of collagenase should be 0.3 mg/mL). Incubate for 2 h at 37 °C on a platform shaker at 160 rpm.

4. Filtrate digested tissue through 150 μm pore polypropylene mesh to remove debris, and transfer the filtrate containing single-cell suspension and erythrocytes to 50 mL polypropylene tubes. Spin down cells of the filtrate as well as of the washing buffer obtained in **step 1** for 10 min at 200 g. Discard the top adipose layer and the supernatant buffer. To remove erythrocytes, resuspend the remaining cell pellet in 20 mL of erythrocyte lysis buffer, and incubate for 3–5 min at room temperature. Next, fill up the tube with PBS until the volume of 50 mL is reached, and pellet the cells by centrifugation for 10 min at 200 g.
5. Wash cells using 10–30 mL of PBS. For cell counting remove a small aliquot of cell suspension (typically 100 μL), mix in a ratio 1:2 with Türk's solution (*see* **Note 7**), and count using a hemocytometer (*see* **Note 8**). Centrifuge cell suspension for 10 min at 200 g, remove the supernatant and resuspend the cell pellet in prewarmed expansion medium (*see* **Note 9**). Seed about 8500 cells/cm^2 (1–2 × 10^6 cells in a T175 flask) in a total volume of 25 mL (T175) or 12.5 mL (T75), respectively, and incubate at 37 °C, 6% CO_2 (*see* **Note 10**). On the next day, aspirate used media, wash cells once with PBS, and add fresh expansion medium. Exchange medium two to three times a week (*see* **Note 11**).
6. Once cells are about 80% confluent, perform cell passaging, aspirate used media. Wash cells twice with PBS. Add about 4 mL of 0.5% trypsin/0.2% EDTA per flask, and incubate at 37 °C, 6% CO_2 up to 10 min until the cells detach. Add 6 mL expansion medium to inhibit trypsin activity. Count cells, centrifuge, remove supernatant, and resuspend the cell pellet in an appropriate volume of expansion medium for seeding of 4.5–5 × 10^3 cells/cm^2. Use cells after passage two or three for differentiation assays.

3.2 Chondrogenic Differentiation via Pellet Culture

1. Harvest cells as for passaging and wash using PBS.
2. Count cells and prepare cell solution of 10^6 cells per 1 mL by resuspending in chondrogenic medium supplemented with TGFβ1 (*see* **Note 12**), BMP6 (*see* **Note 13**) and insulin.
3. Next, transfer 0.5 mL per one 1.5 mL tube, and incubate at 37 °C, 6% CO_2; centrifugation prior to incubation is optional (*see* **Note 14**).
4. Exchange medium above pellets by aspirating 0.5 mL used medium and gently adding 0.5 mL fresh chondrogenic medium three times a week.

5. On day 3–4 after seeding remove medium and transfer pellets into 96-well round bottom plate. Culture cells in a total of 200 μL chondrogenic medium per well and perform media exchange three times a week.
6. Harvest pellets at appropriate time points according to the analysis or freeze at −80 °C until analysis. For histological assessment of cartilage matrix deposition, gently wash pellets with PBS, fix in 4% formaldehyde, section and stain using Safranin O for proteoglycan detection or perform immunohistochemistry of collagen type II.

3.3 Osteogenic and Adipogenic Differentiation of ASC

1. Harvest cells as for passaging, seed 3×10^5 cells per 10 cm^2, in a 6-well or 24-well plate (*see* **Note 15**), and culture in 2 mL of osteogenic or adipogenic induction medium (*see* Subheading 2.2).
2. Culture rest of the cells for 14 days with media exchange twice a week. For control sample, seed cells in expansion medium in a separate plate, and harvest once cells reach confluency.
3. Harvest cells according to the desired analysis method. For histological confirmation of osteogenesis or adipogenesis, gently wash cells with PBS, fix in 4% formaldehyde, and perform Alizarin Red or Oil Red O staining, respectively (*see* **Note 16**).

4 Notes

1. Heat up the indomethacin solution using 37 °C water bath until dissolved completely.
2. Isobutyl-methylxanthine powder is electrostatically charged and therefore needs to be handled with care while weighing.
3. Given the extreme variability of growth factor activity between different companies and batches as well as high variability and undefined composition of FCS, we advise testing multiple growth factors or FCS preparations for their performance and activity, e.g., to support cell expansion and maintenance of ASC differentiation capacity.
4. Adding FGF-2 prior to use and/or to the smaller volumes of medium will help to increase cost-efficiency. Avoid prolonged prewarming of medium once FGF-2 is added.
5. For most cost-efficient use and to reduce contamination risk prepare 40 mL aliquots of basal chondrogenic medium using 50 mL tubes, store at −20 °C, and use within a few weeks. Add insulin and growth factors (TGFβ, BMP6) only after thawing medium, store at 4 °C, and use within a few days.
6. Add BSA stepwise to avoid excessive foaming.

7. Prepare the cell and Türk's solution mixture depending on the macroscopic assessment of cell pellet size upon centrifugation to ensure reliable cell counting (*see* **Note 8**). The ratio may vary from 1:2 to 1:50.

8. Transfer 200 μL of cell suspension into a 0.5 mL tube, transfer 100 μL into a new 0.5 mL tube, and add 200 μL Türk's solution. Apply 10 μL onto the rim of the counting chamber, wait for the fluid to be pulled in by the capillary forces. To ensure reliable cell counting the number of cells counted in four squares should be approximately 100, and the difference between cell numbers in each of the four squares should not exceed ten. The remaining 100 μL of the cell solution can be used to prepare dilution.

9. Cells can also be grown in DMEM medium containing 10% FCS, 100 U/mL penicillin, and 100 μg/mL of streptomycin. We use here described sophisticated expansion medium to increase cell proliferation, which can correlate with successful chondrogenic differentiation outcome.

10. For better growth, cells can be seeded in gelatin-coated flasks. To prepare 0.1% gelatin solution dissolve 1 g sterile gelatin powder in 1 L PBS by heating up in a microwave for a few minutes. Once the solution is clear and cooled down, use approximately 8 mL to cover 175 cm^2 of tissue culture surface for 30 min at room temperature. Treated culture flasks can be stored at 37 °C for a few days. Remove excess of gelatin solution immediately prior to cell seeding.

11. Prior to each medium exchange carefully assess the phenol red indicative coloring in the medium. Increase media exchanges if the media turns yellowish (acidic) suggesting high cell metabolic rate.

12. Generate a number of pellets appropriate for the experiment purposes, i.e., for each time point of RNA or protein analysis a dedicated group of pellets. Note that the longer the pellet culture lasts, the more cells are lost. For instance, for gene expression analysis based on mRNA, prepare therefore pellet excess for harvesting at later time points (day 7 to day 14, 3 pellets; day 21, 4 pellets; day 28, 6 pellets; day 35, 9 pellets; day 42, 9 pellets).

13. In contrast to BMSC, for ASC chondrogenic induction medium requires supplementation with 10 ng/mL BMP6 to support high-quality chondrogenesis [16, 17].

14. To ensure appropriate oxygen and carbon dioxide content the lids of the 1.5 mL microcentrifuge tubes should be perforated using a 25G needle. To support pellet formation, tubes can be centrifuged for no more than 10 min at 200 g.

15. Seed an appropriate number of cells per well and wells for all analytical methods to be performed. We recommend seeding generally approximately 3.5×10^4 cells per 10 cm^2 and two wells in a 6-well plate per time point for RNA isolation and gene expression analysis via qPCR. For histological assessment, we recommend seeding cells in 24-well plate format, in triplicate (three wells per staining per sample or control). To evaluate adipogenesis, samples and control should be stained with Oil Red O staining. For osteogenesis analysis, samples, as well as controls, should be stained using Alizarin Red or von Kossa staining. Control cells can be stained in addition with nuclear Fast Red.
16. Fixed cells can be also covered with PBS containing 0.1% sodium azide, plates secured with paraffin, and stored at 4 °C for up to 2–3 weeks until staining is performed.

References

1. Boeuf S, Richter W (2010) Chondrogenesis of mesenchymal stem cells: role of tissue source and inducing factors. Stem Cell Res Ther 1(4): 31. https://doi.org/10.1186/scrt31
2. Ruetze M, Richter W (2014) Adipose-derived stromal cells for osteoarticular repair: trophic function versus stem cell activity. Expert Rev Mol Med 16:e9. https://doi.org/10.1017/erm.2014.9
3. Estes BT, Diekman BO, Gimble JM et al (2010) Isolation of adipose-derived stem cells and their induction to a chondrogenic phenotype. Nat Protoc 5(7):1294–1311. https://doi.org/10.1038/nprot.2010.81
4. Zuk PA, Zhu M, Mizuno H et al (2001) Multilineage cells from human adipose tissue: implications for cell-based therapies. Tissue Eng 7(2):211–228. https://doi.org/10.1089/107632701300062859
5. Winter A, Breit S, Parsch D et al (2003) Cartilage-like gene expression in differentiated human stem cell spheroids: a comparison of bone marrow-derived and adipose tissue-derived stromal cells. Arthritis Rheum 48(2): 418–429. https://doi.org/10.1002/art.10767
6. Dmitrieva RI, Minullina IR, Bilibina AA et al (2012) Bone marrow- and subcutaneous adipose tissue-derived mesenchymal stem cells: differences and similarities. Cell Cycle 11(2): 377–383. https://doi.org/10.4161/cc.11.2.18858
7. Van RL, Bayliss CE, Roncari DA (1976) Cytological and enzymological characterization of adult human adipocyte precursors in culture. J Clin Invest 58(3):699–704. https://doi.org/10.1172/JCI108516
8. Mitchell JB, McIntosh K, Zvonic S et al (2006) Immunophenotype of human adipose-derived cells: temporal changes in stromal-associated and stem cell-associated markers. Stem Cells 24(2):376–385. https://doi.org/10.1634/stemcells.2005-0234
9. Bourin P, Bunnell BA, Casteilla L et al (2013) Stromal cells from the adipose tissue-derived stromal vascular fraction and culture expanded adipose tissue-derived stromal/stem cells: a joint statement of the International Federation for Adipose Therapeutics and Science (IFATS) and the International Society for Cellular Therapy (ISCT). Cytotherapy 15(6):641–648. https://doi.org/10.1016/j.jcyt.2013.02.006
10. Brocher J, Janicki P, Voltz P et al (2013) Inferior ectopic bone formation of mesenchymal stromal cells from adipose tissue compared to bone marrow: rescue by chondrogenic pre-induction. Stem Cell Res 11(3): 1393–1406. https://doi.org/10.1016/j.scr.2013.07.008
11. Oedayrajsingh-Varma MJ, van Ham SM, Knippenberg M et al (2006) Adipose tissue-derived mesenchymal stem cell yield and growth characteristics are affected by the tissue-harvesting procedure. Cytotherapy 8(2):166–177. https://doi.org/10.1080/14653240600621125
12. Pittenger MF, Mackay AM, Beck SC et al (1999) Multilineage potential of adult human mesenchymal stem cells. Science 284(5411): 143–147. https://doi.org/10.1126/science.284.5411.143
13. Dexheimer V, Frank S, Richter W (2012) Proliferation as a requirement for in vitro chondrogenesis of human mesenchymal stem cells.

Stem Cells Dev 21(12):2160–2169. https://doi.org/10.1089/scd.2011.0670

14. Bothe F, Lotz B, Seebach E et al (2018) Stimulation of calvarial bone healing with human bone marrow stromal cells versus inhibition with adipose-tissue stromal cells on nanostructured beta-TCP-collagen. Acta Biomater 76:135–145. https://doi.org/10.1016/j.actbio.2018.06.026
15. Shafiee A, Seyedjafari E, Soleimani M et al (2011) A comparison between osteogenic differentiation of human unrestricted somatic stem cells and mesenchymal stem cells from bone marrow and adipose tissue. Biotechnol Lett 33(6):1257–1264. https://doi.org/10.1007/s10529-011-0541-8
16. Hennig T, Lorenz H, Thiel A et al (2007) Reduced chondrogenic potential of adipose tissue derived stromal cells correlates with an altered TGFbeta receptor and BMP profile and is overcome by BMP-6. J Cell Physiol 211(3):682–691. https://doi.org/10.1002/jcp.20977
17. Hildner F, Peterbauer A, Wolbank S et al (2010) FGF-2 abolishes the chondrogenic effect of combined BMP-6 and TGF-beta in human adipose derived stem cells. J Biomed Mater Res A 94(3):978–987. https://doi.org/10.1002/jbm.a.32761

Chapter 8

Chondrogenic Differentiation of Human-Induced Pluripotent Stem Cells

Amanda R. Dicks, Nancy Steward, Farshid Guilak, and Chia-Lung Wu

Abstract

The generation of large quantities of genetically defined human chondrocytes remains a critical step for the development of tissue engineering strategies for cartilage regeneration and high-throughput drug screening. This protocol describes chondrogenic differentiation of human-induced pluripotent stem cells (hiPSCs), which can undergo genetic modification and the capacity for extensive cell expansion. The hiPSCs are differentiated in a stepwise manner in monolayer through the mesodermal lineage for 12 days using defined growth factors and small molecules. This is followed by 28 days of chondrogenic differentiation in a 3D pellet culture system using transforming growth factor beta 3 and specific compounds to inhibit off-target differentiation. The 6-week protocol results in hiPSC-derived cartilaginous tissue that can be characterized by histology, immunohistochemistry, and gene expression or enzymatically digested to isolate chondrocyte-like cells. Investigators can use this protocol for experiments including genetic engineering, in vitro disease modeling, or tissue engineering.

Key words Human iPSCs, Chondrogenesis, Stem cells, Tissue-engineered cartilage

1 Introduction

Articular cartilage is the tissue lining the ends of long bones in synovial joints, providing a nearly frictionless surface for joint motion while withstanding millions of cycles of loading per year [1, 2]. The unique mechanical properties of cartilage [3] are due to the composition and structure of the cartilage matrix – predominantly a variety of proteoglycans and collagens as well as hyaluronate and fibronectin [4]. Glycosaminoglycans (GAGs), largely comprising the large aggregating proteoglycan aggrecan (ACAN), make up 4–7% of the tissue [4–6]. Due to their negative charge, GAGs retain water, which composes 65–80% of the tissue weight, contributing to the compressive properties of cartilage [4–6]. Type

Farshid Guilak and Chia-Lung Wu contributed equally with all other contributors.

Martin J. Stoddart et al. (eds.), *Cartilage Tissue Engineering*,
Methods in Molecular Biology, vol. 2598, https://doi.org/10.1007/978-1-0716-2839-3_8,

II collagen, making up approximately 10–20% of cartilage matrix, primarily contributes to the shear and tensile properties of the tissue [4–6].

Chondrocytes, the sole cell type in articular cartilage, are responsible for maintaining the homeostasis of cartilage matrix proteins in response to genetic and environmental signals, such as growth factors or physiologic loading [7, 8]. However, under pathologic, injurious loading, the chondrocytes shift to a degradative and inflammatory phenotype [9]. Since the cartilage is avascular and aneural, it is susceptible to degeneration in an inflammatory environment, while lacking the ability to regenerate, leading to diseases such as arthritis [4, 10].

Arthritis is a family of joint diseases characterized by degradation of the cartilage matrix, which leads to joint space narrowing, along with osteophyte formation, bone remodeling, and synovitis. These progressive changes are associated with pain, swelling, and loss of motion [11, 12]. There is a variety of types of arthritis, affecting approximately 54 million adults in the United States [13], with osteoarthritis (OA) and rheumatoid arthritis (RA) as the most common. OA affects the largest number of people with increasing age, female sex, genetics, joint injury, and obesity as the primary risk factors [11, 14]. RA is an autoimmune disease [12], where the immune system leads to the inflammatory environment and onset of degeneration and symptoms [12]. Unfortunately, current treatment for OA is limited to nonsteroidal, anti-inflammatory drugs for mild-to-moderate cases and joint replacements for severe cases [11]. Disease-modifying antirheumatic drugs have been developed to treat RA; however, they are effective in only a fraction of individuals and have been associated with significant side effects [12]. Therefore, there remains a need for improved therapies for cartilage injury or arthritis. In this regard, cartilage tissue engineering not only provides a potential regenerative therapy for joint diseases but can also provide in vitro model systems for disease modeling and drug development [15–17]. Furthermore, such in vitro models can be used to elucidate mechanisms of and develop treatments for a variety of other cartilage-associated diseases [18–25]. In this regard, the availability of large numbers of genetically defined cells that can be chondrogenically differentiated could serve as a critical component in the development of new cell-based therapies or in vitro disease models for drug screening.

1.1 Development of the Protocol

This protocol was developed based on a series of previous studies that elucidated the sequence of signaling cues required for cartilage development in vivo coupled with various reports of in vitro chondrogenic differentiation of mouse and human pluripotent stem cells [26–29]. In previous studies, we developed a method to derive chondrocytes from hiPSCs in a stepwise manner via the paraxial mesodermal lineage [30] based on the mesodermal roadmap developed by Loh et al. using embryonic stems cells and iPSCs [26]. The next steps toward directed chondrogenic differentiation were induced using bone morphogenic protein 4 (BMP4) and transforming growth factor beta 3 (TGF-β3), given their known roles in precartilaginous mesenchymal condensation and chondrogenic differentiation [29, 31–36]. In this process, hiPSC-derived sclerotome cells were treated with BMP4. The resulting chondroprogenitor cells were then cultured in a conventional 3D pellet system with TGF-β3 to further specify mesodermal cells into the chondrogenic fate. However, despite the robust generation of chondrocytes from hiPSCs using BMP4 and TGF-β3, we and others have shown significant and unpredictable cellular heterogeneity within newly formed cartilaginous tissues [30, 37–40].

To determine the identity of such nonchondrocytic cells, bulk and single-cell RNA sequencing were applied at multiple time points throughout the course of hiPSC chondrogenesis. These studies revealed that the primary off-target cells were of neural and melanocytic lineages [37, 38]. With the analysis of the gene regulatory networks, we identified that *WNTs* and melanocyte inducing transcription factor (*MITF*) are the primary hub genes responsible for off-target neurogenesis and melanogenesis during hiPSC chondrogenic differentiation, respectively. Thus, in the improved protocol, we use small molecules to inhibit Wnt and MITF signaling during chondrogenic pellet culture, significantly enhancing the efficiency and homogeneity of hiPSC chondrogenesis. We further validated our optimized protocol using multiple hiPSC lines, histological and quantitative biochemical analysis of cartilage matrix production, and real-time quantitative polymerase chain reaction (RT-qPCR) and single-cell RNA sequencing techniques to evaluate gene expression (Fig. 1).

1.2 Applications of the Protocol

The hiPSC-derived chondrocytes and tissue-engineered cartilage from this protocol can facilitate the development of patient-specific regenerative approaches for a variety of cartilaginous disorders including, but not limited to, arthritis, osteochondritis dissecans, relapsing polychondritis, chondrocalcinosis, cartilaginous tumors, and arthropathies [41]. Particularly, our protocol can allow for in vitro disease modeling and high-throughput drug screening [15]. Furthermore, in vitro modeling of specific genetic conditions can be established through targeted genome engineering of the cells (e.g., CRISPR-Cas9 genome editing) with isogenic controls or patient-derived iPSCs, in conjunction with simulation of a diseased environment (e.g., inflammatory cytokines) [16, 42, 43]. Furthermore, since the protocol follows the developmental lineage, such models allow investigation of the mechanisms underlying developmental disorders, such as skeletal dysplasias, chondrodysplasias, collagenopathies, and aggrecanopathies [41].

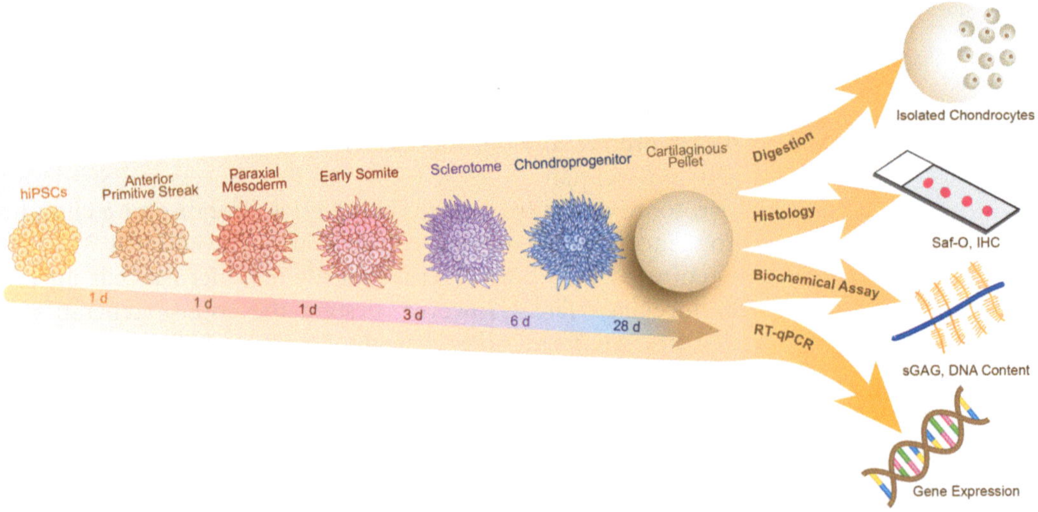

Fig. 1 Overview schematic of the protocol. hiPSCs undergo mesodermal differentiation in monolayer for 12 days. The cells go through the anterior primitive streak, paraxial mesoderm, early somite, sclerotome, and finally chondroprogenitor stage. Cells are then cultured in a 3D pellet culture to become chondrocytes and synthesize cartilaginous matrix. The protocol then has four options to either digest the tissue to isolate single cells or validate chondrogenesis with histology (Saf-O and IHC), biochemical assays (DMMB for sGAG and PicoGreen for dsDNA), and/or RT-qPCR

Additionally, the protocol can be applied to cartilage tissue engineering and other topics in the musculoskeletal field. Studies can be developed to investigate mechanisms that promote or repress chondrogenic differentiation to improve tissue engineering and regenerative strategies. For example, various forms of loading such as osmotic, compressive, and shear forces can be applied at different magnitudes, time points, and regimens to optimize matrix production and mechanical properties [44]. The similarity between hiPSC-derived chondrocytes and primary chondrocytes allows for the study of fundamental questions regarding cell physiology, such as mechanisms of chondrocyte mechanobiology (e.g., role of ion channels, integrins, and other mechanotransduction pathways), the development and physiology of the chondrocyte circadian rhythm, and the cells' responses to metabolic syndrome and inflammation. Furthermore, the rapid expansion of genome engineering in the cartilage field provides the opportunity to apply principals of synthetic biology for the development of "smart," self-regulating cells or mechanogenetic gene circuits in human cells [45–48]. Therefore, the ability to generate a large number of chondrogenically differentiated cells from hiPSCs provides the opportunity for numerous advances in cartilage research, including other types of cartilaginous tissues [49].

1.3 Comparison with Other Methods

To tissue-engineer cartilage, a culture system is needed to differentiate and culture chondrocytes that can synthesize and accumulate cartilaginous matrix, with or without a biomaterial scaffold [6]. A variety of cell types have also been used for this purpose including primary chondrocytes, adult multipotent stem cells, and pluripotent stem cells [6]. While primary chondrocytes can synthesize a cartilaginous matrix in 3D, they are limited in their expansion potential and undergo dedifferentiation with passage in monolayer [50]. Additionally, harvesting of human chondrocytes from a patient results in donor site morbidity, while other sources (e.g., cadaveric, surgical waste) are difficult to obtain [6]. Therefore, adult stem cells, such as bone marrow-derived mesenchymal stem cells [51] and adipose-derived stem cells [52], have been used for cartilage tissue engineering. While these methods have been optimized to successfully produce cartilage-like tissue, adult stem cells represent a heterogeneous population of cells and show significant donor-to-donor variability. Furthermore, they have limited in vitro expansion capacity, making it difficult to perform gene editing and clonal isolation [53–55]. Induced pluripotent stems cells have high proliferation and differentiation capacities, allowing for the study of genetic perturbations. In addition, they can be derived in a patient-specific manner with low or no donor morbidity, and they do not have the ethical concerns associated with embryonic stem cells [42, 43, 56].

Previously published chondrogenesis protocols of stem cells, including ours, have greatly extended our knowledge in chondrocyte biology and cartilage tissue engineering; however, many of these approaches rely on the application of fetal bovine serum (FBS) in culture medium. While it may enhance cell viability, FBS can also lead to off-target differentiation because of its undefined chemical composition. Lot-to-lot variability of FBS may also make the protocols difficult to reproduce in different laboratories [34, 35, 57–61]. Here, we have established a chondrogenic differentiation protocol for hiPSCs, using serum-free, chemically defined medium, that we have validated with 8 hiPSC lines with and without genetic mutations.

1.4 Experimental Design

1.4.1 Cell Source

hiPSC lines derived from foreskin or skin fibroblasts with either retroviral or Sendai viral induction of the Yamanaka factors (i.e., *OCT3/4*, *SOX2*, *KLF4*, and *c-MYC*) [56] have been tested with this protocol. Sendai viral hiPSCs are free from genomic integration of the Yamanaka factors, unlike retrovirally transduced hiPSCs, which have the genes integrated in their genome [62]. The cells are maintained in hiPSC-maintenance medium, avoiding overcrowding and spontaneous differentiation. In some cases, cleaning, either colony picking or scraping of differentiated cells, can be carried out. hiPSC maintenance, expansion, and mesodermal differentiation are performed on vitronectin-coated plates.

1.4.2 Mesodermal Differentiation

Approximately 48 h after passaging, when the hiPSCs have reached 30–40% confluency, the cells are induced for mesodermal differentiation in monolayer. The cells are fed every 24 h with a defined growth factor and small molecule cocktail to guide differentiation. First, basic fibroblast growth factor (FGF) alongside activation of the TGF and Wnt-signaling pathways drive the cells into the anterior primitive streak. Wnt activation and FGF are continued while the TGF/BMP pathway is inhibited on the second day to derive paraxial mesodermal cells. Next, all these pathways (i.e., FGF, Wnt, and TGF/BMP) are inhibited for differentiation into the early somite. Wnt inhibition is continued in combination with sonic hedgehog (SHH) activation for the next three days achieving sclerotome cells. The following 6 days drive the cells into chondroprogenitors by adding BMP4 each day (Fig. 1). Notably, mesodermal differentiation begins along the perimeter of the hiPSC colonies, and cell morphology will change from the small, round, colonized hiPSCs into longer, more fibrotic cells that spread throughout the culture plate.

1.4.3 Chondrogenic Differentiation

After the 12-day mesodermal differentiation in monolayer, chondroprogenitor cells are disassociated with TrypLE into single cells. Cells are resuspended at a concentration of 5×10^5 cells/mL using complete chondrogenic medium containing TGF-β3. These cells are transferred to a 15 mL conical tube and centrifuged to form the 3D chondrogenic pellet culture (Fig. 1). The media is replaced every 3–4 days with complete chondrogenic medium until the time point of interest, typically 28 days after chondrogenic induction.

1.4.4 Validation of Chondrogenic Differentiation

We use five different methods to validate the outcome of the chondrogenic differentiation: isolated single cells (chondrocytes), histology, immunohistochemistry (IHC), biochemical analysis, and RT-qPCR (Fig. 1).

The chondrogenic pellets are digested to obtain hiPSC-derived chondrocytes at the single cell level (Fig. 1). Pellets are washed with DPBS$^{-/-}$ and transferred into a 0.4% (wt/vol) type II collagenase solution in a conical tube. The tubes are vortexed and placed on an orbital shaker in a 37 °C incubator. The tubes are vortexed every 20 min until the pellets are digested. After pellet digestion, the collagenase solution with single cell suspension is neutralized with medium containing 10% FBS. Cells are then resuspended in the appropriate serum-free culture medium for follow-up experiments.

For histological and IHC validation, pellets are fixed in 10% neutral buffered formalin overnight, dehydrated in 70% ethanol at 4 °C, processed following standard histological protocol (including multiple dehydration steps), and embedded in paraffin wax. Wax blocks containing the pellets are cut into 8 μm sections, and sections are placed on microscope slides. Histological slides are stained for the nuclei and proteoglycans with hematoxylin and Safranin-O (Saf-O), respectively. For IHC, slides are rehydrated, prepared for staining with a primary antibody of interest; visualized with a secondary antibody, enzyme conjugation, and a chromogen substrate, and then counterstained for the nucleus. Slides are mounted with paramount and cover slipped for imaging and storage. IHC can be semiquantified using the previously published ImageJ protocol [63].

For a more quantitative analyses and validation of our protocol, we recommend two additional methods. The first is the biochemical assay using dimethylmethylene blue (DMMB) and PicoGreen for quantitative measures of sulfated GAG (sGAG) and DNA contents, respectively [52]. Chondrogenic pellets are washed in DPBS$^{-/-}$, transferred into a 1.7 mL tube containing 200 μL of 125 μg/mL papain, vortexed, and digested at 65 °C overnight to release DNA and sGAG content. The following day, the digested samples are vortexed again and can be frozen until needed. This protocol measures the sGAG content, which can then be normalized to DNA. The second method is to quantify gene expression using RT-qPCR [52, 64]. Each pellet is washed with PBS$^{-/-}$ and transferred into a 2 mL tube to be snap frozen in liquid nitrogen and stored at −80 °C until the RNA isolation.

2 Materials

2.1 hiPSC Culture

1. hiPSC lines: This protocol has been validated with hiPSCs derived from fibroblasts reprogrammed using Sendai virus or retrovirus. It has also been successful with CRISPR-Cas9-edited hiPSCs.
2. Culture dishes: 6-well cell culture plate or T75 or T225 cm^2 cell culture flask.
3. Matrix substrate: 1 μg/mL vitronectin recombinant human protein, truncated (Invitrogen) in PBS. Coat plate at 0.5 μg/cm^2, distribute by rocking plate to ensure surface is coated, and incubate at room temperature 1 h. Plates can be used immediately or stored at 2–8 °C wrapped in plastic film. If stored, warm for 1 h at room temperature before using. Aspirate vitronectin solution and discard before culturing cells.
4. hiPSC culture medium: Essential 8 flex media (E8; Gibco). Prepare according to manufacturer instruction and store at 4 °C for up to 2 weeks. Add 10 μM Y-27632 (STEMCELL Technologies) to E8 medium for the first 24 h after thawing or passaging.
5. Disassociation solution: ReLeSR (STEMCELL Technologies).
6. Cryomedium: PSC cryopreservation kit (Gibco).
7. 2 ml cryotubes.

2.2 Mesodermal Differentiation

1. Differentiation medium (500 mL): 242.5 mL Iscove's Modified Dulbecco's Medium, glutaMAX (IMDM; Gibco), 242.5 mL Ham's F-12 nutrient mix, glutaMAX (F12; Gibco), 5 mL penicillin-streptomycin (P/S; Gibco), 5 mL insulin-transferrin-selenium (ITS+; Gibco), 19.5 μL 1-thioglycerol (Millipore Sigma), 5 mL chemically defined concentrated lipids (Thermo Fisher Scientific). Store at 4 °C for up to 2 weeks.
2. Wash medium (500 mL): 248.5 mL IMDM, 248.5 mL F12, 5 mL P/S. Store at. 4 °C for up to 2 weeks.
3. Anterior primitive streak differentiation medium (day 1): 30 ng/mL activin A protein (R&D Systems), 20 ng/mL recombinant human fibroblast growth factor basic protein (FGF2; R&D Systems), 4 μM CHIR99021 (Reprocell) in warmed differentiation medium (37 °C water bath).
4. Paraxial mesoderm differentiation medium (day 2): 20 ng/mL FGF2, 3 μM CHIR99021, 2 μM SB505124 (Tocris Bioscience), 4 μM dorsomorphin (Reprocell) in warmed differentiation medium.
5. Early somite differentiation medium (day 3): 2 μM SB505124, 4 μM dorsomorphin, 500 nM PD173074 (Tocris Bioscience), 1 μM Wnt-C59 (Cellagen Technologies) in warmed differentiation medium.
6. Sclerotome differentiation medium (days 4–6): 1 μM Wnt-C59, 2 μM purmorphamine (Reprocell) in warmed differentiation medium.
7. Chondroprogenitor differentiation medium (days 7–12): 20 ng/mL recombinant human bone morphogenic protein 4 (BMP4, R&D Systems) in warmed differentiation medium.
8. Disassociation reagent: TrypLE Select enzyme (Gibco).
9. Neutralization medium: 494 mL Dulbecco's Modified Eagle Medium/F12, glutaMAX (DMEM/F12; Gibco), 5 mL fetal bovine serum (FBS; Atlanta Biologicals).
10. Trypan blue stain (Invitrogen).
11. Cell counter.
12. 50 mL conical tubes.
13. Cryomedium (20 mL): 16 mL FBS, 2 mL DMEM, 2 mL dimethylsulfoxide (DMSO).
14. 2 ml cryotubes.

2.3 Chondrogenic Differentiation

1. Dexamethasone (100 μM): Add 19.62 mg dexamethasone powder (Millipore Sigma) to 1 mL absolute ethanol. Transfer 0.8 mL of the ethanol solution to 39.2 mL DMEM/F12 for a 1 mM solution. Transfer 4 mL of the 1 mM solution to 36 mL DMEM/F12 to make a 100 μM solution. Store aliquots at −80 °C up to 1 year.

2. Ascorbic acid solution: 50 mg/mL ascorbic acid (Millipore Sigma) in DMEM/F12. Store aliquots at −80 °C for up to 3 months.
3. Proline: 40 mg/mL proline (Millapore Sigma) in DMEM/F12. Store aliquots at −80 °C for up to 3 months.
4. Chondrogenic medium (500 mL): 483 mL DMEM/F12, 5 mL P/S, 5 mL ITS+, 5 mL Modified Eagle Medium (MEM) with nonessential amino acids (Gibco), 0.5 mL dexamethasone, 0.5 mL 2-Mercaptoethanol (Gibco). Store at 4 °C for up to 2 weeks.
5. Complete chondrogenic medium: 10 ng/mL recombinant human transforming growth factor beta 3 protein (TGF-β3; R&D Systems), 1 μM Wnt-C59, 1 μM ML329 (Cayman Chemical), 0.1% ascorbic acid, 0.1% proline in warmed chondrogenic medium.
6. 15 mL conical tubes.

2.4 Chondrogenic Validation

2.4.1 Digestion of Chondrogenic Pellets

1. DPBS$^{-/-}$.
2. Neutralization medium (500 mL): 494 mL Dulbecco's Modified Eagle Medium/F12, glutaMAX (DMEM/F12; Gibco), 5 mL fetal bovine serum (FBS; Atlanta Biologicals).
3. Digestion medium: 0.4% wt/vol type II collagenase (Worthington Biochemical, activity 225 units/mL) in warmed neutralization medium. Sterile filter and use immediately.
4. 15 mL or 50 mL conical tubes.
5. Orbital shaker.

2.4.2 Histology Preparation

1. DPBS$^{-/-}$.
2. 20 mL glass scintillation vials.
3. 10% neutral buffered formalin.
4. Paraffin.
5. Ethanol, 200 proof.
6. Xylenes.
7. Microscope slides.
8. Microtome.

2.4.3 Safranin-O Staining

1. Safranin-O solution (Millipore Sigma).
2. Harris hematoxylin with glacial acetic acid (Poly Scientific).
3. Ethanol, 200 proof.
4. Xylenes.
5. Differentiation solution (Millipore Sigma).
6. Permount mounting media.
7. Glass cover slips.

2.4.4 Immunohistochemistry

1. Ethanol, 200 proof.
2. Xylenes.
3. COL1A1 antibody (Abcam, cat. no. ab90395).
4. COL2A1 antibody (Iowa Hybridoma Bank, cat. no. II-II6B3-s).
5. COL6A1 antibody (Fitzgerald Industries, cat. no. 70F-CR009X).
6. COL10A1 antibody (Millipore Sigma, cat. no. C7974).
7. Goat anti-mouse antibody (Abcam, cat. no. ab97021).
8. Goat anti-rabbit antibody (Abcam, cat. no. ab6720).
9. Hydrogen peroxide.
10. Methanol.
11. 10% goat serum (Thermo Fisher Scientific).
12. Pepsin (Thermo Fisher Scientific).
13. Proteinase K (Millipore Sigma): 0.5% wt/vol in TE buffer. Prepare fresh.
14. Quenching solution: 3% hydrogen peroxide in methanol.
15. Histostain Plus Kit (Thermo Fisher Scientific).
16. AEC substrate solution (Abcam).
17. Vector hematoxylin QS counterstain (Vector Laboratories).
18. VectaMount AQ aqueous mounting medium (Vector Laboratories).
19. Glass cover slips.
20. Aluminum foil.

2.4.5 Biochemical Analysis Preparation

1. $DPBS^{-/-}$.
2. Papain solution (125 mg/L papain, pH 6.5): Weigh 125 mg of papain (Millipore Sigma), 13.8 g of sodium phosphate, 1.46 g ethylenediaminetetraacetic acid (EDTA), and 0.79 g cysteine hydrochloric acid (HCl). Mix reagents using a stir bar and plate at room temperature with 1 L ultrapure distilled water. Reagents may take 1.5–2 h to dissolve. Adjust pH to 6.5 using approximately 38 mL of 1 N NaOH. Aliquot and store at −20 °C for up to 3 months.
3. 1.7 mL tubes.

2.4.6 RT-qPCR Preparation

1. $DPBS^{-/-}$.
2. 2 ml screw-top tubes.
3. Liquid nitrogen.
4. Primers. See Table 1 for suggested chondrogenic primers.

Table 1
RT-qPCR primers

Target gene	Forward primer (5′-3′)	Reverse primer (5′-3′)
ACAN	CACTTCTGAGTTCGTGGAGG	ACTGGACTCAAAAAGCTGGG
COL1A1	TGTTCAGCTTTGTGGACCTC	TTCTGTACGCAGGTGATTGG
COL2A1	GGCAATAGCAGGTTCACGTA	CTCGATAACAGTCTTGCCCC
COL10A1	CATAAAAGGCCCACTACCCAAC	ACCTTGCTCTCCTCTTACTGC
SOX9	CGTCAACGGCTCCAGCAAGAACAA	GCCGCTTCTCGCTCTCGTTCAGAAGT
TBP	AACCACGGCACTGATTTTCA	ACAGCTCCCCACCATATTCT

3 Methods

3.1 hiPSC Culture

≥ 1 h for coating plate(s), 30 min for plating/passaging, 30 min for feeding, 2–7 days for culture

1. Coat a 6-well plate with vitronectin.
2. Thaw a vial of hiPSCs in a 37 °C water bath.
3. Transfer the cells using sterile serological plastic pipettes into 10 mL of room-temperature hiPSC maintenance medium in a 15 mL conical tube.
4. Centrifuge cells at 200 g for 5 min at 23 °C.
5. Aspirate the supernatant.
6. Add 13 mL hiPSC maintenance medium containing 10 μM of Y-27632 to the cell pellet. Do not pipette up and down; instead gently rock the tube back and forth twice. This will prevent breaking up the colonies into single cells.
7. Aspirate the vitronectin-PBS$^{-/-}$ solution from plate.
8. Add 2 mL of cell solution to each well. Gently tilt the plate in circular and back-and-forth motions to distribute the cells evenly throughout the wells. Failure to do so will cause the cells to cluster in one area, preventing proper proliferation, which may lead to spontaneous differentiation.
9. Incubate the plate at 37 °C.
10. After 24 h, aspirate the medium from the plate and feed with hiPSC maintenance medium without Y-27632.
11. Continue cell culture, feeding every day until cells reach approximately 80% confluency. Cells can be frozen in cryopreservation medium. We typically freeze cells sufficient for one 6-well plate in 1 mL of cryopreservation medium for future passaging. The split ratio may vary depending on hiPSC lines.

12. Before passaging, coat the appropriate number of 6-well plates, T75, and/or T225 flasks with vitronectin.
13. Aspirate medium from cell plate.
14. Wash each well with 2 mL of $PBS^{-/-}$.
15. Add 1.5 mL of ReLeSR for 1 min at RT.
16. Aspirate ReLeSR.
17. Incubate plate for 2 min at 37 °C. If the cells are ≥85% confluent or if there is spontaneous differentiation, the time can be shortened to 1 min.
18. Tape the plate several times and pipette 2 mL of hiPSC maintenance medium onto the bottom of each well, you should see the cells lift off with the medium, and transfer into a conical tube using sterile serological plastic pipettes. Do not pipette up and down.
19. Centrifuge cells at 200 g for 5 min at 23 °C.
20. Aspirate the supernatant.
21. Resuspend cells in hiPSC maintenance medium containing 10 μM of Y-27632. Do not pipette up and down, instead gently rock the tube back and forth twice. This will prevent breaking up the colonies into single cells.
22. Aspirate the vitronectin-$PBS^{-/-}$ solution from plate(s)/flask(s).
23. Add cell solution to plate(s)/flask(s). Gently tilt the plate(s)/flask(s) in a circular and back-and-forth motions to distribute the cells throughout the wells. Failure to do so will cause the cells to cluster, preventing proper proliferation and leading to spontaneous differentiation
24. Check cell density in microscope. If too confluent, remove some of the cell solution and add up volume with maintenance medium containing Y-27632.
25. Culture for at least 48 h until 30–40% confluent. Too high cell density inhibits mesodermal differentiation. Too low cell density prevents adhesion and increases cell death (*see* **Notes 1–3**).

3.2 Mesodermal Differentiation

30 min, 1 h for feeding, 12 days for culture

1. Warm appropriate volume of wash and differentiation medium in 37 °C water bath. For days 2–12: if medium changes from orange to yellow in color after 24 h, suggesting high metabolic activity of the cells, increase differentiation medium volume by 1 mL per well and/or 5 mL per flask.
2. Make the appropriate differentiation medium for the corresponding day by adding the correct growth factors. Day 1: anterior primitive streak; day 2: paraxial mesoderm; day 3: early somite; days 4–6: sclerotome; days 7–12: chondroprogenitor.

3. Aspirate maintenance media from plate(s)/flask(s).
4. Rinse plate(s)/flask(s) with wash medium.
5. Add complete differentiation medium.
6. Incubate plate at 37 °C.
7. Feed the cells every 24 h with mesodermal differentiation medium supplemented with the appropriate growth factor and small molecule cocktails. **Note that more-than-usual cell death may be observed after day 3 due to the inhibition of several major signaling pathways.** When switching to a new set of the growth factors and small molecules (i.e., days 1, 2, 3, 4, and 7), it is critical to feed at the 24 h time points to ensure proper lineage specification. Cells can be harvested after days 1, 2, 3, 6, and 12 if the specific lineage stages are of interest to the experiment (e.g., RT-qPCR) (*see* **Notes 4–7**).
8. On day 13, aspirate medium from cell plate(s)/flask(s).
9. Wash with DPBS$^{-/-}$.
10. Pipette TrypLE disassociation reagent onto plate(s)/flask(s).
    2 mL per well of 6-well plate and 20–25 mL per T225 flask.
11. Incubate for 3 min at 37 °C.
12. Gently tap plate(s)/flask(s) several times to disassociate cells. You should see cells floating in medium (*see* **Note 8**).
13. Add slightly more than an equal volume of neutralization media.
14. Pipette half of the medium into a conical tube.
15. Pipette up and down the other half in the wells/flask(s) twice to lift any remaining cells. Transfer to a conical tube.
16. Rinse the wells/flask(s) twice with neutralization medium and transfer to tubes each time.
17. Centrifuge cell tubes at 300 g for 5 min at 23 °C.
18. Aspirate supernatant.
19. Chondroprogenitor cells can be immediately used for chondrogenesis or be frozen in cryomedium for future chondrogenesis.

3.3 Chondrogenic Differentiation

1–3 h for pelleting, 28 days for differentiation

1. Warm neutralization and chondrogenic medium in 37 °C water bath.
2. Add TGF-β3, ascorbate, proline, Wnt-C59, and ML329 to make complete chondrogenic medium.
3. Resuspend chondroprogenitor cells in complete chondrogenic medium. Combine cells if you have multiple tubes.

4. Stain 10 μL of the cell solution with 10 μL of trypan blue to count cell number on an automated cell counter or hemocytometer.
5. Using the cell count, calculate the volume needed for a concentration of 5×10^5 cells per 1 mL of complete chondrogenic medium (*see* **Note 9**).
6. Pipette the cell solution to make pellet cultures—1 mL per 15 mL conical tube.
7. Centrifuge tubes at 300 g for 5 min at 23 °C.
8. Loosen the caps of the tubes in the biosafety cabinet. The tubes must be loosened to provide oxygen supply to the cells. The cap should however still be screwed on so that it cannot be lifted off to prevent contamination.
9. Incubate the tubes at 37 °C.
10. Check tubes to confirm pellet formation after 24 h (*see* **Note 10**).
11. Feed pellets every 3–4 days.
12. Warm incomplete chondrogenic medium in 37 °C water bath.
13. Add TGF-β3, ascorbate, proline, Wnt-C59, and ML329 to make complete chondrogenic medium.
14. Aspirate medium from the conical tube with a 9" Pasteur pipette.
15. Use at 12 mL stereological pipette to feed 6 tubes at a time with 2 mL per pellet. Ensure lids remain loose (*see* **Notes 11–12**).
16. Chondrogenic pellets can be harvested at various time points as desired. Proceed to Subheading 3.4. We recommend weekly/ bi-weekly harvests on days 7, 14, 28, and 42. In general, cells start to deposit matrix 14 days postchondrogenic induction (i.e., apparent pellet enlargement). Additionally, chondrocytes usually can be observed approximately 14–21 days postchondrogenic induction.
17. 28 days after chondrogenic induction, most of the cells should have differentiated into chondrocytes and formed cartilaginous matrix. We recommend digesting the chondrogenic pellet to retrieve the chondrocytes at a single cell level (Subheading 3.4.1) for further experimentation or using histology/IHC (Subheadings 3.4.2, 3.4.3, and 3.4.4), biochemical assays (Subheading 3.4.5), and/or RT-qPCR (Subheading 3.4.6) to confirm and study chondrogenic differentiation.

3.4 Chondrogenic Validation

3.4.1 Digestion of Chondrogenic Pellets

3 h

1. Warm neutralization and desired medium for the experiment. If plating the cells, we recommend using the neutralization medium.
2. Prepare digestion medium.
3. Aspirate medium from pellets.
4. Wash each pellet with 2–3 mL of DPBS$^{-/-}$.
5. Transfer pellets into a tube with digestion medium. Use 1 mL of digestion medium per pellet being digested (± 1 mL). A 15 mL or 50 mL tube should not exceed 8 mL or 25 mL of digestion medium, respectively.
6. Vortex the tube(s) and shake manually then place on an orbital shaker (80 RPMs) in a 37 °C incubator.
7. Every 20 min remove the tube(s) from the incubator to vortex and check digestion progress.
8. After approximately 2 h, the matrix of chondrogenic pellets should be mostly digested. The length of digestion time needed depends on harvest time point and size of the tube. For example, pellets harvested prior to 28 days postchondrogenic induction may only require 1 h to achieve full digestion. Additionally, we have observed faster digestion using 50 mL conical tubes (*see* **Note 13**).
9. Add neutralization medium to the tube(s). 4 mL for a 15 mL conical tube or 20 mL for a 50 mL conical tube.
10. Centrifuge tube(s) at 300 g for 5 min at 23 °C.
11. Aspirate supernatant.
12. Resuspend in neutralization medium.
13. Stain 10 μL of the cell solution with 10 μL of trypan blue to count on an automated cell counter or hemocytometer.
14. Centrifuge at 300 g for 5 min at 23 °C.
15. Resuspend the cells at the desired concentration. The number of cells retrieved per pellet may depend on hiPSC lines. On average, we get approximately 6.5×10^5 cells per pellet. The cell recovery rate may be increased if multiple pellets from the same group are pooled and digested together. If plating the chondrocytes, we recommend adding the cells to desired dish(es) and incubating for 6–8 h. This provides sufficient time for the cells to adhere without dedifferentiating and losing their phenotype.

3.4.2 Histology Preparation

At least 3 days

1. Transfer pellet(s) into a scintillation vial with 10% (vol/vol) neutral buffered formalin.
2. Store at 4 °C overnight.
3. Remove formalin and add 70% (vol/vol) ethanol. Pellets can be stored in 70% (vol/vol) ethanol at 4 °C long term.
4. Transfer pellet into a plastic cassette for processing. Biopsy foam pads can be used to sandwich the chondrogenic pellet within the cassette to prevent it from falling out. Processing can be done by hand or using a tissue processor.
5. Dehydrate the chondrogenic pellet for 30 min in 80% (vol/vol) ethanol followed by 30 min in 100% (vol/vol) ethanol. Exchange the 100% (vol/vol) ethanol for another 30 min. Pellets can be stored in 100% (vol/vol) ethanol overnight.
6. Clear the pellet for 30 min with a 1:1 solution of ethanol and xylene followed by 30 min of 100% (vol/vol) xylene. Exchange the 100% (vol/vol) xylene for another 30 min.
7. Begin embedding the pellet for 1 h with a 1:1 solution of xylene and paraffin wax at 60 °C followed by 1 h with 100% (vol/vol) paraffin at 60 °C. Exchange the 100% (vol/vol) paraffin for another 1 h at 60 °C.
8. Transfer pellet into an embedding tray, place cassette on top, and fill with paraffin wax.
9. Store for a few hours or overnight at 4 °C to harden wax. Pellets can be stored in wax long term at RT.
10. Cut the wax blocks in 8 μm-thick sections and place a short ribbon of sections in a 42 °C water bath.
11. Remove a ribbon of sections from the water bath by allowing it to attach to a microscope slide. In general, there should be 3–5 sections per ribbon on each slide. Slides can be stored long term at RT.
12. Dry slides in a warmer at 37 °C for staining or in a drying rack at RT for storage overnight.
13. Perform desired histology, such as staining for sGAGs with Safranin-O (Subheading 3.4.3) or labeling of collagenous proteins with immunohistochemistry (Subheading 3.4.4).

3.4.3 Safranin-O and Hematoxylin

1–2 h

1. Remove paraffin wax with 3 rounds of soaking slide(s) in xylene for 5 min.
2. Rehydrate tissue with 100% (vol/vol) ethanol for 2 min followed by 50% (vol/vol) ethanol for 2 min.
3. Rinse slide(s) in tap water for 2 min.

4. Remove all excess water.
5. Stain slide(s) with Harris hematoxylin for 3 min (nuclei will be stained purple). Filter Harris hematoxylin staining solution before each use to remove potential precipitations and avoid deposition of particulates on slide. Do not reuse stain more than ten times.
6. Rinse slide(s) in tap water for 3 min.
7. Differentiate slide(s) in differentiation solution (acid alcohol) for 15 s.
8. Rinse slide(s) in tap water for 3 min.
9. Stain slide(s) with Safranin-O for 3–5 min (sGAGs will be stained pink/red). Do not reuse stain more than 10 times.
10. Rinse slide(s) multiple times with 100% (vol/vol) ethanol until no excess stain remains.
11. Let slide(s) partially dry before rinsing with xylene for 30 s.
12. Mount slide(s) with Permount, coverslip, let dry, and image (*see* **Note 14**).

3.4.4 Immunohistochemistry

4–5 h

1. Remove paraffin wax with 3 rounds of soaking slide(s) in xylene for 5 min.
2. Rehydrate tissue by washing slide(s) with 100% (vol/vol) ethanol for 5 min twice, 95% (vol/vol) ethanol for 5 min twice, 70% (vol/vol) ethanol for 5 min, and 50% (vol/vol) ethanol for 5 min, and then tap water for 5 min. Do not let the slides dry out. Place slide(s) in a container with a hydrated paper towel.
3. Perform antigen retrieval by adding 100 μL of proteinase K or pepsin on the slide and incubating (*see* Table 2 for recommended reagent, timing, and temperature).
4. Wash the slide(s) with $DPBS^{-/-}$ for 5 min twice.
5. Quench the slide(s) for 1 h in 3% (vol/vol) hydrogen peroxide in methanol.
6. Wash the slide(s) with $DPBS^{-/-}$ for 5 min three times.
7. Use a PAP-pen to circle each piece of tissue on the slide(s) to prevent solution spreading.
8. Perform blocking by adding a couple drops (enough to cover the tissue) of 2.5% (vol/vol) goat serum to each piece of tissue for 1 h at RT.
9. Blot excess serum from bottom of inclined slide. Do not rinse.
10. Stain tissue with primary antibody for 1 h at RT in hydrated container. *See* Table 2 for antibodies and dilutions. Do not add primary antibody to one piece of tissue. Instead add more goat serum from step viii to serve as negative control.

Table 2
IHC antibodies

Antibody	Host	Manufacturer	Cat. No.	Retrieval (time, temp)	Dilution	AEC Time
COL1A1	Mouse	Abcam	ab90395	Pepsin (5 min, RT)	1:800	2 min
COL2A1	Mouse	Iowa Hybridoma Bank	II-II6B3-s	Proteinase K (3 min, 37 °C)	1:10	2.5 min
COL6A1	Rabbit	Fitzgerald Industries	70F-CR009X	Proteinase K (3 min, 37 °C)	1:1000	2.5 min
COL10A1	Mouse	Millipore Sigma	C7974	Pepsin (5 min, RT)	1:200	2 min
Goat anti-mouse IgG, biotin	Goat	Abcam	ab97021		1:500	
Goat anti-rabbit IgG, biotin	Goat	Abcam	ab6720		1:500	

11. Wash the slide(s) with DPBS$^{-/-}$ for 5 min three times.
12. Add a couple drops (enough to cover the tissue) of the proper secondary antibody at a 1:500 dilution for 30 min at RT (Table 2).
13. Wash the slide(s) with PBS for 5 min three times.
14. Combine 5 mL of ImmPACT AEC diluent, 2 drops of AEC reagent 1, 3 drops of AEC reagent 2, and 2 drops of AEC reagent 3. Mix well.
15. Add AEC solution to the slide(s) and incubate at RT. See Table 2 for timing (*see* **Note 15**).
16. Wash the slide(s) with distilled water for 5 min.
17. Rinse the slide(s) in tap water.
18. Cover tissue sections with Vector Hematoxylin QS counterstain for 45 s.
19. Rinse the slide(s) in tap water for 10 s. Do not dehydrate slides.
20. Mount the slide(s) with VectaMount AQ aqueous mounting medium, coverslip, let dry, and image.

3.4.5 Biochemical Assay Preparation

1 day for sample preparation, 2–3 h for assay

1. Wash chondrogenic pellets with 2–3 mL of DPBS$^{-/-}$.
2. Transfer each pellet into a 1.7 mL tube with 200 μL of papain and vortex.
3. Digest tissue at 65 °C overnight and vortex. Use of caplock on the tubes may help securely seal the tube at high temperatures. Samples can be stored at −20 °C until analysis.
4. Follow previously published protocol to determine sGAG/DNA ratio using DMMB and PicoGreen Assays [52].

3.4.6 RT-qPCR Preparation

1–2 days

1. Clean work area with RNaseZap solution.
2. Wash chondrogenic pellets with 2–3 mL DPBS$^{-/-}$.
3. Transfer pellet into a tube and snap freeze in liquid nitrogen. Samples can be frozen at −80 °C until analysis (*see* **Note 16**).
4. Follow previously published protocol to determine gene expression using RT-qPCR [64, 52].

3.5 Anticipated Results

During regular culture, the hiPSCs have a small, round phenotype, and they grow in dense, round colonies (Fig. 2a). When passaging and plating, caution must be taken to prevent from breaking up the colonies too much as the hiPSCs tend to trigger cell death or spontaneously differentiate if cultured as single cells. In our laboratory, most hiPSC lines had optimal survival and growth on vitronectin-coated 6-well plates while there was lower viability when cultured on 10 cm dishes. Some hiPSC lines that were originally maintained on other substrates (e.g., Matrigel) were able to be switched onto vitronectin substrate by culturing cells on vitronectin-coated plates for several passages before mesodermal differentiation. After mesodermal induction, hiPSCs begin to differentiate at the edges of the colonies. As the cells become more elongated and spread out, the center of the colony begins to differentiate as well (Fig. 2b–e). Some cell death may be observed after the third day of mesodermal induction due to the inhibition of multiple essential pathways; however, the remaining cells should recover and differentiate as expected. Excess cell death may occur if hiPSCs are induced at too low of a cell density. In our previous study, we have demonstrated a decrease in pluripotent genes *OCT4* and *NANOG* throughout the mesodermal differentiation with an upregulation of *MIXL1* indicating anterior primitive streak cells after day 1, *MSGN1* indicating the paraxial mesoderm after day 2, and *PARAXIS* indicating the early somite state after day 3 (30). After day 6, the cells are committed into the sclerotome

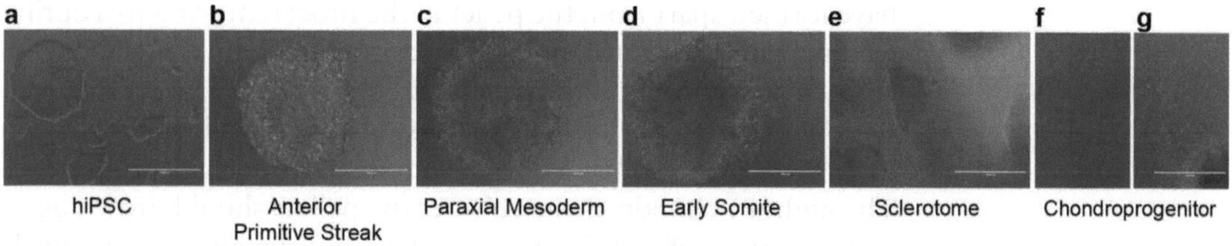

Fig. 2 Phase contrast images of cells throughout mesodermal differentiation. (**a**) Induce cells when they are 30–40% confluent. (**b–f**) As cells differentiate, from (**b**) anterior primitive streak, (**c**) paraxial mesoderm, (**d**) early somite, (**e**) sclerotome, to (**f**) chondroprogenitor, they spread and become more spindled. (**g**) If cells are induced at too high of a density, they may not fully differentiate and form nodules in the center of the colonies. Scale bar = 1 mm

lineage with an upregulation of *SOX9*, *PDGFRA*, and *PDGFRB*. The upregulation of *PDGFRB* continues through the chondroprogenitor stage alongside *COL2A1* expression after day 12 of mesodermal induction (30). We have also demonstrated that the chondroprogenitors express the surface markers CD166, CD146, and PDGFRβ, but not CD45, using flow cytometric analysis (37). Cells in the chondroprogenitor stage should be fully differentiated and spread out throughout the dish with an elongated phenotype (Fig. 2f). If the mesodermal differentiation is induced at too high of a cell density; however, one might observe formation of nodules in the center of cell colonies (Fig. 2g) and/or formation of a cell "sheet" that often spontaneously lifts off the culture plates, resulting into failed mesodermal lineage commitment.

Chondroprogenitor cells in monolayer culture (i.e., day 12 of mesodermal differentiation) should be disassociated into single cells to form 3D, chondrogenic pellets for chondrogenesis. Cells should lift off the bottom of the culture dish in the TrypLE solution after a 3 min incubation and gentle tapping. If cells do not lift off, increase the time of incubation. We have tried using cell scrapers in instances where cells do not lift after 8 min of incubation; however, the cells had poor viability and failed to form chondrogenic pellets. We average a yield of 6×10^7 chondroprogenitor cells per T225 flask; however, this value may vary due to different hiPSC lines and plating density used. Approximately 24 h after chondrogenic induction, we observe the formation of a spherical pellet in the bottom of the conical tube. Occasionally, the pellet may need additional time before the proper shape is visible. Pellets that do not form after 48 h should be discarded. This may be caused by improper mesodermal differentiation or incompatible disassociation reagent (cell line specific). In some cell lines, we have found disassociating chondroprogenitor cells in 0.05 mM EDTA and/or differentiating cells on Matrigel has improved pelleting and chondrogenesis. As pellet cultures approach 28 days, the medium may begin to turn yellow due to increased cellular metabolic activity, which can be prevented by increasing the volume of chondrogenic medium during feeding. In other cases, some excess matrix may have formed apart from the pellet in the tube; removing this during feeding will also reduce the rate the culture medium is metabolized. If black dots appear on the pellets, suggestive of off-target differentiation and presence of melanocytes, increasing the concentration of ML329 will inhibit MITF and potential melanin production. Throughout chondrogenic culture, the pellets should grow due to accumulation of cartilaginous matrix but maintain a relatively spherical shape. We have reported a significant increase in chondrogenic genes *SOX9*, *COL2A1*, and *ACAN* from the chondroprogenitor stage through day 42 of culture, while fibrocartilage and hypertrophic cartilage markers *COL1A1* and *COL10A1*, respectively, remain relatively low until later time points (Fig. 3a–e) [30, 37].

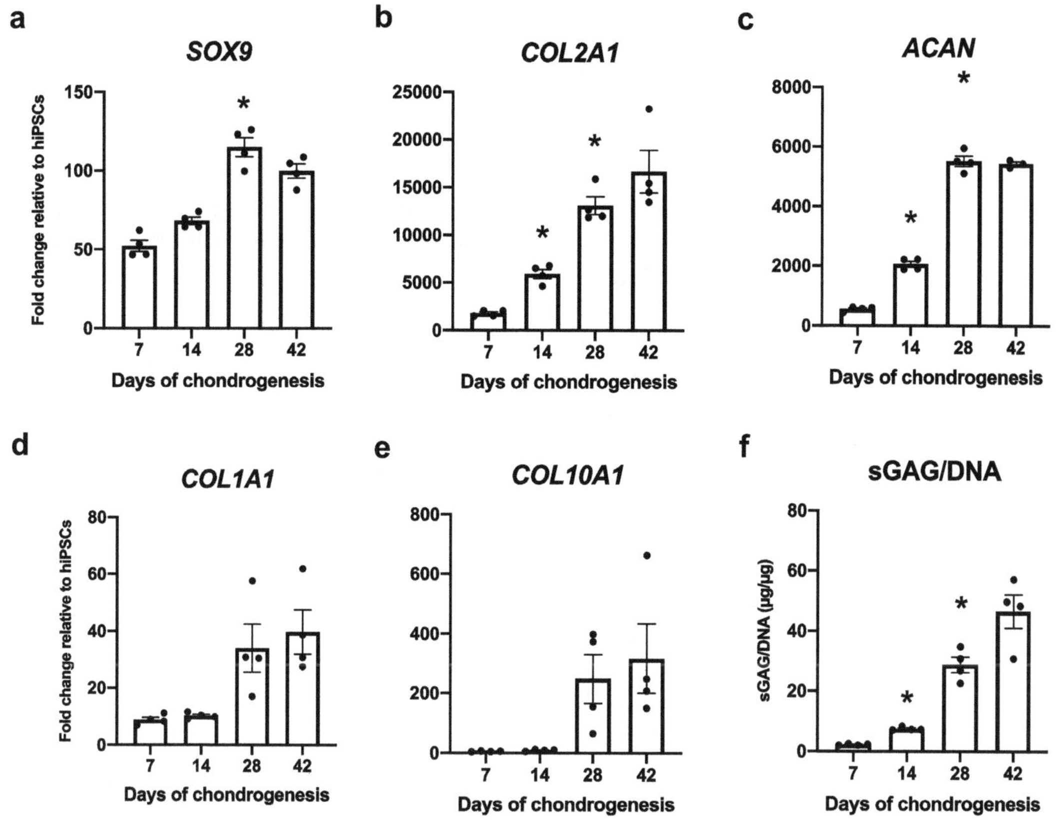

Fig. 3 Anticipated results – gene expression and matrix quantification. (**a**–**c**) Chondrogenic transcription factor *SOX9* gene expression should increase early, followed by matrix genes *COL2A1* and *ACAN*. (**d**–**e**) Relative to the expression of *COL2A1*, lower gene expression of fibrocartilage and hypertrophic cartilage markers *COL1A1* and *COL10A1*, respectively, was observed. (**f**) sGAG/DNA ratio should increase throughout chondrogenesis, reaching 20–30 ng/ng at day 28 and over 40 ng/ng at day 42. Mean ± SEM, $n = 4$. *$p < 0.05$ compared to previous time point

To obtain hiPSC-derived chondrocytes, day 28 chondrogenic pellets can be digested with type 2 collagenase. Vortexing the tubes every 20 min will facilitate the breakdown of the tissue matrix, thus increasing digestion efficiency. However, if the chunks of pellets remain intact after a lengthy period of digestion, "smashing" the pellets using a tool with a flat surface (e.g., spatula with shovel head, or lid of an Eppendorf tube) in a sterile Petri dish may help break up the tissue matrix. We do not recommend digesting for longer than 2.5 h as it may decrease the cell viability. Occasionally, small pieces of tissue may still be visible after the digestion process; however, most of the cells will be dissociated from the matrix. We average a cell viability of 85% and approximately 6.5×10^5 cells per digested pellet. We found an increase in viability and efficiency when more pellets are digested at a time and 50 mL conical tubes are used.

To validate and visualize the cartilage matrix produced by hiPSC-derived chondrocytes, we use Safranin-O staining to reveal the presence of sGAGs (Fig. 4a–c), IHC to label various types of collagens (Fig. 4d–f), biochemical DMMB assay to quantify sGAG production (Fig. 3f), and RT-qPCR to analyze gene expression (Fig. 3a–e). We normally observe round chondrocyte-like cells within cartilaginous matrix with robust, homogenous Safranin-O staining (Fig. 4a–c). If cellular heterogeneity is observed, concentration of Wnt-C59 and ML329 can be increased to further prevent off-target differentiation. If there is little or no sGAG staining, differentiation was unsuccessful, which can be attributed to several issues addressed in Notes. If hiPSCs have successfully differentiated into chondrocytes, there should also be significant labeling for COL2A1 (Fig. 4d), with minimal labeling of COL1A1 (Fig. 4e) and COL10A1 (Fig. 4f). Biochemical analysis and RT-qPCR for these genes can further support successful chondrogenesis. Chondrogenic pellets typically have an sGAG/DNA ratio between 20 and 30 ng/ng in day-28 pellet samples, which can continue to increase with more time in culture (Fig. 3f). A significant upregulation of *SOX9*, *COL2A1*, and *ACAN* gene expression should be detected in the pellets at day 28 versus day-0 pellet samples (Fig. 3a–c), while the expression levels of *COL1A1* and *COL10A1* remain low until later time points (Fig. 3d–e). The hiPSC-derived chondrocytes and cartilage matrix generated using this protocol can be used for a variety of experiments including genetic engineering, in vitro disease modeling, and tissue engineering.

4 Notes

1. If hiPSCs are confluent in 2–3 days, the cell concentration of the frozen stock is too high. Plate one vial to more than one 6-well plate (e.g., 1.5 or 2 × 6-well plates).
2. If there is hiPSC death, cells were plated at too low of a density, media was changed too soon, Y-27632 was not added, cells were washed and fed too aggressively, or the cell line is not compatible with vitronectin. Reduce number of wells and/or flasks you are plating, make sure media is changed 24 h after plating (or try extending the time slightly longer than 24 h), make sure to add Y-27632 to media for the first 24 h, pipette media slowly and onto the side of the dish (not directly on the cells), or try other matrix substrates, such as Matrigel.
3. If there is spontaneous differentiation, colonies were not maintained but broken into single cells or cells were plated at an incorrect density. Do not pipette up and down when plating and passaging hiPSCs or change the number of wells you plate per cell vial/plate. You can scrape differentiated cells to clean the wells.

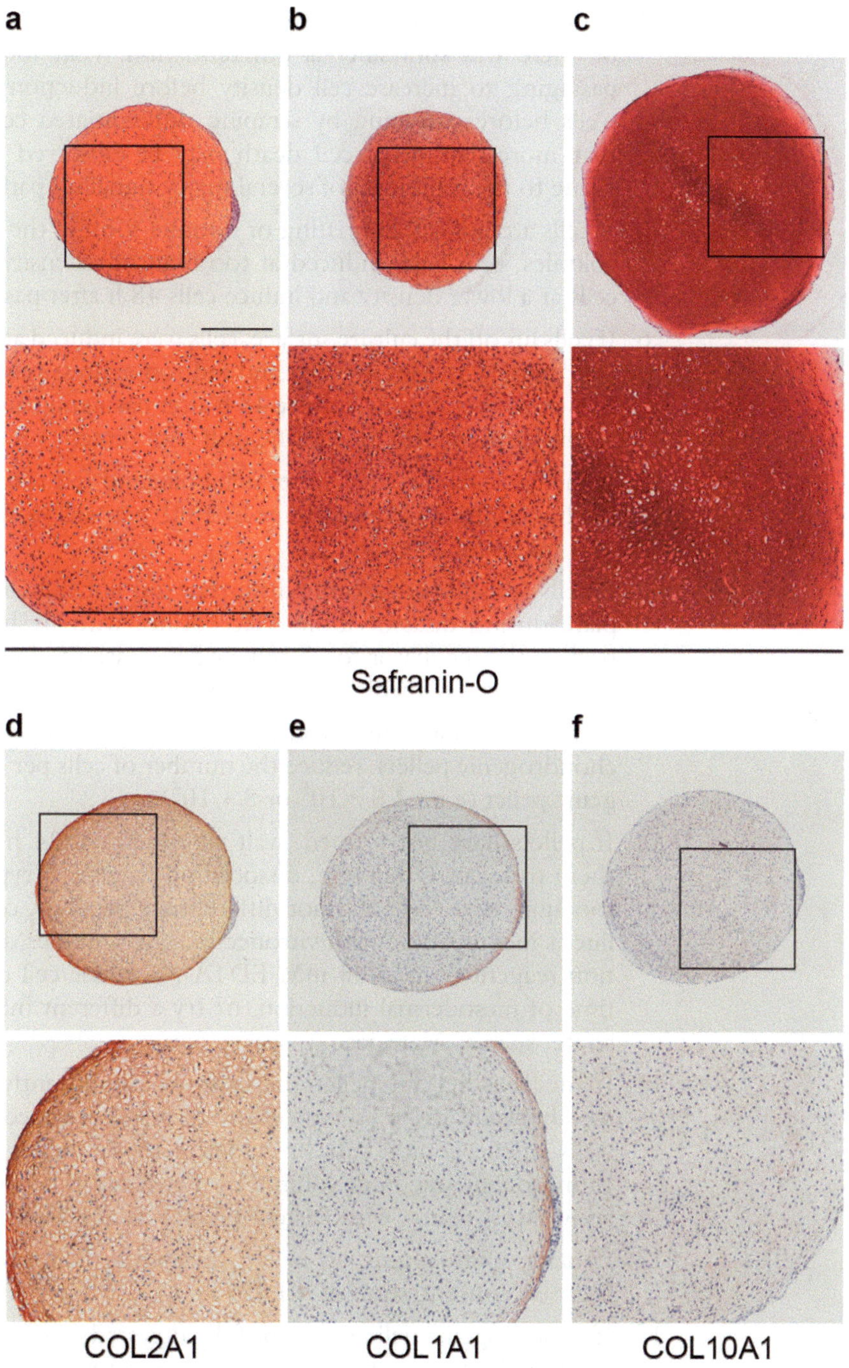

Fig. 4 Anticipated results – histology. (**a**–**c**) Robust, homogenous safranin-O staining for sGAGs in three different cell lines. (**d**–**f**) Pellet in panel a with IHC labeling of COL2A1 (**d**), COL1A1 (**e**), and COL10A1 (**f**). Scale bar = 500 μm

4. If there is cell death, cells were induced at too low of a density or there was spontaneous differentiation. Wait longer after passaging to increase cell density before induction or clean wells before passaging by scraping differentiated cells. Note that more-than-usual cell death may be observed after day 3 due to the inhibition of several major signaling pathways.
5. If cells are not differentiating or nodules form in the center of colonies, cells were induced at too high of a density. Passage cells at a lower density and induce cells 48 h after passaging.
6. If cells lift off the culture surface, cells were induced at too high of a density or were washed and fed too aggressively. Passage cells at a lower density, induce 48 h after passaging, and pipette media slowly on the side of the dish (not directly on cells).
7. If media is yellow, cells were metabolic and fed too low of a media volume. Increase volume by 1 mL per well (6-well plate) or 5 mL per T225 flask.
8. If cells are still stuck to the dish, they did not fully lift off the plate during disassociation. Place dishes with fresh TrypLE back in the incubator for 2 min, and increase the force of slap without causing media to splash on top of dish.
9. If you do not have enough cells to create the desired number of chondrogenic pellets, reduce the number of cells per chondrogenic pellet (e.g., 2.5×10^5 or 3×10^5).
10. If pellets have not formed, wait an additional 24 h to allow them to form. Otherwise, dissociation reagent (TrypLE) was too aggressive, cells did not differentiate properly, or the cell line is incompatible with vitronectin. Use a milder disassociation reagent (e.g., 0.05 mM EDTA), decrease cell density at time of mesodermal induction, or try a different matrix substrate, such as Matrigel.
11. If media is yellow, pellets have grown significantly and are metabolically active or extra matrix is being produced around walls of the conical tube in addition to the pellet. Increase chondrogenic medium volume by 1 mL per tube or aspirate away excess matrix with Pasteur pipette during aspiration.
12. If black spots appear on pellet, melanin is being produced. Increase concentration of ML329.
13. If chondrogenic pellets are not fully digested, the collagenase solution is not strong enough or the matrix is too dense. Ensure proper ratio of collagenase to medium was used and test different lots of collagenase. Remove pellets from tube and place in an empty well of 6-well plate. Use a closed, sterile 1.7 mL Eppendorf to gently "smash" pellets. Use collagenase solution to rinse the well and transfer back to tube. Digest an additional 20 min.

14. If there is cell and matrix heterogeneity in chondrogenic pellet, Wnt and MITF signaling is occurring and causing off-target differentiation. Increase concentrations of Wnt-C59 and ML329.
15. If the positive control is not showing stain, development time is too short. Increase time of development with AEC.
16. We generally use screw-top microcentrifuge tubes with O-ring cap.

Acknowledgements

This work was supported by the Shriners Hospitals for Children, Nancy Taylor Foundation, Arthritis Foundation, NIH (AG46927, AG15768, AR075899, AR072999, AR074992, AR073752, AR080902, T32 DK108742, T32 EB018266), and Taiwan GSSA Scholarship.

Conflicts of Interest FG is an employee of Cytex Therapeutics, Inc. The other authors declare that they have no competing interests.

Author Contribution Statement ARD, NS, CLW, and FG were involved in the development, testing, and troubleshooting of these protocols and the writing of this paper.

References

1. Mansour JM (2003) Biomechanics of Cartilage. In: Kinesiology: the mechanics and pathomechanics of human movement, vol 2e, pp 66–75
2. Guilak F (2011) Biomechanical factors in osteoarthritis. Best Pract Res Clin Rheumatol 25(6):815–823. https://doi.org/10.1016/j.berh.2011.11.013
3. Antons J, Marascio MGM, Nohava J et al (2018) Zone-dependent mechanical properties of human articular cartilage obtained by indentation measurements. J Mater Sci Mater Med 29(5):57. https://doi.org/10.1007/s10856-018-6066-0
4. Sophia Fox AJ, Bedi A, Rodeo SA (2009) The basic science of articular cartilage: structure, composition, and function. Sports Health 1(6):461–468. https://doi.org/10.1177/1941738109350438
5. Xia Y, Momot KI, Chen Z et al (2017) Chapter 1 Introduction to Cartilage. In: Biophysics and biochemistry of Cartilage by NMR and MRI. The Royal Society of Chemistry, pp 1–43. https://doi.org/10.1039/9781782623663-00001
6. Zhang L, Hu J, Athanasiou KA (2009) The role of tissue engineering in articular cartilage repair and regeneration. Crit Rev Biomed Eng 37(1–2):1–57. https://doi.org/10.1615/critrevbiomedeng.v37.i1-2.10
7. Guilak F, Hung C (2005) Physical regulation of cartilage metabolism. In: Mow V, Huiskes R (eds) Basic orthopaedic biomechanics and mechanobiology, 3rd edn. Lippincott-Raven Publishers, Philadelphia, pp 179–207
8. Shieh AC, Athanasiou KA (2003) Principles of cell mechanics for cartilage tissue engineering. Ann Biomed Eng 31(1):1–11. https://doi.org/10.1114/1.1535415
9. Sanchez-Adams J, Leddy HA, McNulty AL et al (2014) The mechanobiology of articular cartilage: bearing the burden of osteoarthritis. Curr Rheumatol Rep 16(10):451. https://doi.org/10.1007/s11926-014-0451-6
10. Berenbaum F, Griffin TM, Liu-Bryan R (2017) Review: metabolic regulation of inflammation

in osteoarthritis. Arthritis Rheumatol 69(1): 9–21. https://doi.org/10.1002/art.39842

11. Breedveld FC (2004) Osteoarthritis--the impact of a serious disease. Rheumatology (Oxford) 43(Suppl 1):i4–i8. https://doi.org/10.1093/rheumatology/keh102
12. Smolen JS, Aletaha D, McInnes IB (2016) Rheumatoid arthritis. Lancet 388(10055): 2023–2038. https://doi.org/10.1016/S0140-6736(16)30173-8
13. Barbour KE, Moss S, Croft JB et al (2018) Geographic variations in arthritis prevalence, health-related characteristics, and management - United States, 2015. MMWR Surveill Summ 67(4):1–28. https://doi.org/10.15585/mmwr.ss6704a1
14. Palazzo C, Nguyen C, Lefevre-Colau MM et al (2016) Risk factors and burden of osteoarthritis. Ann Phys Rehabil Med 59(3):134–138. https://doi.org/10.1016/j.rehab.2016.01.006
15. Willard VP, Diekman BO, Sanchez-Adams J et al (2014) Use of cartilage derived from murine induced pluripotent stem cells for osteoarthritis drug screening. Arthritis Rheumatol 66(11):3062–3072. https://doi.org/10.1002/art.38780
16. Adkar SS, Brunger JM, Willard VP et al (2017) Genome engineering for personalized arthritis therapeutics. Trends Mol Med 23(10): 917–931. https://doi.org/10.1016/j.molmed.2017.08.002
17. O'Connor SK, Katz DB, Oswald SJ et al (2020) Formation of Osteochondral organoids from murine induced Pluripotent stem cells. Tissue Eng Part A. https://doi.org/10.1089/ten.TEA.2020.0273
18. Saitta B, Passarini J, Sareen D et al (2014) Patient-derived skeletal dysplasia induced pluripotent stem cells display abnormal chondrogenic marker expression and regulation by BMP2 and TGFbeta1. Stem Cells Dev 23(13):1464–1478. https://doi.org/10.1089/scd.2014.0014
19. Nur Patria Y, Stenta T, Lilianty J et al (2020) CRISPR/Cas9 gene editing of a SOX9 reporter human iPSC line to produce two TRPV4 patient heterozygous missense mutant iPSC lines, MCRIi001-A-3 (TRPV4 p.F273L) and MCRIi001-A-4 (TRPV4 p.P799L). Stem Cell Res 48:101942. https://doi.org/10.1016/j.scr.2020.101942
20. Liu H, Yang L, Yu FF et al (2017) The potential of induced pluripotent stem cells as a tool to study skeletal dysplasias and cartilage-related pathologic conditions. Osteoarthr Cartil 25(5):616–624. https://doi.org/10.1016/j.joca.2016.11.015
21. Sanjurjo-Rodriguez C, Castro-Vinuelas R, Pineiro-Ramil M et al (2020) Versatility of induced Pluripotent stem cells (iPSCs) for improving the knowledge on musculoskeletal diseases. Int J Mol Sci 21(17). https://doi.org/10.3390/ijms21176124
22. Yamashita A, Morioka M, Kishi H et al (2014) Statin treatment rescues FGFR3 skeletal dysplasia phenotypes. Nature 513(7519): 507–511. https://doi.org/10.1038/nature13775
23. Matsumoto Y, Hayashi Y, Schlieve CR et al (2013) Induced pluripotent stem cells from patients with human fibrodysplasia ossificans progressiva show increased mineralization and cartilage formation. Orphanet J Rare Dis 8: 190. https://doi.org/10.1186/1750-1172-8-190
24. Miller EE, Kobayashi GS, Musso CM et al (2017) EIF4A3 deficient human iPSCs and mouse models demonstrate neural crest defects that underlie Richieri-Costa-Pereira syndrome. Hum Mol Genet 26(12):2177–2191. https://doi.org/10.1093/hmg/ddx078
25. Okada M, Ikegawa S, Morioka M et al (2015) Modeling type II collagenopathy skeletal dysplasia by directed conversion and induced pluripotent stem cells. Hum Mol Genet 24(2): 299–313. https://doi.org/10.1093/hmg/ddu444
26. Loh KM, Chen A, Koh PW et al (2016) Mapping the pairwise choices leading from Pluripotency to human bone, heart, and other mesoderm cell types. Cell 166(2):451–467. https://doi.org/10.1016/j.cell.2016.06.011
27. Umeda K, Zhao J, Simmons P et al (2012) Human chondrogenic paraxial mesoderm, directed specification and prospective isolation from pluripotent stem cells. Sci Rep 2:455. https://doi.org/10.1038/srep00455
28. Oldershaw RA, Baxter MA, Lowe ET et al (2010) Directed differentiation of human embryonic stem cells toward chondrocytes. Nat Biotechnol 28(11):1187–1194. https://doi.org/10.1038/nbt.1683
29. Diekman BO, Christoforou N, Willard VP et al (2012) Cartilage tissue engineering using differentiated and purified induced pluripotent stem cells. Proc Natl Acad Sci U S A 109(47): 19172–19177. https://doi.org/10.1073/pnas.1210422109
30. Adkar SS, Wu CL, Willard VP et al (2019) Step-wise chondrogenesis of human induced Pluripotent stem cells and purification via a reporter Allele generated by CRISPR-Cas9 genome editing. Stem Cells 37(1):65–76. https://doi.org/10.1002/stem.2931

31. Lim J, Tu X, Choi K et al (2015) BMP-Smad4 signaling is required for precartilaginous mesenchymal condensation independent of Sox9 in the mouse. Dev Biol 400(1):132–138. https://doi.org/10.1016/j.ydbio.2015.01.022
32. Patil AS, Sable RB, Kothari RM (2011) An update on transforming growth factor-beta (TGF-beta): sources, types, functions and clinical applicability for cartilage/bone healing. J Cell Physiol 226(12):3094–3103. https://doi.org/10.1002/jcp.22698
33. Tang QO, Shakib K, Heliotis M et al (2009) TGF-beta3: a potential biological therapy for enhancing chondrogenesis. Expert Opin Biol Ther 9(6):689–701. https://doi.org/10.1517/14712590902936823
34. Craft AM, Rockel JS, Nartiss Y et al (2015) Generation of articular chondrocytes from human pluripotent stem cells. Nat Biotechnol 33(6):638–645. https://doi.org/10.1038/nbt.3210
35. Yamashita A, Morioka M, Yahara Y et al (2015) Generation of scaffoldless hyaline cartilaginous tissue from human iPSCs. Stem Cell Rep 4(3):404–418. https://doi.org/10.1016/j.stemcr.2015.01.016
36. Johnstone B, Hering TM, Caplan AI et al (1998) In vitro chondrogenesis of bone marrow-derived mesenchymal progenitor cells. Exp Cell Res 238(1):265–272. https://doi.org/10.1006/excr.1997.3858
37. Dicks A, Wu CL, Steward N et al (2020) Prospective isolation of chondroprogenitors from human iPSCs based on cell surface markers identified using a CRISPR-Cas9-generated reporter. Stem Cell Res Ther 11(1):66. https://doi.org/10.1186/s13287-020-01597-8
38. Wu CL, Dicks A, Steward N et al (2021) Single cell transcriptomic analysis of human pluripotent stem cell chondrogenesis. Nat Commun 12(1):362. https://doi.org/10.1038/s41467-020-20598-y
39. Yamashita A, Tamamura Y, Morioka M et al (2018) Considerations in hiPSC-derived cartilage for articular cartilage repair. Inflamm Regen 38:17. https://doi.org/10.1186/s41232-018-0075-8
40. Murphy C, Mobasheri A, Tancos Z et al (2018) The potency of induced Pluripotent stem cells in Cartilage regeneration and osteoarthritis treatment. Adv Exp Med Biol 1079:55–68. https://doi.org/10.1007/5584_2017_141
41. Krishnan Y, Grodzinsky AJ (2018) Cartilage diseases. Matrix Biol 71-72:51–69. https://doi.org/10.1016/j.matbio.2018.05.005
42. Musunuru K (2013) Genome editing of human pluripotent stem cells to generate human cellular disease models. Dis Model Mech 6(4):896–904. https://doi.org/10.1242/dmm.012054
43. Yumlu S, Stumm J, Bashir S et al (2017) Gene editing and clonal isolation of human induced pluripotent stem cells using CRISPR/Cas9. Methods 121-122:29–44. https://doi.org/10.1016/j.ymeth.2017.05.009
44. O'Conor CJ, Case N, Guilak F (2013) Mechanical regulation of chondrogenesis. Stem Cell Res Ther 4(4):61. https://doi.org/10.1186/scrt211
45. Brunger JM, Zutshi A, Willard VP et al (2017) Genome engineering of stem cells for autonomously regulated, closed-loop delivery of biologic drugs. Stem Cell Rep 8(5):1202–1213. https://doi.org/10.1016/j.stemcr.2017.03.022
46. Pferdehirt L, Ross AK, Brunger JM et al (2019) A synthetic gene circuit for self-regulating delivery of biologic drugs in engineered tissues. Tissue Eng Part A 25(9–10):809–820. https://doi.org/10.1089/ten.TEA.2019.0027
47. Nims RJ, Pferdehirt L, Ho NB et al (2021) A synthetic mechanogenetic gene circuit for autonomous drug delivery in engineered tissues. Sci Adv 7(5). https://doi.org/10.1126/sciadv.abd9858
48. Guilak F, Pferdehirt L, Ross AK et al (2019) Designer stem cells: genome engineering and the next generation of cell-based therapies. J Orthop Res 37(6):1287–1293. https://doi.org/10.1002/jor.24304
49. Tang R, Jing L, Willard VP et al (2018) Differentiation of human induced pluripotent stem cells into nucleus pulposus-like cells. Stem Cell Res Ther 9(1):61. https://doi.org/10.1186/s13287-018-0797-1
50. Thirion S, Berenbaum F (2004) Culture and phenotyping of chondrocytes in primary culture. Methods Mol Med 100:1–14. https://doi.org/10.1385/1-59259-810-2:001
51. Nasrabadi D, Rezaeiani S, Eslaminejad MB et al (2018) Improved protocol for chondrogenic differentiation of bone marrow derived mesenchymal stem cells -effect of PTHrP and FGF-2 on TGFbeta1/BMP2-induced chondrocytes hypertrophy. Stem Cell Rev Rep 14(5):755–766. https://doi.org/10.1007/s12015-018-9816-y
52. Estes BT, Diekman BO, Gimble JM et al (2010) Isolation of adipose-derived stem cells and their induction to a chondrogenic

phenotype. Nat Protoc 5(7):1294–1311. https://doi.org/10.1038/nprot.2010.81

53. Caplan AI (2017) Mesenchymal stem cells: time to change the name! Stem Cells Transl Med 6(6):1445–1451. https://doi.org/10.1002/sctm.17-0051

54. Yang YK, Ogando CR, Wang See C et al (2018) Changes in phenotype and differentiation potential of human mesenchymal stem cells aging in vitro. Stem Cell Res Ther 9(1):131. https://doi.org/10.1186/s13287-018-0876-3

55. Drela K, Stanaszek L, Nowakowski A et al (2019) Experimental strategies of mesenchymal stem cell propagation: adverse events and potential risk of functional changes. Stem Cells Int 2019:7012692. https://doi.org/10.1155/2019/7012692

56. Takahashi K, Tanabe K, Ohnuki M et al (2007) Induction of pluripotent stem cells from adult human fibroblasts by defined factors. Cell 131(5):861–872. https://doi.org/10.1016/j.cell.2007.11.019

57. Nejadnik H, Diecke S, Lenkov OD et al (2015) Improved approach for chondrogenic differentiation of human induced pluripotent stem cells. Stem Cell Rev Rep 11(2):242–253. https://doi.org/10.1007/s12015-014-9581-5

58. Lee J, Taylor SE, Smeriglio P et al (2015) Early induction of a prechondrogenic population allows efficient generation of stable chondrocytes from human induced pluripotent stem cells. FASEB J 29(8):3399–3410. https://doi.org/10.1096/fj.14-269720

59. Suchorska WM, Augustyniak E, Richter M et al (2017) Comparison of four protocols to generate chondrocyte-like cells from human induced Pluripotent stem cells (hiPSCs). Stem Cell Rev Rep 13(2):299–308. https://doi.org/10.1007/s12015-016-9708-y

60. Lian Q, Zhang Y, Zhang J et al (2010) Functional mesenchymal stem cells derived from human induced pluripotent stem cells attenuate limb ischemia in mice. Circulation 121(9):1113–1123. https://doi.org/10.1161/CIRCULATIONAHA.109.898312

61. Xu M, Shaw G, Murphy M et al (2019) Induced Pluripotent stem cell-derived mesenchymal stromal cells are functionally and genetically different from bone marrow-derived mesenchymal stromal cells. Stem Cells 37(6):754–765. https://doi.org/10.1002/stem.2993

62. Chen IP, Fukuda K, Fusaki N et al (2013) Induced pluripotent stem cell reprogramming by integration-free Sendai virus vectors from peripheral blood of patients with craniometaphyseal dysplasia. Cell Reprogram 15(6):503–513. https://doi.org/10.1089/cell.2013.0037

63. Crowe AR, Yue W (2019) Semi-quantitative determination of protein expression using immunohistochemistry staining and analysis: an integrated protocol. Bio Protoc 9(24):10.21769/BioProtoc.3465

64. Nolan T, Hands RE, Bustin SA (2006) Quantification of mRNA using real-time RT-PCR. Nat Protoc 1(3):1559–1582. https://doi.org/10.1038/nprot.2006.236

Chapter 9

The Dimethylmethylene Blue Assay (DMMB) for the Quantification of Sulfated Glycosaminoglycans

Yann D. Ladner, Mauro Alini, and Angela R. Armiento

Abstract

The 1,9-dimethylmethylene blue (DMMB) assay enables the detection of sulfated glycosaminoglycans (sGAGs). This assay can be used to quickly quantify the sGAG content in a large number of samples using spectrophotometry. While this widespread assay appears straightforward, there are certain pitfalls that need to be considered.

Key words Dimethylmethylene blue assay, DMMB, Sulfated glycosaminoglycans, sGAG, Cartilage

1 Introduction

The 1,9-dimethyl-methylene blue (DMMB) assay finds wide application in the field of cartilage tissue engineering. First introduced by Taylor and Jeffree (1969), it quantifies sulphated glycosaminoglycans (sGAGs), a major component of cartilage. Even though there exist more precise techniques that give insights into the composition of sGAGs [1, 2], the 96-well plate format of the DMMB assay has the advantage that it can be used for a large number of samples. The DMMB assay is mostly used to measure sGAGs (predominantly chondroitin sulphate) in both samples and culture media of tissues and engineered cellular constructs.

When the dye forms a complex with polyanions such as sGAGs, a metachromatic shift in absorbance is detected using spectrophotometry [3, 4]. The metachromatic change is clearly visible as the dye changes from blue to violet and further to pink with increasing concentration of polyanions. Usually, the absorbance is measured either at a wavelength of around 525 nm (A525) or 595 nm (A595). At A525, the absorbance increases with increased sGAG content (Fig. 1). On the other hand, at A595, the absorbance decreases with increased sGAG content (Fig. 1). The change in absorbance is linearly correlated to the sGAG content. However,

Martin J. Stoddart et al. (eds.), *Cartilage Tissue Engineering*,
Methods in Molecular Biology, vol. 2598, https://doi.org/10.1007/978-1-0716-2839-3_9,

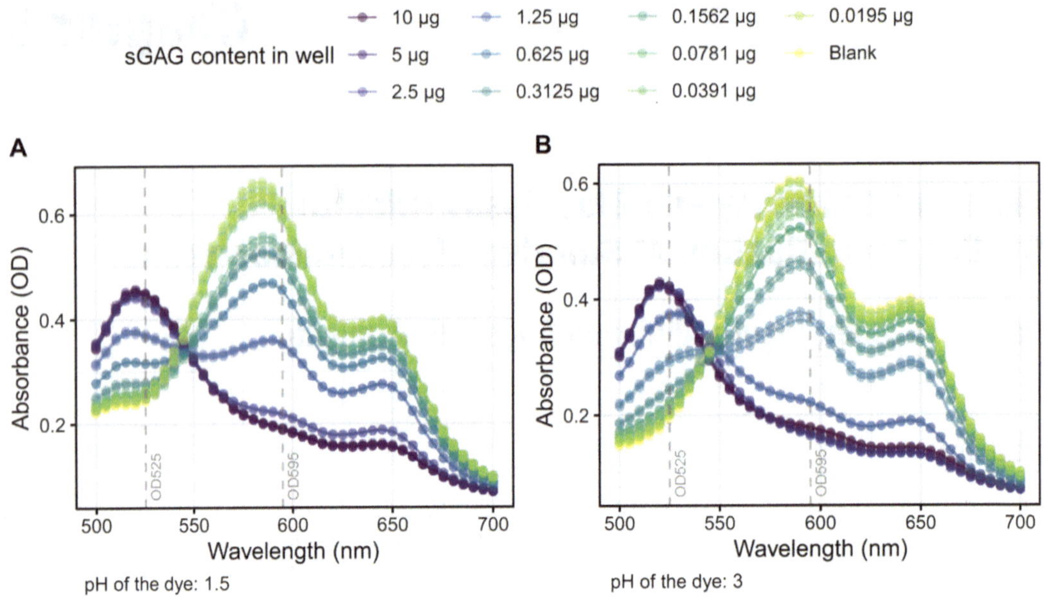

Fig. 1 Absorbance curves for samples with different chondroitin sulfate (CS) contents per well for the DMMB color reagent (dye) at a pH 1.5 (**a**) and pH 3 (**b**). Samples were prepared in Dulbecco's modified Eagle medium 4.5 g/L glucose (DMEM) (the blank consists of only DMEM)

there are parameters such as pH, salt content, presence of polyanions (*e.g.*, DNA), and the duration of the reaction that interfere with the spectrophotometric reading—for a more complete understanding of what can affect the DMMB assay, readers are advised to read Zheng and Levenston (2015) [5]. While the DMMB assay is commonly used at a pH of 3, it has been shown that decreasing the pH to 1.5 and subtracting the absorbance values at A595 from the values at A525 reduce the influence of artifacts.

The DMMB assay measures total sGAG content and cannot discriminate between different sGAG types (*e.g.*, keratan *vs.* chondroitin sulphate). Should the reader be interested in the sGAG composition, there exist methods such as immunological approaches (*e.g.*, immunohistochemistry), fluorophore-assisted carbohydrate electrophoresis (FACE) [6], or liquid chromatography-mass spectrometry [2]. However, DMMB really shines thanks to its simplicity and ease of use, which even allows for automation [7, 8].

In this chapter, we describe how we quantify sGAGs using the DMMB assay and note the tips we find useful for a successful and accurate measurement.

2 Materials

1. Plate reader capable of measuring absorbance at 525 nm and 595 nm (*e.g.*, Tecan Infinite PRO Micro-Plate Reader).
2. Well assay plate: (*e.g.*, 96-well assay plates (Corning, #3590)).
3. Multichannel pipette.
4. Stock standard solution: 1 mg/mL chondroitin sulfate (CS – *e.g.*, chondroitin sulfate A sodium salt from bovine trachea, Sigma-Aldrich, C9819) in distilled H_2O. Store in aliquots at −20 °C.
5. Glycine.
6. Sodium chloride (NaCl).
7. Hydrochloric acid (HCl).
8. Brown 1 L bottle or transparent bottle wrapped in aluminum foil. The DMMB reagent is stable in a brown bottle (or an aluminum-wrapped transparent bottle) at room temperature for at least 3 months (*see* **Note 1**).
9. DMMB color reagent (dye): 16 mg DMMB (1,9-dimethyl-methylene blue) in 1 L distilled H_2O containing 3.04 g glycine and 2.37 g sodium chloride. Stir overnight at room temperature, in a brown bottle or in a transparent bottle wrapped in aluminum.

   On the next day, adjust the pH (*see* **Note 2** and **Note 3**).
10. Solvents for standard preparation and dilution of samples (if necessary): depends on the experiment (*see* **Note 4**).

3 Methods

Carry out the procedure at room temperature. Prepare enough solvent for standard preparation and dilution of samples.

1. Prepare standards using a serial dilution (*see* Table 1 for an example of a serial dilution for the standards and **Note 5** for advice on the concentration of the first standard).
2. For the blanks, use appropriate solvent (*see* **Note 4**).
3. Pipet 20 μL (or up to 50 μL) of standards, samples, or blanks according to your predefined plate layout into designated wells of a 96-well assay plate. Measure your standards, samples, and blanks in duplicates (triplicates are preferred) (*see* **Note 6**).
4. Add 200 μL of DMMB color reagent to each well using a multichannel pipette.
5. Tap plate gently and immediately read the absorbance at 525 nm (A525) (*see* **Note 7**).

Table 1
Example of a serial dilution for standards used in 50 μL of sample volume (conditioned medium assay, *see* Note 5)

Dilution (standard)	Volume and source of CS	Volume of solvent (μL)	CS content in well (μg) within 50 μL	CS concentration (μg/mL)
1	10 μL of 1 mg/mL CS stock	390	1.25	25
2	200 μL of Dilution 1	200	0.625	12.5
3	200 μL of Dilution 2	200	0.3125	6.25
4	200 μL of Dilution 3	200	0.15625	3.125
5	200 μL of Dilution 4	200	0.078125	1.5625
6	200 μL of Dilution 5	200	0.0390625	0.78125
7	200 μL of Dilution 6	200	0.01953125	0.390625
Blank	–	200	0	0

6. Subtract the average A525 value of the blanks from the A525 values of the standards and samples.
7. Prepare a linear standard curve A525 *vs.* sGAG content (μg/well), and use the slope to calculate the sGAG content or concentration of the samples.

4 Notes

1. It is not recommended to store the color reagent at 4 °C because its color fades, likely due to aggregation of dye molecules. Make up new dye solution if precipitation occurs.
2. For pH 3, adjust pH with 1 M HCl (ca. 9.5 mL) to give an absorbance of around 0.2 at 525 nm. For pH 1.5, adjust pH with 12 M (concentrated) HCl (ca. 7.5 mL) (*see* **Note 3**). 1 M HCl can be used to fine-adjust pH.
3. Polyanions, such as DNA and RNA and polyanionic extracellular matrix (ECM) components like hyaluronic acid interfere (overestimate sGAG content) with reading at pH 3. For analysis of sGAG content from samples with high DNA content and from cells grown in polyanionic scaffolds, such as, alginate or hyaluronic acid, adjust the pH to 1.5 (Fig. 2). The pH adjustment from pH 3 to pH 1.5 results in a loss of sensitivity (slope of standard curve decreases—Fig. 3d *vs.* Fig. 3c). This loss of sensitivity can be counteracted by subtracting blank corrected A595 values from blank corrected A525 values ([5], Fig. 4).

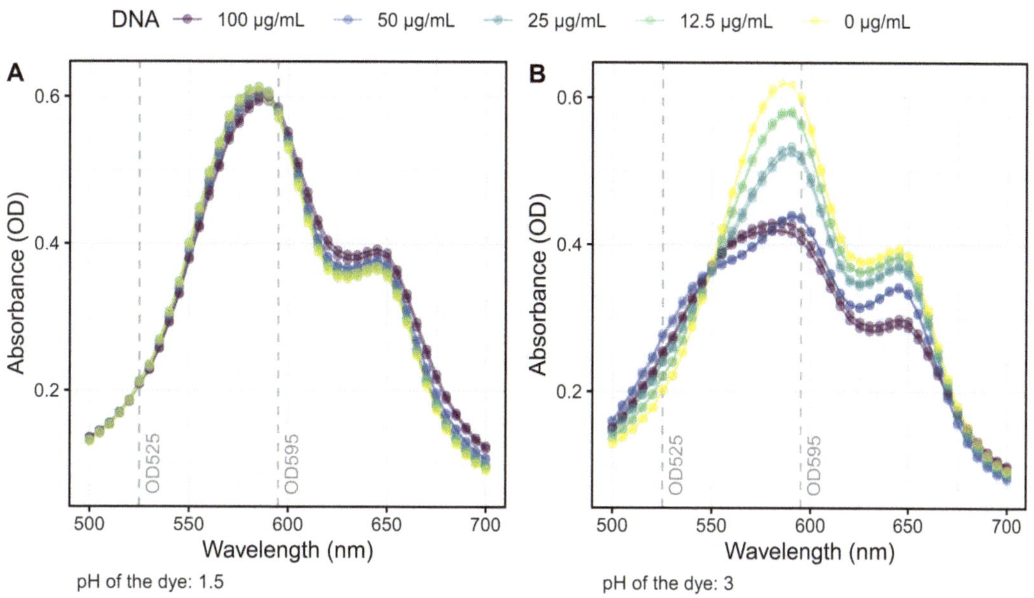

Fig. 2 Interference of DNA at different concentrations on the DMMB assay with DMMB color reagent (dye) at pH 1.5 (**a**) and pH 3 (**b**)

4. The standards should be prepared in the same base solvent as the samples as the content within the solvent itself can affect the reading. Samples need to be diluted within the same base solvent as well, should the concentration of the samples not be within the linear part of the standard curve.
5. The concentration of the highest CS standard should not be higher than 25 μg/mL for conditioned medium assays and 62.5 μg/mL for lysed cell assays. This corresponds to a CS content that should be no greater than 1.25 μg per well. Above the mentioned concentrations, the standard curve will leave its linear range of sensitivity (Fig. 3a, b). The DMMB assay would not be able to accurately discern between a CS content of 5 μg and 10 μg within a well. Therefore, we strongly advise the reader not to extrapolate if samples do not fall within the standard curve. Samples need to be diluted so they can fit within a range of 0–1.25 μg CS within a well.
6. Should the sGAG content within the sample be low (*e.g.*, conditioned medium samples from a small number of cells), one can use a higher sample to dye ratio (*e.g.*, 50 μL of sample to 150 μL of dye solution).
7. Precipitates can form shortly (within minutes) after addition of the dye reagent; therefore, reading should be performed immediately [9].

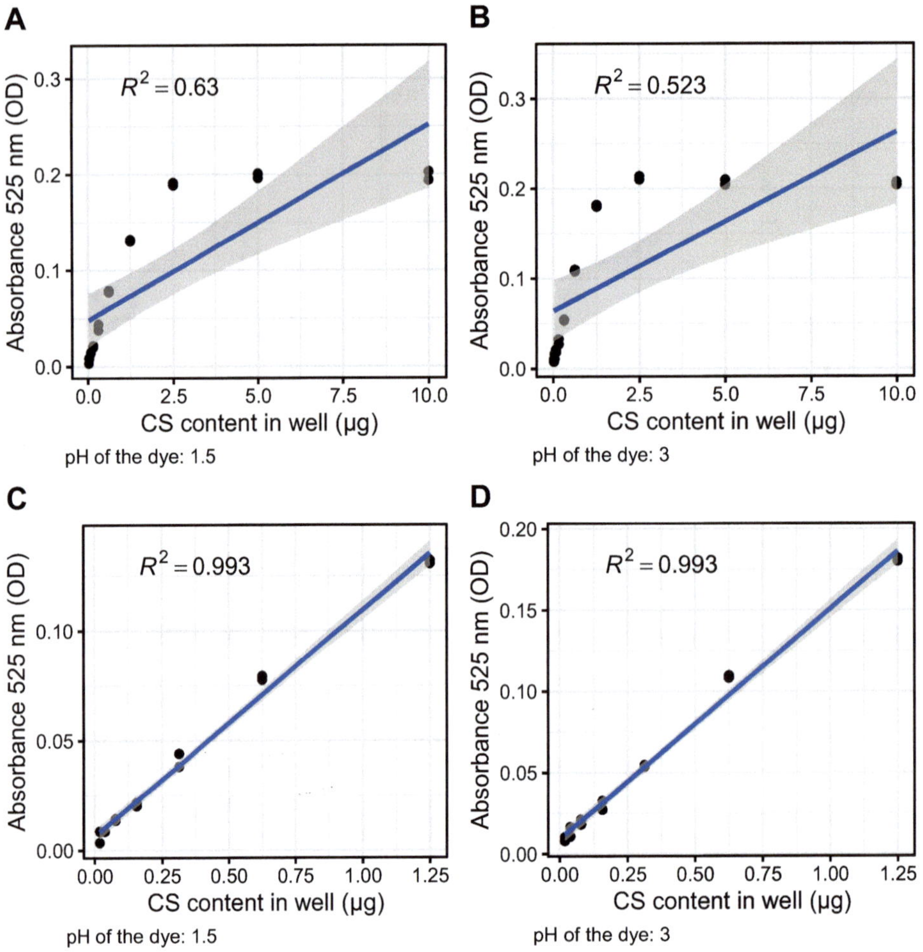

Fig. 3 Standard curves (blue line with shaded gray area covering the 95% confidence interval) for CS standards prepared in DMEM with DMMB color reagent (dye) at pH 1.5 and pH 3. The chondroitin sulfate (CS) content is measured at A525. Above a CS content of ca. 1.25 μg within a well, the standard curve is neither linear at pH 1.5 (**a**) nor pH 3 (**b**). Below a CS content of ca. 1.25 μg within a well, the standard curve is linear at both pH 1.5 (**c**) and pH 3 (**d**). A higher R^2 value indicates better fit of the curve and should be close to 1

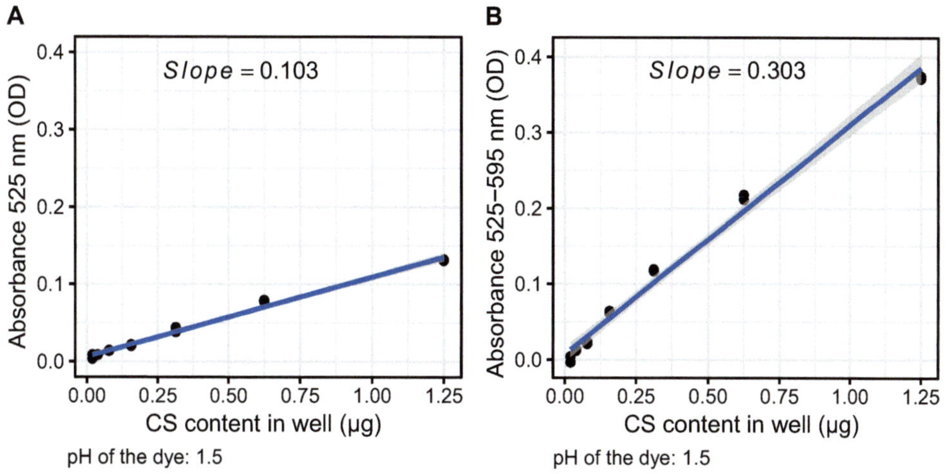

Fig. 4 Standard curves (blue line with shaded gray area covering the 95% confidence interval) for CS standards prepared in DMEM with DMMB color reagent (dye) at pH 1.5. For a single reading at A525 (**a**), the linear standard curve has a flatter slope than for the double reading A525-A595 (**b**), which indicates inferior sensitivity of the single reading compared to the double reading

References

1. Hjerpe A, Antonopoulos CA, Engfeldt B (1979) Determination of sulphated disaccharides from chondroitin sulphates by high-performance liquid chromatography. J Chromatogr 171:339–344. https://doi.org/10.1016/s0021-9673(01)95313-0
2. Liu X, Krishnamoorthy D, Lin L et al (2018) A method for characterising human intervertebral disc glycosaminoglycan disaccharides using liquid chromatography-mass spectrometry with multiple reaction monitoring. Eur Cell Mater 35:117–131. https://doi.org/10.22203/eCM.v035a09
3. Farndale RW, Sayers CA, Barrett AJ (1982) A direct spectrophotometric microassay for sulfated glycosaminoglycans in cartilage cultures. Connect Tissue Res 9(4):247–248. https://doi.org/10.3109/03008208209160269
4. Taylor KB, Jeffree GM (1969) A new basic metachromatic dye, I:9-dimethyl methylene blue. Histochem J 1(3):199–204. https://doi.org/10.1007/BF01081408
5. Zheng CH, Levenston ME (2015) Fact versus artifact: avoiding erroneous estimates of sulfated glycosaminoglycan content using the dimethylmethylene blue colorimetric assay for tissue-engineered constructs. Eur Cell Mater 29:224–236; discussion 236
6. Kuiper NJ, Sharma A (2015) A detailed quantitative outcome measure of glycosaminoglycans in human articular cartilage for cell therapy and tissue engineering strategies. Osteoarthr Cartil 23(12):2233–2241. https://doi.org/10.1016/j.joca.2015.07.011
7. Whitley CB, Spielmann RC, Herro G et al (2002) Urinary glycosaminoglycan excretion quantified by an automated semimicro method in specimens conveniently transported from around the globe. Mol Genet Metab 75(1):56–64. https://doi.org/10.1006/mgme.2001.3271
8. Afyoncu E, Yilmaz G, Yilmaz FM et al (2013) Performance of different screening methods for the determination of urinary glycosaminoclycans. Clin Chem Lab Med 51(2):347–350. https://doi.org/10.1515/cclm-2012-0471
9. Mort JS, Roughley PJ (2007) Measurement of glycosaminoglycan release from cartilage explants. Methods Mol Med 135:201–209. https://doi.org/10.1007/978-1-59745-401-8_12

Chapter 10

Extracellular Vesicle Isolation and Characterization for Applications in Cartilage Tissue Engineering and Osteoarthritis Therapy

Alexander Otahal, Andrea De Luna, Ali Mobasheri, and Stefan Nehrer

Abstract

Extracellular vesicles (EVs) have the capacity for use in cartilage tissue engineering by stimulating tissue repair and microenvironmental reprogramming. This makes them ideal candidates for treating focal cartilage defects and cartilage degeneration in osteoarthritis (OA). Observational studies have reported beneficial biological effects of EVs, such as inhibition of inflammation, enhanced extracellular matrix deposition, and reduced cartilage degradation. Isolation of EVs derived from different source materials such as conditioned cell culture media or biofluids is essential to attribute observed biological effects to EVs as genuine effectors. This chapter presents a density- and a size-based method as well as a combination of both for isolation of EVs from conditioned cell culture media or biofluids. In addition, three methods for characterization of isolated EVs are suggested based on physical properties, protein profiling, and ultrastructural morphology.

Key words Osteoarthritis, Cartilage defect, Chondrocyte, Extracellular vesicle, Regeneration, Immunomodulation

1 Introduction

Articular chondrocytes are the only cell type present in the cartilage that lines the articulating surfaces of load-bearing synovial joints. Despite its simple tissue architecture, cartilage has very limited capacity for intrinsic regeneration following injury and is prone to degeneration as seen in osteoarthritis (OA), the most common form of arthritis. Cell-based therapies that focus on rebuilding degraded or damaged cartilage employ primary chondrocytes, mesenchymal/stromal stem cells (MSCs), or genetically modified cells. MSCs are multipotent stem cells that are found in various adult and perinatal tissues including the skin, adipose tissue, bone marrow, dental pulp, as well as the amnion or umbilical cord [1]. MSCs can differentiate into chondrocytes due to their multi-lineage potential,

Martin J. Stoddart et al. (eds.), *Cartilage Tissue Engineering*,
Methods in Molecular Biology, vol. 2598, https://doi.org/10.1007/978-1-0716-2839-3_10,

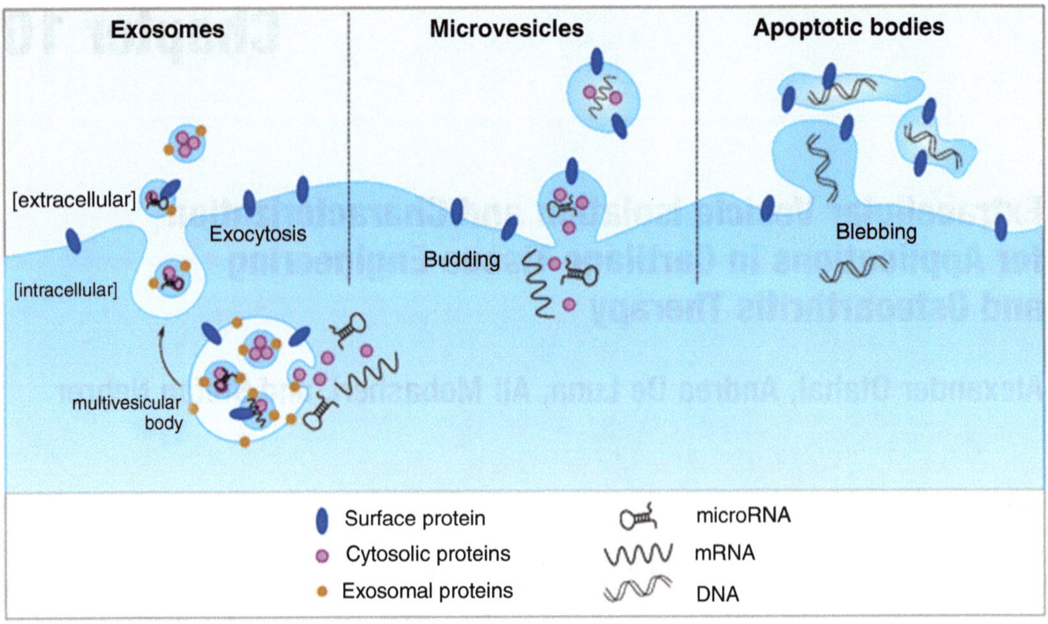

Fig. 1 Synthesis and cargos of EV subtypes

together with their low-immunogenicity MSCs which were frequently used in the treatment musculoskeletal conditions, in particular cartilage degradation [2–4]. Drawbacks of cell therapies include failure of engraftment, erroneous differentiation, and uncontrolled proliferation at the target site [5–7], even tumorigenic side effects of MSC therapies are discussed in the literature due to the replicative potential of MSCs [8]. However, mounting evidence suggests that the MSC secretome, in particular MSC-derived extracellular vesicles (EVs), rather than cell-mediated effects is the main driver of a therapeutic outcome [9–11]. EVs are a heterogeneous group of membrane nanoparticles with a size between 50 and 5000 nm [12]. They are synthesized by cells either in the cytosol or at the cell surface. Based on the site of origin, EVs are classified into exosomes, microvesicles, and apoptotic bodies (Fig. 1). Exosomes are 50–150 nm in size and are generated intracellularly within multivesicular bodies (MVBs) and are released via exocytosis. Microvesicles are considered to be larger than 100 nm, smaller than 1000 nm, and are formed via budding at the plasma membrane upon rising intracellular Ca^{2+} concentration [13]. Apoptotic bodies are remnants of apoptotic cells up to 5000 nm in size [14, 15]. It is currently not possible to identify unequivocally an EV based on marker proteins, unique lipids, or specific cargos as soon as an EV has been released from a cell and the general term EV should be used [16]. Nevertheless, small EVs tend to be enriched in TSG101, Alix and the tetraspanins CD9, CD63, or CD81, as well as heat shock proteins 70 and 90 (Hsp70, Hsp90)

[17–21], while large EVs are characterized by the exposure of negatively charged lipids such as phosphatidylserine (PS) [22]. To describe an EV population, current recommendations state that the presence of EV marker proteins and absence of contaminating components such as lipoproteins should be determined and details about the source material, cellular origin, size, and/or density range should be provided [16]. Other methods for EV characterization include nanoparticle tracking analysis (NTA) or electron microscopic techniques such as cryo-electron microscopy (cryo-EM) or transmission electron microscopy (TEM) [22].

EVs have been initially discovered as "platelet dust" in platelet-free plasma [23], before it became evident that these particles are not only released from platelets [24], but also reticulocytes [25, 26]. Nowadays, a variety if not all cells have been shown or are considered to be capable of EV release, including chondrocytes [27, 28]. Interestingly, neither CD9, CD63, CD81, nor Hsp70 and Hsp90 were detected in articular cartilage derived EVs [29]. EVs have been detected in all body fluids including saliva, urine, milk, synovial fluid, or blood [30–35]. However, the paradigm of EVs as means of waste disposal [36] has changed, because EV-associated signal molecules were found to be biologically active in immunomodulation, angiogenesis, and metastasis, and their active role in cartilage homeostasis has been acknowledged [37–40]. There is evidence that EVs are also involved in OA disease progression, as IL1beta stimulated synoviocytes release EVs that reduce chondrocytic gene expression in articular chondrocytes [41]. This highlights the intercellular communication that is mediated by EVs in physiologic and pathologic conditions.

Nevertheless, EVs can promote cartilage recovery as shown for MSC-derived EVs in vitro and in vivo [42–44]. Chondrocytes take up bone marrow MSC-derived EVs in vitro and respond with increased aggrecan and type II collagen expression concomitant with decreased IL-1, IL6, IL8, and IL17, as well as COX2 expression as a result of decreased NF-κB activity [42]. Adipose MSC-derived EVs downregulate COX2 and reduce TNF-a, IL6, prostaglandin E2, nitric oxide (NO) release, and MMP-13 activity, while anti-inflammatory IL-10 release and type II collagen expression increased by EV treatment [45]. Similarly, EVs enhanced cartilage matrix deposition and reduced matrix degradation via ADAMTS5 or MMP-2 repression in vivo [46, 47]. So far, therapeutic EV-based approaches for cartilage recovery focus on EVs derived from MSCs, and little is known about regenerative potentials of EV fractions in body fluids such as blood or blood-derived products, which are frequently applied as intra-articular injections [48–50]. However, EVs are only a small fraction of submicron particles present in body fluids and typical cell culture media. Therefore, proper isolation is required to reliably attribute observed therapeutic effects to EVs and to avoid confounded experimental results by co-purified components such as lipoproteins. Residual oxidized low-density lipoprotein (ox-LDL) binds to Toll-like receptor 4 (TLR4) [51–53], which could induce IL1beta release and aggravate OA and cartilage degradation counteracting potential regenerative effects mediated by EVs. This might be relevant in autologous blood product therapy in elderly patients [54], although there were no differences in anti-inflammatory cytokine profiles between OA patients and healthy volunteers [55]. Similarly, residual high-density lipoprotein (HDL), which can bind potentially pro-inflammatory miRNAs [56, 57], could adversely affect chondrocytes or joint tissues upon intracellular injection of EV preparations.

Ultracentrifugation (UC) is still considered a gold standard technique to enrich EVs from cell culture supernatants and body fluids [58]. The method is scalable and easy to implement. Disadvantages include the cost of the equipment and the long duration up to several hours of the preparation of an EV batch [59]. Extravesicular protein aggregates or HDL are usually co-enriched depending on the sample type, and the integrity of EVs could be compromised by internal friction or aggregation [60–62]. Size-exclusion chromatography (SEC) emerged as method to separate small soluble components from EVs. SEC is considered a gentler method of EV isolation and requires low-cost equipment; however, it could co-isolate large (lipo-)proteins [63].

The optimal EV isolation procedure might be dependent on the type of source material. EV recovery and yield might also be influenced by isolation procedure [64]. Cell culture supernatant EVs might be isolated via SEC only methods, in case the EV production protocols was performed under serum-free conditions. Therefore, low amounts of non-EV particles can be expected that would end up in EV preparations. However, body fluids or cell culture systems supplemented with EV-depleted sera are more complex mixtures with components that might be co-enriched or co-purified together with EVs such as protein aggregates or lipoproteins. Therefore, more sophisticated approaches are required for isolation of pure EVs such as combinations of methods employing different physical principles including size and density to yield purified EVs. This chapter focuses on the isolation and characterization of EVs.

2 Materials

2.1 Devices and Software

- Ultracentrifuge OptimaMax.
- MLA-80 fixed angle rotor.
- UC-tubes (Beckman, #355647).
- SEC columns qEVoriginal (iZON).
- ZetaView (ParticleMetrix).

- Synergy 2 plate reader/Gen5 software.
- Buffer chamber.
- SDS-PAGE 4–12% precast gels (Invitrogen, # NP0321).
- iBlot semidry transfer device (Invitrogen).
- hemiDoc/PDQuest 7.4.
- Cryo-electron microscope, e.g., Glacios (Thermo Scientific).

2.2 Buffers, Reagents, and Media

- PBS (GIBCO, #70011-036).
- NTA-standard 100 nm (Applied Microspheres).
- RIPA buffer (Pierce, #89900).
- DC assay (Bio-Rad, #5000112).
- MES (Invitrogen, #NP0002)
- MOPS (Invitrogen, #NP0001).
- 1 M DTT stock solution.
- LDS sample buffer (Invitrogen, #NP0007).
- Nonfat dry milk powder.
- Molecular weight marker (Invitrogen, #LC5800 or #LC5699).
- WesternBright ECL reagent (Advansta, #K-12045).
- CD9 (System-Biosciences, #EXOAB-CD9A-1).
- CD63 (BioLegend, #353005, clone H5C9).
- Alix (Cell Signaling, #2171).
- ApoAl (Santa Cruz, #sc-376818).
- ApoB100 (Santa Cruz, #sc-393636).
- Liquid ethane.

2.3 Consumables

- 0.22 μm filter for PBS (Sartorius, #16534).
- 1 ml syringes.
- Transfer stacks (Invitrogen, #IB401031).
- Copper grids (1 μm mesh) for cryo-electron microscopy.

3 Methods

3.1 Isolation of EVs

EV isolation from body fluids or cell culture supernatants is started by depletion of residual cells and cell debris via low-speed centrifugation (2500 × g; 15 min; room temperature). The pellet is discarded, and the precleared supernatant is used for subsequent isolation steps. Optionally, the preclearing can be repeated. The preclearing step is essential to avoid contamination of EV preparations with large non-EV particles.

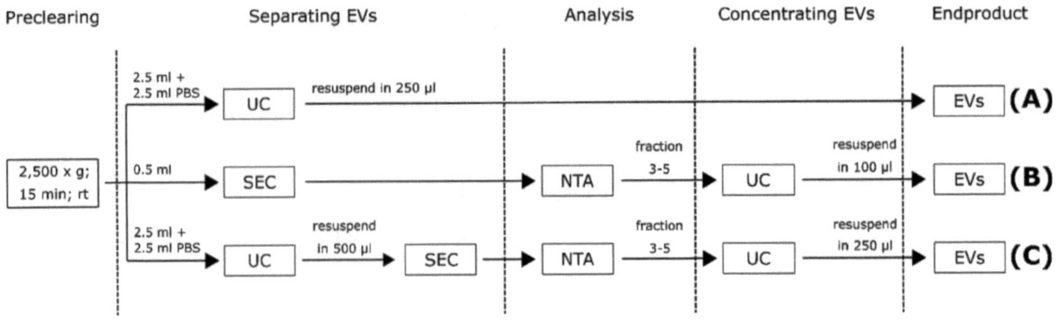

Fig. 2 Overview of presented EV isolation strategies

3.2 Ultracentrifugation (UC)

Isolation or enrichment of EVs via UC is still considered a gold standard technique to obtain EVs from various body fluids [58] (Fig. 2a). Predilution of body fluids such as plasma, serum, or synovial fluid is advised as EVs might experience damage due to internal friction in viscous media [61]. Cell culture media might not need predilution, as they behave similarly to water [65]. EV preparation from synovial fluid, for example, might require pretreatments with enzymes such as hyaluronidase to optimize EV yield [67].

1. Put the fixed angle UC rotor (MLA80; k-factor 29; 45,000 rpm) in the fridge 30 min before a UC run for precooling.
2. Prepare particle depleted PBS (fPBS) devoid of Mg^{2+} and Ca^{2+} by filtration through a 0.2 µm pore size filter (Sartorius, #16534).
3. Label UC tubes (Beckman-Coulter, #355647), and mark the side of the tube which will face outward in the fixed angle UC rotor (MLA80; k-factor 29; 45,000 rpm) (*see* **Note 1**). Put a balance with at least 0.001 g precision inside a laminar flow hood.
4. Place 2.5 ml precleared blood product in the UC tubes under sterile conditions.
5. Add 2.5 ml fPBS.
6. Weigh all filled tubes, and adjust weights with a precision of at least 0.01 g with fPBS.
7. Spin the samples at 100,000 × g for 120 min at 4 °C.
8. For body fluids such as plasma or serum, three phases will be present in the tubes: (1) an opaque fat layer floating on top; (2) a cleared supernatant below; and (3) a gel-like EV pellet on the lower part of the outward facing tube wall. The pellet might not be visible before removing phase 1 and 2. Cell culture supernatants typically do not show a phase (1).

9. Carefully aspirate the fat layer, which contains lipoproteins such as LDL that account for more than 95% of all particles detectable in the precleared blood product.
10. After complete removal of fat, aspirate the remaining supernatant while progressively tilting the tube. Thereby, the fluid flows away from the pellet and can be aspirated easily without coming into vicinity of the pellet with the tip.
11. Aspirate any drops of liquid that might still adhere to the tube wall.
12. Add 250 μl fPBS onto the pellet (*see* **Note 2**).
13. Set the micropipette to 180 μl, and resuspend the pellet by letting the buffer flow over the pellet at least 20–40 times. Complete resuspension should be checked visually (*see* **Note 3**).
14. Store the EV suspension at −80 °C or proceed to characterization.

3.3 Size-Exclusion Chromatography (SEC)

SEC separates EVs and EV-sized particles from soluble proteins. Particles elute first, because components smaller than the cut-off size of the SEC matrix are retained within the porous matrix and have a longer residence time within the column than particles that cannot enter the pores.

1. Filtrate PBS devoid of Mg^{2+} and Ca^{2+} a 0.2 μm pore size filter to prepare particle depleted PBS (fBPS) (Sartorius, #16534). Presence of Mg^{2+} or Ca^{2+} in a phosphate buffer could promote formation of phosphate nanocrystals that could mistakenly be identified as EVs by NTA.
2. Place the prefilled SEC column (iZON, qEVoriginal) in a column holder, open the top cap, and attach the reservoir funnel.
3. Add 14 ml fPBS, open the bottom cap to replace the storage buffer by gravity flow.
4. Shortly before the frit runs dry add up to 500 μl sample.
5. Let it settle inside the gel matrix; then immediately add gently 14 ml fPBS for elution.
6. Start taking 1 ml fractions by holding 1.5 ml microcentrifuge tubes to catch the drops eluting from the column. About 20 drops sum up to 1 ml (*see* **Note 4**).

After taking 20 fractions, the sample should have completely eluted, and the column should be washed. Therefore, add 14 ml fPBS twice to wash out any remaining material. Columns can be reused several times. When elution time increases, columns should be regenerated according to the manufacturer's recommendations (*see* **Note 5**).

When taking 1 ml fraction, fraction 3 to 5 should be the EV-rich fractions; however, it is always advised to perform NTA of fraction 2 to 7 to affirm their particle concentration before proceeding. Starting from fraction 6, soluble proteins elute from the column. Successful separation of EVs from soluble proteins should be checked by determining protein concentrations in the individual fraction by an appropriate total protein assay, for example, DC assay (BioRad). Particle-rich fractions might show a protein concentration of around zero at this point.

As SEC dilutes the sample, EVs have to be concentrated afterwards. This can be done via ultracentrifugation (Fig. 2b).

7. Pool up to five particle-rich fractions as determined via NTA in UC tubes (Beckman-Coulter, #355647), and mark the side of the tube which will face outward in the fixed angle UC rotor (MLA80; k-factor 29; 45,000 rpm). If less than five fractions are pooled, add fPBS up to 5 ml, and adjust weights with fPBS as described above.
8. Centrifuge at 100,000 × g for 120 min at 4 °C.
9. Pellets might be very transparent or invisible.
10. Aspirate the supernatant while progressively tilting the tube.
11. Aspirate any drops of liquid that might still adhere to the tube wall.
12. Add 250 μl fPBS onto the pellet.
13. Set the micropipette to 180 μl, and resuspend the pellet by letting the buffer flow over the pellet at least 20–40 times.
14. Store EVs at −80 °C or proceed to characterization.

3.4 Combined UC and SEC

As enrichment of EVs via UC might result in the co-enrichment of non-EV proteins such as HDL [60] and isolating EVs from blood plasma, serum or cell culture media supplemented with EV-depleted serum via SEC co-isolates large lipoprotein particles [66], a combination method for EV isolation such as UC and SEC in tandem can yield purified EVs (Fig. 2c).

1. Perform EV enrichment via UC as described above (*see* Subheadings 3.3, **steps 3–14**).
2. Resuspend the EV pellet in 500 μl fPBS.
3. Transfer the EV suspension onto a SEC column.
4. Perform SEC as described above (*see* Subheading 3.3, **steps 4–11**).
5. Add 100 μl fPBS onto the pellet. Attention! The pellet after UC+SEC can be very small compared to an UC pellet. Take special care to not overlook the pellet. Set the micropipette to 80 μl and resuspend the pellet by letting the buffer flow over the pellet at least 20–40 times.
6. Store EVs at −80 °C or proceed to characterization.

3.5 Characterization of EVs

Current methods of EV isolation yield a heterogeneous population of EVs and potentially co-isolated particles and proteins. To assess purity or the extent of non-EV components in a preparation, various methods exist including nanoparticle tracking analysis (NTA), western blot (WB), or cryo-EM [22].

3.5.1 Nanoparticle Tracking Analysis (NTA)

The method is able to characterize the concentration of submicron particles present in samples like cell culture supernatants, but also body fluids including urine, cerebrospinal fluid, synovial fluid, blood, and blood derivates.

Particle concentrations are usually too high in biological samples, therefore appropriate dilutions have to be prepared. The following considerations should be kept in mind:

- To start with a new sample type, a 1:1000 dilution is recommended as a starting point to find the most suitable dilution.
- The test sample should yield 100–200 particles per field of view to allow optimal particle tracking.
- Higher particle numbers make it difficult to track individual particles, which could result in wrong particle sizes determined from a wrongly determined Brownian motion.
- Low particle numbers decrease the reliability of the measurement as a significant proportion of detected signals might be background rather than actual sample.
- Avoid diluting biological samples with water as this might result in aggregation of EVs.

Follow these steps to prepare a dilution of the sample preferentially in 1 ml of PBS.

1. Put 1000 - x μl PBS in a microcentrifuge tube, where x is the volume of sample added in the next step.
2. Add x μl sample.
3. Mix by gently pipetting up and down several times and stirring with the tip.
4. Alternatively, close the cap, and gently invert several times to ensure mixing, especially when small sample volumes are added (*see* **Note 6**).
5. Take up dilution with a 1 ml syringe, and release it three times to improve the mixing.
6. Inject the dilution and check the number of particles in the field of view.
7. If OK, proceed to setting preacquisition camera settings:
   (a) Determine the optimal camera sensitivity. This ensures that detected signals reflect light refracted from particles

rather than background. Higher sensitivity also collects signals from smaller particles, while setting sensitivity too low might lead to overlooking particles (*see* **Note 7**).

(b) Adjust the shutter to avoid overexposure. Refracted light from overexposed large particles might superimpose signals from smaller particles.

8. Wait until sample has equilibrated in the flow chamber and particle drift is within an acceptable range.
9. Always measure samples in duplicate or triplicate because inefficient mixing can cause large errors.
10. Multiply the concentration with the dilution factor to obtain the original particle concentration.
11. If interested in particle size, it is recommended to investigate mode size rather than average or median size of particles in the sample, because the size distribution is often skewed in biological samples.
12. As a control, measure an aliquot of diluent (PBS) to determine the amount of background signals that are introduced by the diluent. The particle concentration of the diluent should be in the range of 10^6 particles per ml or lower, which is the detection limit of the NTA device. Determining the particle size in the diluent is usually not possible due to too low number of trackable particles.

3.5.2 Western Blot Analysis

Protein profiling is recommended by the MISEV 2018 guidelines to check the presence of EV markers and absence of contaminants [16].

Determine total protein content of EV preparations via DC assay, which is compatible with detergents required for lysis of EVs prior to total protein analysis.

1. Mix 5–10 μl EV suspension as well as 5–10 μl BSA standard dilutions with 5–10 μl RIPA buffer in 96-well plates.
2. Incubate at 4 °C for 10 min.
3. Add 25 μl reagent A′ and 200 μl reagent B as described in the manufacturer's protocol.
4. Incubate for at least 15 min at room temperature in the dark.
5. Measure absorption at 670 nm on a plate reader.

Use an appropriate volume of EV suspension carrying 10 μg of total protein per sample. While this is enough for UC-enriched EVs, SEC- or UC+SEC-, enriched EVs might require higher total protein up to 80 μg to be able to detect EV marker proteins (*see* **Note 8**). When aiming to detect tetraspanins like CD9 or CD63, avoid reducing the protein samples via DTT or other reducing agents, as this destroys the epitopes and completely prevents detection.

6. Mix the determined volume of EV suspension and 8 μl 4X LDS sample buffer in a 1.5 ml test tube. Optional: Add 2 μl 1 M DTT in H_2O for reducing disulfide bonds.
7. Add PBS to a final volume of 32 μl.
8. Denature the sample at 95 °C for 10 min.
9. Thaw prestained molecular weight marker (Invitrogen, #LC5800 or #LC5699).
10. Assemble gel chamber for protein separation. Use 4–12% Bis-Tris gradient gels when separation and detection of proteins larger than 150 kD is desired. Alternatively, 10% Bis-Tris gels can be used for better separation of smaller proteins.
11. Choose MES- or MOPS running buffer to achieve more efficient separation in the lower or upper size range, respectively.
12. Gently load the samples and molecular weight marker, and avoid overflowing of the slots.
13. Run the gel at 150 V until the lowest protein marker band reaches the lower quarter of the gel. Increasing the voltage shortens the run time but increases heat generation and decreases separation efficiency.
14. After separation, disassemble the gel cartridge, cut away the slot ridges, and equilibrate the gel in running buffer.
15. In the meanwhile, place the lower electrode of iBlot transfer stack into the iBlot transfer device.
16. Always place the PVDF membrane in pure methanol for 1–5 min to assure optimal protein adhesion.
17. Rinse PVDF membrane with distilled water, and place it back on the electrode.
18. Place the gel onto the membrane. Be careful when placing a 4–12% gradient gel as the upper gel part can easily rupture.
19. Put the filter paper onto the gel and remove air bubbles with a roller.
20. Distribute 1–2 ml of distilled water on the filter paper. This reduces heat development and oxidation of the copper electrodes during transfer.
21. Finish transfer stack assembly by adding the upper electrode, and run program P3 (20 V, 13 min) twice for optimal protein transfer. The extended transfer time allows efficient transfer of very large proteins such as ApoB100/48 (~550 kD) but did not result in considerable signal loss of smaller proteins like ApoA1 (~30 kD).
22. Disassemble the transfer stack and rinse membrane in distilled water.

23. Optionally check for residual protein in the gel with simply blue total protein stain.

Stain the membrane with Ponceau dye to check successful protein transfer and to detect inconsistencies such as air bubbles which could interfere with interpretation of results after detection.

24. Place the membrane in 10 ml Ponceau dye for 5 min on a shaker.
25. Recover the dye solution for reuse. Dye can be reused up to 10 times.
26. Destain the membrane with distilled water until bands appear on a shaker.
27. Take a photo.

Detection of proteins is performed via enhanced chemiluminescence (ECL).

28. Block membrane in 5% skimmed milk in PBST (PBS + 0.1% Tween 20) for at least 1 h or overnight at 4 °C.
29. In case blocking was done overnight, place membrane on a shaker for 30 min to equilibrate to room temperature.
30. Wash membrane three times in PBST for 5 min on a shaker.
31. Add primary antibody directed against CD9, CD63, Alix, or TSG101 as EV markers or ApoA1, ApoB100, GM130, or albumin as contamination markers diluted in 1% BSA in PBST.
32. Incubate for at least 1 h at room temperature on a shaker or (preferentially) over night at 4 °C. This could improve detection of proteins such as CD9 or CD63 on nonreduced membranes.
33. Wash three times with PBST for 5 min on a shaker.
34. Add secondary antibody conjugated to horseradish peroxidase diluted in PBST.
35. Incubate for 1 h at room temperature on a shaker.
36. Wash three times with PBST for 5 min on a shaker.
37. Prepare 1.5 ml working reagent per membrane for chemiluminescent detection by mixing reagent A and B 1:1 (Advansta, #K-12045-D20).
38. Place membrane on a flat surface after removing excess buffer.
39. Add 1.5 ml working reagent per membrane.
40. Incubate for 2 min.
41. Wrap in a transparent plastic sheet, and place it in the Chemidoc device (Bio-Rad).
42. Detect for 30 min in 30 s intervals in high sensitivity mode via PDQuest 7.4 software.

43. Take a photo in Epi White mode to image the prestained molecular weight marker before removing the membrane from the device. This step can be omitted if the marker is luminescent itself.
44. Wash membrane in PBST before storing at 4 °C or reprobing with other primary antibodies.

3.5.3 Cryo-electron Microscopy (cryoEM)

The MISEV guidelines recommend that three methods are used to characterize EV preparations. To assess the structural properties of EVs in a preparation, electron microscopy can be used. Frequently, TEM is used; however, TEM sample preparation involves a drying step that results in collapse of EVs due to water removal. In contrast, cryo-EM retains the structural integrity of EVs due to snap freezing in liquid ethane. This allows visualization of the morphological appearance of isolated EVs (Fig. 3).

1. Place 10 μl sample onto 1 μm copper grid. Recommended particle amount is 1 × 10^9 to 1 × 10^10 as determined via NTA. Less particles can be used.
2. Blot the grid for 1–4 s onto filter paper.
3. Plunge the grid in liquid ethane.
4. Proceed to visualization on a Glacios device.

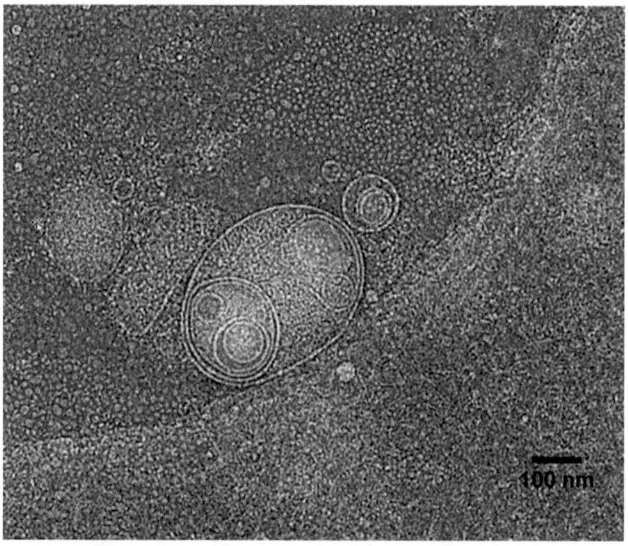

Fig. 3 Cryo-electron micrograph of EVs isolated from blood plasma via combined UC and SEC (*see* Fig. 2C). Note the presence of small EVs and large multilamellar EVs indicating the heterogeneity of EV populations

4 Notes

1. Different source materials for EV isolation might have different viscosity; therefore, time and speed of ultracentrifugation should be adjusted as a compensation for optimal EV recovery [61].
2. IMPORTANT! The EV pellet will appear as very small gel-like transparent material; the tube marking will tell the area where the pellet should be located after centrifugation. Otherwise, the resuspension of EVs might fail.
3. IMPORTANT! AVOID BUBBLES! Bubbles will disrupt EVs.
4. There is a void volume which does not contain sample material, neither EVs nor soluble proteins. The void volume can be skipped if the volume is known. Taking fractions too late might result in EV loss, which elute first.
5. A maximum reuse of five times is recommended for the SEC columns, but no decrease in separation efficiency was observed after extended use. To store columns for reuse, make sure all sample material was completely eluted. If unsure, wash with one additional column volume (10 ml) elution buffer (fPBS). Store at 4 °C with closed caps in PBS; never let to dry out the column matrix. If a decrease in elution velocity is noticed, i.e., when drops hang longer than approximately 4 s before abscission, clean the column with one-column volume 0.5 M NaOH and three-column volumes PBS.
6. IMPORTANT! NEVER VORTEX THE DILUTION FOR MIXING! This creates nanobubbles which are detected by the NTA device as particles! Similarly, dropping sample tubes onto the bench or the floor obscures actual particles with loads of nanobubbles.
7. Although each sample might have an individual optimal camera sensitivity, it is recommended to agree on a constant sensitivity to allow comparison across samples.
8. EV populations can differ substantially in their EV protein marker sets. While CD9, CD63, or CD81 were not found in chondrocyte EVs, syntenin could be used instead [29].

References

1. Heldring N, Mäger I, Wood MJ et al (2015) Therapeutic potential of multipotent mesenchymal stromal cells and their extracellular vesicles. Hum Gene Ther 26(8):506–517. https://doi.org/10.1089/hum.2015.072
2. Mancuso P, Raman S, Glynn A et al (2019) Mesenchymal stem cell therapy for osteoarthritis: the critical role of the cell secretome. Front Bioeng Biotechnol 7:9–9. https://doi.org/10.3389/fbioe.2019.00009
3. Kong L, Zheng L-Z, Qin L et al (2017) Role of mesenchymal stem cells in osteoarthritis treatment. J Orthop Transl 9:89–103. https://doi.org/10.1016/j.jot.2017.03.006

4. Song Y, Zhang J, Xu H et al (2020) Mesenchymal stem cells in knee osteoarthritis treatment: a systematic review and meta-analysis. J Orthop Transl 24:121–130. https://doi.org/10.1016/j.jot.2020.03.015
5. O'Shea C, Hynes SO, Shaw G et al (2008) 271. Adverse effects of mesenchymal stem cell transplantation in a denuded Rabbit Carotid artery. Mol Ther 16:S102. https://doi.org/10.1016/S1525-0016(16)39674-5
6. Wyles CC, Houdek MT, Behfar A et al (2015) Mesenchymal stem cell therapy for osteoarthritis: current perspectives. Stem Cells Clon Adv Appl 8:117–124. https://doi.org/10.2147/sccaa.s68073
7. Lukomska B, Stanaszek L, Zuba-Surma E et al (2019) Challenges and controversies in human mesenchymal stem cell therapy. Stem Cells Int 2019:9628536. https://doi.org/10.1155/2019/9628536
8. Musiał-Wysocka A, Kot M, Majka M (2019) The pros and cons of mesenchymal stem cell-based therapies. Cell Transplant 28(7):801–812. https://doi.org/10.1177/0963689719837897
9. Bruno S, Deregibus MC, Camussi G (2015) The secretome of mesenchymal stromal cells: role of extracellular vesicles in immunomodulation. Immunol Lett 168(2):154–158. https://doi.org/10.1016/j.imlet.2015.06.007
10. Bian S, Zhang L, Duan L et al (2014) Extracellular vesicles derived from human bone marrow mesenchymal stem cells promote angiogenesis in a rat myocardial infarction model. J Mol Med (Berl) 92(4):387–397. https://doi.org/10.1007/s00109-013-1110-5
11. ter Huurne M, Schelbergen R, Blattes R et al (2012) Antiinflammatory and chondroprotective effects of intraarticular injection of adipose-derived stem cells in experimental osteoarthritis. Arthritis Rheum 64(11):3604–3613. https://doi.org/10.1002/art.34626
12. van der Pol E, Böing AN, Harrison P et al (2012) Classification, functions, and clinical relevance of extracellular vesicles. Pharmacol Rev 64(3):676–705. https://doi.org/10.1124/pr.112.005983
13. Rubin O, Canellini G, Delobel J et al (2012) Red blood cell microparticles: clinical relevance. Transfus Med Hemother 39(5):342–347. https://doi.org/10.1159/000342228
14. De Jong OG, Van Balkom BWM, Schiffelers RM et al (2014) Extracellular vesicles: potential roles in regenerative medicine. Front Immunol 5(608). https://doi.org/10.3389/fimmu.2014.00608
15. Yáñez-Mó M, Siljander PRM, Andreu Z et al (2015) Biological properties of extracellular vesicles and their physiological functions. J Extracell Vesicl 4:27066–27066. https://doi.org/10.3402/jev.v4.27066
16. Thery C, Witwer KW, Aikawa E et al (2018) Minimal information for studies of extracellular vesicles 2018 (MISEV2018): a position statement of the International Society for Extracellular Vesicles and update of the MISEV2014 guidelines. J Extracell Vesicl 7(1):1535750. https://doi.org/10.1080/20013078.2018.1535750
17. Wyciszkiewicz A, Kalinowska-Łyszczarz A, Nowakowski B et al (2019) Expression of small heat shock proteins in exosomes from patients with gynecologic cancers. Sci Rep 9(1):9817. https://doi.org/10.1038/s41598-019-46221-9
18. Chanteloup G, Cordonnier M, Isambert N et al (2020) Membrane-bound exosomal HSP70 as a biomarker for detection and monitoring of malignant solid tumours: a pilot study. Pilot Feasibil Stud 6(1):35. https://doi.org/10.1186/s40814-020-00577-2
19. Andreu Z, Yáñez-Mó M (2014) Tetraspanins in extracellular vesicle formation and function. Front Immunol 5:442–442. https://doi.org/10.3389/fimmu.2014.00442
20. Iavello A, Frech VS, Gai C et al (2016) Role of Alix in miRNA packaging during extracellular vesicle biogenesis. Int J Mol Med 37(4):958–966. https://doi.org/10.3892/ijmm.2016.2488
21. Raposo G, Stoorvogel W (2013) Extracellular vesicles: exosomes, microvesicles, and friends. J Cell Biol 200(4):373
22. Tissot J-D, Canellini G, Rubin O et al (2013) Blood microvesicles: from proteomics to physiology. Transl Proteom 1(1):38–52. https://doi.org/10.1016/j.trprot.2013.04.004
23. Wolf P (1967) The nature and significance of platelet products in human plasma. Br J Haematol 13(3):269–288. https://doi.org/10.1111/j.1365-2141.1967.tb08741.x
24. O'Brien JR (1955) The platelet-like activity of serum. Br J Haematol 1(2):223–228
25. Harding C, Stahl P (1983) Transferrin recycling in reticulocytes: pH and iron are important determinants of ligand binding and processing. Biochem Biophys Res Commun 113(2):650–658. https://doi.org/10.1016/0006-291x(83)91776-x
26. Pan BT, Johnstone RM (1983) Fate of the transferrin receptor during maturation of

sheep reticulocytes in vitro: selective externalization of the receptor. Cell 33(3):967–978. https://doi.org/10.1016/0092-8674(83)90040-5
27. Anderson HC (1969) Vesicles associated with calcification in the matrix of epiphyseal cartilage. J Cell Biol 41(1):59–72. https://doi.org/10.1083/jcb.41.1.59
28. Derfus BA, Kurtin SM, Camacho NP et al (1996) Comparison of matrix vesicles derived from normal and osteoarthritic human articular cartilage. Connect Tissue Res 35(1-4):337–342. https://doi.org/10.3109/03008209609029209
29. Rosenthal AK, Gohr CM, Ninomiya J et al (2011) Proteomic analysis of articular cartilage vesicles from normal and osteoarthritic cartilage. Arthritis Rheum 63(2):401–411. https://doi.org/10.1002/art.30120
30. Iwai K, Minamisawa T, Suga K et al (2016) Isolation of human salivary extracellular vesicles by iodixanol density gradient ultracentrifugation and their characterizations. J Extracell Vesicl 5(1):30829. https://doi.org/10.3402/jev.v5.30829
31. Merchant ML, Rood IM, Deegens JKJ et al (2017) Isolation and characterization of urinary extracellular vesicles: implications for biomarker discovery. Nat Rev Nephrol 13(12):731–749. https://doi.org/10.1038/nrneph.2017.148
32. van Herwijnen MJC, Zonneveld MI, Goerdayal S et al (2016) Comprehensive proteomic analysis of human milk-derived extracellular vesicles unveils a novel functional proteome distinct from other milk components*. Mol Cell Proteomics 15(11):3412–3423. https://doi.org/10.1074/mcp.M116.060426
33. Foers AD, Chatfield S, Dagley LF et al (2018) Enrichment of extracellular vesicles from human synovial fluid using size exclusion chromatography. J Extracell Vesicl 7(1):1490145. https://doi.org/10.1080/20013078.2018.1490145
34. Serrano-Pertierra E, Oliveira-Rodríguez M, Rivas M et al (2019) Characterization of plasma-derived extracellular vesicles isolated by different methods: a comparison study. Bioengineering (Basel) 6(1). https://doi.org/10.3390/bioengineering6010008
35. Onódi Z, Pelyhe C, Terézia Nagy C et al (2018) Isolation of high-purity extracellular vesicles by the combination of iodixanol density gradient ultracentrifugation and Bind-Elute chromatography from blood plasmA. Front Physiol 9:1479–1479. https://doi.org/10.3389/fphys.2018.01479
36. Johnstone RM, Mathew A, Mason AB et al (1991) Exosome formation during maturation of mammalian and avian reticulocytes: evidence that exosome release is a major route for externalization of obsolete membrane proteins. J Cell Physiol 147(1):27–36. https://doi.org/10.1002/jcp.1041470105
37. Miyaki S, Lotz MK (2018) Extracellular vesicles in cartilage homeostasis and osteoarthritis. Curr Opin Rheumatol 30(1):129–135. https://doi.org/10.1097/BOR.0000000000000454
38. Todorova D, Simoncini S, Lacroix R et al (2017) Extracellular vesicles in angiogenesis. Circ Res 120(10):1658–1673. https://doi.org/10.1161/circresaha.117.309681
39. Becker A, Thakur BK, Weiss JM et al (2016) Extracellular vesicles in cancer: cell-to-cell mediators of metastasis. Cancer Cell 30(6):836–848. https://doi.org/10.1016/j.ccell.2016.10.009
40. Silva AM, Teixeira JH, Almeida MI et al (2017) Extracellular vesicles: immunomodulatory messengers in the context of tissue repair/regeneration. Eur J Pharm Sci 98:86–95. https://doi.org/10.1016/j.ejps.2016.09.017
41. Kato T, Miyaki S, Ishitobi H et al (2014) Exosomes from IL-1beta stimulated synovial fibroblasts induce osteoarthritic changes in articular chondrocytes. Arthritis Res Ther 16(4):R163. https://doi.org/10.1186/ar4679
42. Vonk LA, van Dooremalen SFJ, Liv N et al (2018) Mesenchymal stromal/stem cell-derived extracellular vesicles promote human Cartilage regeneration in vitro. Theranostics 8(4):906–920. https://doi.org/10.7150/thno.20746
43. Kim M, Steinberg DR, Burdick JA et al (2019) Extracellular vesicles mediate improved functional outcomes in engineered cartilage produced from MSC/chondrocyte cocultures. Proc Natl Acad Sci U S A 116(5):1569–1578. https://doi.org/10.1073/pnas.1815447116
44. Cosenza S, Ruiz M, Toupet K et al (2017) Mesenchymal stem cells derived exosomes and microparticles protect cartilage and bone from degradation in osteoarthritis. Sci Rep 7(1):16214. https://doi.org/10.1038/s41598-017-15376-8
45. Tofino-Vian M, Guillen MI, Perez Del Caz MD et al (2018) Microvesicles from human Adipose tissue-derived mesenchymal stem cells as a new protective strategy in osteoarthritic chondrocytes. Cell Physiol Biochem 47(1):11–25. https://doi.org/10.1159/000489739
46. Wang Y, Yu D, Liu Z et al (2017) Exosomes from embryonic mesenchymal stem cells alleviate osteoarthritis through balancing synthesis and degradation of cartilage extracellular

matrix. Stem Cell Res Ther 8(1):189. https://doi.org/10.1186/s13287-017-0632-0
47. Xiang C, Yang K, Liang Z et al (2018) Sphingosine-1-phosphate mediates the therapeutic effects of bone marrow mesenchymal stem cell-derived microvesicles on articular cartilage defect. Transl Res 193:42–53. https://doi.org/10.1016/j.trsl.2017.12.003
48. Filardo G, Kon E, Roffi A et al (2015) Platelet-rich plasma: why intra-articular? A systematic review of preclinical studies and clinical evidence on PRP for joint degeneration. Knee Surg Sports Traumatol Arthrosc 23(9): 2459–2474. https://doi.org/10.1007/s00167-013-2743-1
49. Filardo G, Previtali D, Napoli F et al (2020) PRP injections for the treatment of knee osteoarthritis: a meta-analysis of randomized controlled trials. Cartilage 194760352093117. https://doi.org/10.1177/1947603520931170
50. Kon E, Mandelbaum BR, Buda RE et al (2011) Platelet-rich plasma intra-articular injection versus hyaluronic acid viscosupplementation as treatments for cartilage pathology: from early degeneration to osteoarthritis. Arthroscopy J Arthroscop Relat Surg 27(11): 1490–1501
51. Sillat T, Barreto G, Clarijs P et al (2013) Toll-like receptors in human chondrocytes and osteoarthritic cartilage. Acta Orthop 84(6): 585–592. https://doi.org/10.3109/17453674.2013.854666
52. Miller YI (2005) Toll-like receptors and atherosclerosis: oxidized LDL as an endogenous Toll-like receptor ligand. Future Cardiol 1(6): 785–792. https://doi.org/10.2217/14796678.1.6.785
53. Bobacz K, Sunk IG, Hofstaetter JG et al (2007) Toll-like receptors and chondrocytes: the lipopolysaccharide-induced decrease in cartilage matrix synthesis is dependent on the presence of toll-like receptor 4 and antagonized by bone morphogenetic protein 7. Arthritis Rheum 56(6):1880–1893. https://doi.org/10.1002/art.22637
54. O'Donnell C, Migliore E, Grandi FC et al (2019) Platelet-Rich Plasma (PRP) from older males with knee osteoarthritis depresses chondrocyte metabolism and upregulates inflammation. J Orthop Res 37(8):1760–1770. https://doi.org/10.1002/jor.24322
55. O'Shaughnessey K, Matuska A, Hoeppner J et al (2014) Autologous protein solution prepared from the blood of osteoarthritic patients contains an enhanced profile of anti-inflammatory cytokines and anabolic growth factors. J Orthop Res 32(10):1349–1355. https://doi.org/10.1002/jor.22671
56. Vickers KC, Palmisano BT, Shoucri BM et al (2011) MicroRNAs are transported in plasma and delivered to recipient cells by high-density lipoproteins. Nat Cell Biol 13(4):423–433. https://doi.org/10.1038/ncb2210
57. Vickers KC, Remaley AT (2012) Lipid-based carriers of microRNAs and intercellular communication. Curr Opin Lipidol 23(2):91–97. https://doi.org/10.1097/MOL.0b013e328350a425
58. Thery C, Amigorena S, Raposo G et al (2006) Isolation and characterization of exosomes from cell culture supernatants and biological fluids. Curr Protoc Cell Biol Chapter 3:Unit 3.22. https://doi.org/10.1002/0471143030.cb0322s30
59. Pourakbari R, Khodadadi M, Aghebati-Maleki A et al (2019) The potential of exosomes in the therapy of the cartilage and bone complications; emphasis on osteoarthritis. Life Sci 236: 116861. https://doi.org/10.1016/j.lfs.2019.116861
60. Otahal A, Kramer K, Kuten-Pella O et al (2020) Characterization and chondroprotective effects of extracellular vesicles from plasma- and serum-based autologous blood-derived products for osteoarthritis therapy. Front Bioeng Biotechnol 8(1114). https://doi.org/10.3389/fbioe.2020.584050
61. Momen-Heravi F, Balaj L, Alian S et al (2012) Impact of biofluid viscosity on size and sedimentation efficiency of the isolated microvesicles. Front Physiol 3:162. https://doi.org/10.3389/fphys.2012.00162
62. Linares R, Tan S, Gounou C et al (2015) High-speed centrifugation induces aggregation of extracellular vesicles. J Extracell Vesicl 4(1): 29509. https://doi.org/10.3402/jev.v4.29509
63. Monguió-Tortajada M, Gálvez-Montón C, Bayes-Genis A et al (2019) Extracellular vesicle isolation methods: rising impact of size-exclusion chromatography. Cell Mol Life Sci 76(12):2369–2382. https://doi.org/10.1007/s00018-019-03071-y
64. Zonneveld MI, Brisson AR, van Herwijnen MJC et al (2014) Recovery of extracellular vesicles from human breast milk is influenced by sample collection and vesicle isolation procedures. J Extracell Vesicl 3(1):24215. https://doi.org/10.3402/jev.v3.24215
65. Poon C (2020) Measuring the density and viscosity of culture media for optimized computational fluid dynamics analysis of in vitro devices. bioRxiv:2020.2008.2025.266221.

https://doi.org/10.1101/2020.08.25.266221

66. Otahal A, Kuten-Pella O, Kramer K et al (2021) Functional repertoire of EV-associated miRNA profiles after lipoprotein depletion via ultracentrifugation and size exclusion chromatography from autologous blood products. Sci Rep 11(1):5823. https://doi.org/10.1038/s41598-021-84234-5

67. Boere J, van de Lest CH, Libregts SF et al (2016) Synovial fluid pretreatment with hyaluronidase facilitates isolation of CD44+ extracellular vesicles. J Extracell Vesicles 5:31751. https://doi.org/10.3402/jev.v5.31751

Chapter 11

Metabolomic Profiling to Understand Chondrocyte Metabolism

Priyanka P. Brahmachary, Hope D. Welhaven, and Ronald K. June

Abstract

Metabolism has long been recognized as a critical physiological process necessary to maintain homeostasis in all types of cells including the chondrocytes of articular cartilage. Alterations in metabolism in disease and metabolic adaptation to physiological stimuli such as mechanical loading are increasingly recognized as important for understanding musculoskeletal systems such as synovial joints. Metabolomics is an emerging technique that allows quantitative measurement of thousands of small molecule metabolites that serve as both products and reactants to myriad reactions of cellular biochemistry. This protocol describes procedures to perform metabolomic profiling on chondrocytes and other tissues and fluids within the synovial joint.

Key words Metabolomics, Metabolomic profiling, Metabolite extraction, Cartilage, Synovial fluid, Osteoarthritis

1 Introduction

Articular cartilage (AC) is a highly specialized form of cartilage located at joint interfaces. The main role of AC is to provide near-frictionless articulation in synovial joints. Unlike most cartilage and tissues, AC is avascular and aneural. Therefore, repair and healing are limited. AC is composed of extracellular matrix (ECM) and chondrocytes which are the sole resident cell type of AC. Chondrocytes are metabolically active and highly specialized cells that help generate, maintain, and restore the ECM [1]. ECM is primarily composed of water, collagen, and proteoglycans [2]. There are many interactions between the ECM and chondrocytes, and chondrocytes create a specialized local microenvironment termed the pericellular matrix (PCM). Therefore,

Priyanka P. Brahmachary and Hope D. Welhaven contributed equally with all other contributors.

Martin J. Stoddart et al. (eds.), *Cartilage Tissue Engineering*,
Methods in Molecular Biology, vol. 2598, https://doi.org/10.1007/978-1-0716-2839-3_11,

depending on loads and pressure exerted on the joint, chondrocytes respond and take action to maintain the ECM and AC homeostasis.

Chondrocytes can have spatially heterogeneous behavior. AC can be divided into four zones from the surface to the bone termed: the superficial, middle, deep, and calcified zone [1]. In each zone, chondrocyte shape, size, and frequency differ. For example, there are fewer and flatter chondrocytes located in the superficial zone but deeper into AC, chondrocytes are rounder, and more are present in comparison to those located directly at the joint interface. Together, chondrocytes and ECM provide AC with the ability to provide smooth joint articulation.

Chondrocytes must be metabolically active to maintain homeostasis. Yet several unanswered questions regarding the metabolism of chondrocytes and AC remain. Glucose is the most studied form of energy for chondrocytes. Therefore, there is great interest in analyzing metabolic pathways associated with glucose metabolism including glycolysis, the pentose phosphate pathway (PPP), and the tricarboxylic acid cycle (TCA cycle) [3]. One method to study chondrocyte metabolism is metabolomic profiling.

Metabolomics is the study of small molecule intermediates, which are called metabolites. Metabolomic profiling involves analyzing patterns of metabolite expression to better understand cell physiology and pathophysiology. Metabolites "act as a spoken language, broadcasting signals from the genetic architecture and microenvironment" [4]. A metabolomic profile provides information that encapsulates various factors including genetics, environment, stress, and toxic exposure. Chondrocytes respond to physiological compression with large-scale changes in metabolomic profiles [5–7]. Metabolomic profiles of joints capture large-scale changes in response to injury [8, 9], and metabolomic profiles of synovial fluid provide insight into pathophysiology [10, 11].

The field of metabolomics is the newest edition to the "omics" field, and it has been pivotal to many fields of research. For example, metabolomics has been applied to study *Escherichia coli* under stress conditions such as pH, heat, toxic exposure, and genetic manipulation to better understand key metabolic information such as involved biological pathways and associated metabolites. Beyond *E. coli*, metabolomics has been applied to many fields including musculoskeletal research to study various systems including chondrocytes and synovial fluid.

2 Materials

2.1 Chondrocyte Harvest

1. Conical centrifuge tubes, 50 mL, sterile.
2. #22 scalpel blades.
3. #4 scalpel handles.
4. 100 mm surface treated tissue culture petri dish, sterile.
5. Hemostats-Rochester-Ochsner forceps.
6. Long drawn-out capillary tube glass Pasteur pipettes, sterile.
7. Millipore Sigma™ Steriflip™ sterile disposable vacuum filter unit.
8. Sterile cell strainers, 70 μm.
9. Pen-Strep solution: penicillin-streptomycin (10,000 U/mL) solution.
10. Cartilage wash: 1× phosphate buffered saline (pH 7.4), sterile, 1% Pen-Strep solution.
11. Harvest media: Gibco™ DMEM, high glucose, 1% pen-strep solution, sterile.
12. Complete media: Gibco™ DMEM, high glucose, 1% pen-strep solution, 10% fetal bovine serum, sterile.
13. Gibco™ collagenase, type I powder.

2.2 Alginate Encapsulation and Culture

1. Sodium alginate: alginic acid sodium salt, from brown algae, low viscosity, Sigma-Aldrich.
2. PBS: 1× phosphate buffered saline (pH 7.4), sterile.
3. 10 mL syringe with Luer-Lok tips, sterile.
4. Hypodermic needle, 22 × 1 1/2 gauge, sterile.
5. 100 mm and 60 mm polystyrene petri dishes, sterile.
6. 0.22-micron syringe filters.
7. Conical centrifuge tubes: 15 mL, 50 mL, sterile.
8. 10 mL and 25 mL pipettes, sterile.
9. 150 mM NaCl: Weigh 8.76 g of NaCl, and transfer to a beaker. Add 900 mL of deionized water. Mix and make up the volume to 1lt. Autoclave at 121 °C for 20 min for a sterile solution.
10. 102 mM $CaCl_2$: Weigh 11.31 g of $CaCl_2$, and transfer to a beaker. Add 900 mL of deionized water. Mix and make up the volume to 1 L. Autoclave at 121 °C for 20 min for a sterile solution.
11. Trypsin-EDTA (0.05%), phenol red.
12. Complete media: Gibco™ DMEM, high glucose, 1% pen-strep solution, 10% fetal bovine serum, sterile.

2.3 Alginate Release and Agarose Encapsulation

1. Conical centrifuge tubes: 15 mL, 50 mL, sterile.
2. 2 mL and 10 mL pipettes, sterile.
3. Hemocytometer.
4. Trypan blue, 0.4%.
5. Anodized aluminum mold with multiple holes (i.e., one per agarose construct) of diameter of 7 mm and height of 12.7 mm, sterile.
6. Glass rod with a diameter of 2–3 mm, sterile.
7. Surgical blade, sterile.
8. 24-well flat bottom cell culture plates, sterile.
9. Water bath set at 40 °C.
10. PBS: 1× phosphate buffered saline (pH 7.4), sterile.
11. Alginate lysis buffer: 150 mM NaCl, 55 mM sodium citrate, 50 mM EDTA made in 1× PBS. Weigh 0.87 g of NaCl, 1.61 g of sodium citrate, and 1.46 g of EDTA. Add 90 mL of 1× PBS. Mix and make up the volume to 100 mL. Adjust the pH of the solution to 7. Autoclave the solution at 121 °C for 20 min.
12. Agarose, low gelling temperature, Sigma-Aldrich.
13. Harvest media: Gibco™ DMEM, high glucose, 1% pen-strep solution, sterile.
14. Complete media: Gibco™ DMEM, high glucose, 1% pen-strep solution, 10% fetal bovine serum, sterile.

2.4 Metabolite Extraction

1. Sterile heat sink (cold block) to pulverize samples.
2. Steel pounding rods for pulverization, forceps, narrow spatula, all sterile.
3. Hammer.
4. Aluminum foil squares, cut in 5 in × 5 in, sterile.
5. 1.5 mL microcentrifuge tubes.
6. 2 lbs of liquid nitrogen.
7. Insulated bucket filled with wet ice.
8. Microcentrifuge tubes (1.5 mL, *see* **Note 1**).
9. Precipitation solution: Prepare 10 mL of 8:2 (v/v) solution composed of 8 mL of methanol and 2 mL of HPLC grade water.
10. Resuspension solution: Prepare 10 mL of a 1:1 (v/v) solution composed of 5 mL of acetonitrile and 5 mL of HPLC grade water.
11. Vacuum concentrator capable of centrifugation under a vacuum of ~20–50 mbar.
12. MS snap cap SureStop Screw Vials and AVCS caps.
13. Extraction solvent: Prepare 10 mL of 70:30 (v/v) methanol/acetone. Keep at −80 °C till further use.

2.5 Liquid Chromatography–Mass Spectrometry

1. Depending on accessibility and availability, extracted metabolites can be analyzed using HPLC-MS or MS/MS. Specific to this work, an Agilent 6538 Q-TOF mass spectrometer was utilized.

2.6 Data Analysis

1. Access to XCMS, MetaboAnalyst, and analysis platforms such as R and MATLAB.

3 Methods

3.1 Chondrocyte Harvest

1. Add 30 mL cartilage wash solution to a 50 mL sterile conical centrifuge tube (*see* **Note 2**).
2. Attach the scalpel blade to the scalpel handle. Using the Hemostats-Rochester-Ochsner Forceps, pick the donor joint sample and place on a sterile petri dish. Holding the joint in place with the forceps, score cartilage on the donor joint with the blade using a crisscross pattern. Scrape, harvest the small pieces of cartilage into the conical tube containing the cartilage wash solution (*see* **Note 3**).
3. Centrifuge the tube at 28 × g for 2 min. Carefully aspirate the supernatant using a Pasteur pipette connected to a vacuum flask containing 10% liquid bleach. Be careful not to suck up the small pieces of cartilage. Add another 30 mL cartilage wash solution to the harvested cartilage, and repeat the centrifugation process. Perform a final wash with 30 mL of sterile harvest media. Aspirate the supernatant.
4. Aliquot 30 mL of harvest media to a fresh tube and add collagenase, type I at 2 mg/mL. Filter sterilize the solution using sterile disposable vacuum filter unit. Add the filter sterilized harvest media containing collagenase to the washed cartilage.
5. Tape the conical tube to a rotary shaker and place the shaker in an incubator. Digest the cartilage at 37 °C for 14–16 h (*see* **Note 4**).
6. After 14–16 h, take the tube out from the shaker. Working in a sterile hood, place a sterile 70 μm cell strainer on a new sterile 50 mL conical tube. Slowly, add the contents of the digested cartilage solution onto the cell strainer and collect the filtrate. This step removes the undigested cartilage and matrix debris (*see* **Note 5**).
7. Discard the cell strainer. Centrifuge the collected filtrate at 1000 × g for 5 min. (Place a counter tube on the opposite side of the centrifuge to balance it).
8. Gently aspirate the supernatant, taking care not to disturb the cell pellet at the bottom of the tube.

9. Resuspend the cells in 30 mL sterile 1× PBS. Centrifuge at 800 × g for 5 mins. Aspirate the supernatant, and repeat the wash with 30 mL sterile 1× PBS, followed by centrifugation.
10. Aspirate the supernatant after the final wash with PBS, and resuspend the cell pellet in 20 mL complete media (Gibco™ DMEM, high glucose, 1% pen-strep solution, 10% fetal bovine serum).
11. Place 10 mL each of the resuspended media containing the chondrocytes in two 100 mm sterile surface treated tissue culture petri dishes. Place the petri dishes in a humidified incubator at 37 °C with 5% CO_2 and let the cells grow.
12. Harvest the chondrocytes once they reach 70–80% confluency for alginate encapsulation.

3.2 Alginate Encapsulation

1. Sodium alginate solution, 1.2%: Weigh 1.2 g of sodium alginate on a square piece of aluminum foil. Wrap the foil and autoclave at 121 °C for 20 min.
2. Place a magnetic stirrer into an autoclavable 250 mL glass bottle, and sterilize at 121 °C for 20 min.
3. Working under sterile conditions in a laminar flow cabinet, pipette 100 mL of sterile 1× PBS into the autoclaved glass bottle. Slowly add the sterile sodium alginate to the PBS solution in the bottle.
4. Place the bottle on a magnetic stir plate, and leave the solution stirring at a low speed until the solids dissolve. For best results, allow the solution to sit overnight with gentle stirring for a homogenous solution (*see* **Note 6**).
5. Resuspend the harvested cells at a density of 4×10^6 cells/mL in sodium alginate solution in a 50 mL conical tube.
6. Transfer the cell-alginate solution to a 10 mL syringe, and slowly express through a 22-gauge needle in a dropwise fashion into a petri dish containing 102 mM $CaCl_2$ (20 mL).
7. Let the polymerization take place at room temperature for 10 mins.
8. Using a 25 mL pipette, collect the beads into a 50 mL conical tube, and centrifuge at 500 × g for 5 min. Aspirate the supernatant. Be careful not to disturb the alginate beads at the bottom of the tube.
9. Resuspend the beads in 10 mL of 150 mM NaCl, and centrifuge at 500 × g for 5 min. Aspirate the supernatant. Repeat the NaCl wash thrice.
10. Wash the alginate beads in 10 mL of complete media after the last NaCl wash. Aspirate the supernatant.
11. Using a 10 mL pipette, resuspend the alginate encapsulated cells in 10 mL of complete media.

12. Place 5 mL each of the complete media containing alginate encapsulated cells in two sterile 60 mm petri dishes, and culture the beads in a humidified incubator at 37 °C with 5% CO_2. Cell encapsulation must be carried out under sterile conditions to avoid contamination to the encapsulated cells. If culturing the alginate encapsulated beads for a longer period, replace the media every 2–3 days. Collect the alginate beads in a 15 mL conical tube and centrifuge at 500 × g for 5 min. Carefully aspirate the supernatant, and resuspend the alginate beads in 5 mL of complete media and plate in 60 mm petri dish, and incubate at 37 °C in a humidified incubator with 5% CO_2.

3.3 Alginate Release, Agarose Encapsulation, and Tissue Culture

1. Under aseptic conditions, pipette the alginate beads with the culture media into a 15 mL conical tube. Centrifuge at 500 × g for 5 min. Carefully aspirate the supernatant, and resuspend the alginate beads in 10 mL of sterile 1× PBS. Centrifuge at 500 × g for 5 min. Aspirate the supernatant.
2. Resuspend the alginate beads containing the chondrocytes in 5 mL of sterile alginate lysis buffer. Place the tube at 37 °C in a humidified incubator for 3–4 mins, till the solution turns clear, indicating degradation of alginate gel.
3. Centrifuge at 800 × g for 10 min. Aspirate the supernatant.
4. Resuspend the cells in sterile 1× PBS and repeat the centrifugation step.
5. Aspirate the supernatant and resuspend the cells in 1 mL of harvest media.
6. Determine the cell density using a hemocytometer and trypan blue before proceeding with agarose encapsulation.
7. Sterilize the aluminum gel mold and glass rod at 121 °C for 20 min. Unwrap and let cool in the laminar flow hood before using to make the agarose gels. Keep 10 mL and 2 mL sterile pipettes in a 37 °C incubator to keep warm for use with agarose.
8. Make a 4.95% agarose (w/v) in 1× PBS. Sterilize the solution by autoclaving at 121 °C for 20 min. Keep in a 40 °C water bath to ensure the agarose does not solidify.
9. Transfer 9 mL of 4.95% w/v sterile agarose to a 50 mL falcon tube. Working quickly add 1 mL of cells to the agarose to give a final concentration of 1×10^6 cells/mL in 4.5% agarose. Mix the solution by swirling and vortexing at a low speed for 5 s, taking care not to introduce bubbles in the agarose mixture.
10. Using a 2 mL sterile warm pipette, aspirate the agarose cell mixture, and add to a well of the aluminum mold. Let the agarose overflow slightly over the top of the well. Repeat the process to fill up the wells with the agarose-cell mixture. Let the gels solidify. Once the gelling is complete, slide the sterile

surgical blade horizontally across the top of the gel to remove excess agarose gel.

11. Use the sterile glass rod to move over the surface of the gel, trying to release it from the mold. Once the gel is loose, push through with the glass rod into a single well of a 24-well flat bottom cell culture plate, taking care not to break the gel.
12. Fill the well with 2 mL of complete media, ensuring that the gel is completely immersed in the media. Place the plate in a humidified incubator at 37 °C with 5% CO_2.

3.4 Metabolite Extraction (Figs. 1 and 2)

3.4.1 Alginate Released Agarose-Embedded Chondrocytes

1. Autoclave the tombstone, steel pounding rods, and the aluminum foil squares at 121 °C for 20 min.
2. Place the tombstone and pounding rods at 4 °C until use.
3. Fill the ice bucket with ice and place the tombstone in ice.

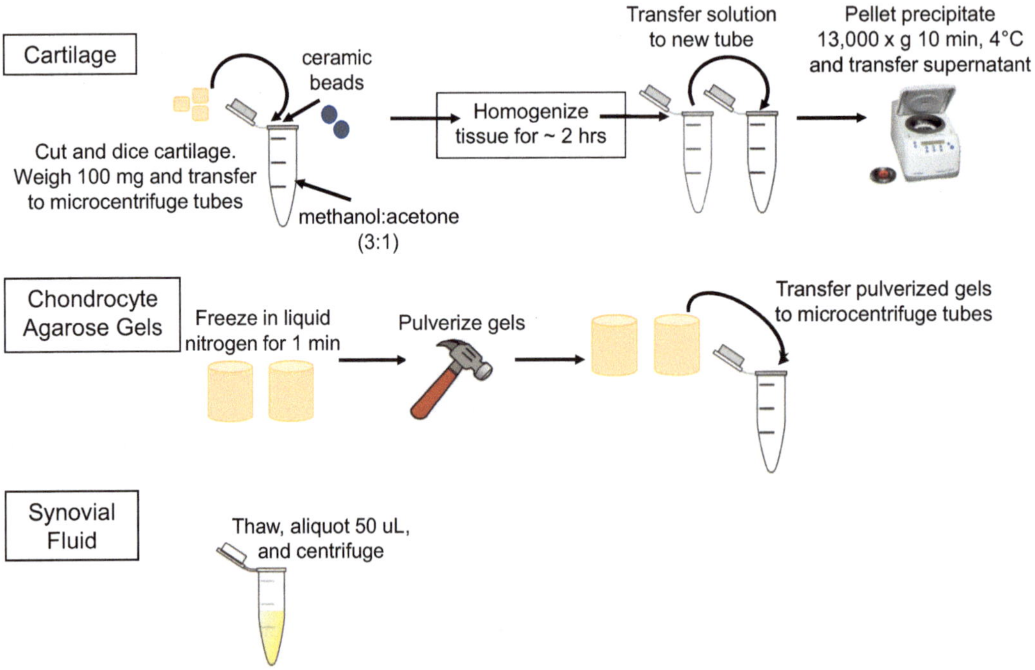

Fig. 1 Metabolite extraction preparation for various types of samples. Cartilage: Begin by cutting and dicing cartilage. Weigh 100 mg of cartilage and transfer to microcentrifuge tubes. Suspend cartilage with 3:1 methanol acetone and add ceramic beads. Homogenize tissue for 2 h. Transfer milky solution to a new microcentrifuge tubes and centrifuge sample. Chondrocyte agarose gels: Flash freeze gels for 1 min, and then pulverize gels. Transfer pulverized gels to microcentrifuge tubes. Synovial fluid: Thaw fluid and aliquot 50 uL to microcentrifuge tubes. Centrifuge samples to remove cells and debris

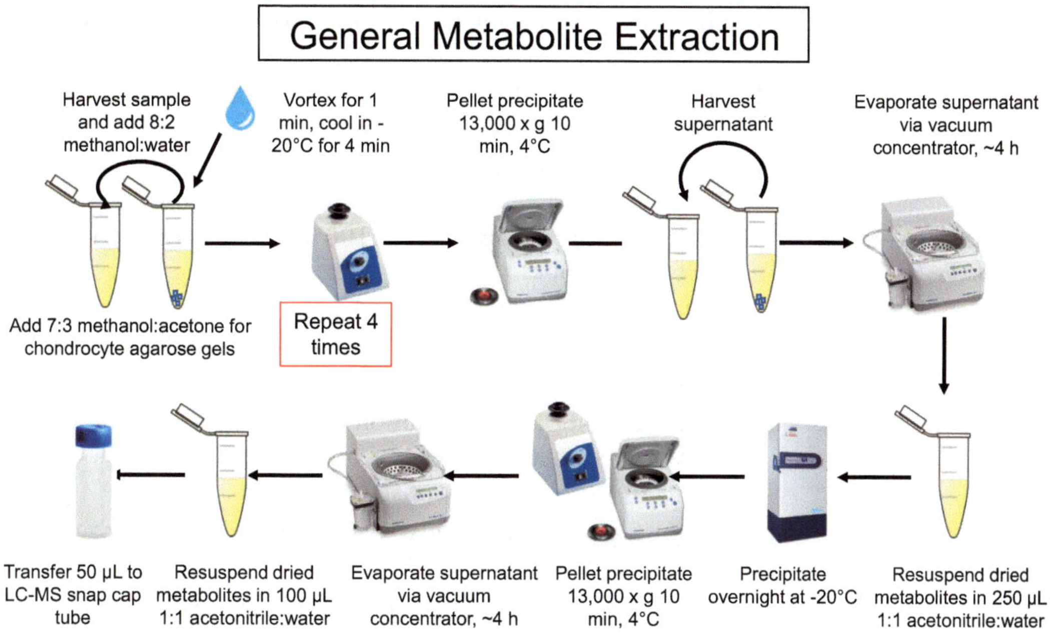

Fig. 2 Experimental procedure for extracting metabolites from various tissue types. After specific extraction protocols for individual sample types (Fig. 1), the protocol to extract metabolites is uniform. The protocol displayed applies to all three sample types discussed and could be extended to other sample types. Protocol may need alteration for sample types not discussed in this chapter

4. Wash chondrocyte agarose gels with 1× PBS, and then wrap a single gel in a sterile square of aluminum foil. Place in liquid nitrogen for approximately 1 min.
5. Remove the flash frozen gel using forceps. Unwrap and place into the well of the tombstone, making sure the gel is on its side.
6. Place the pounding rod on top of the gel and smash it with the hammer.
7. Carefully remove the smashed gel pieces with a spatula, and place in a 1.5 mL microcentrifuge tube (the gel pieces can be stored at −80 °C till further analysis).
8. For metabolite extraction from agarose gels, add 1 mL of cold 70:30 methanol/acetone solution to each gel sample.
9. Vortex vigorously every 5 min for a total of 20 min. When not vortexing, place the tubes at −20 °C.
10. Place the samples at −20 °C overnight.
11. Centrifuge the samples at 19,200 × g at 4 °C for 10 min.
12. Carefully transfer the supernatant to a fresh 1.5 mL microcentrifuge tube, taking care not to aspirate the agarose chunks at the bottom of the tube.

13. Dry the supernatant in a vacuum concentrator for approximately 3 h.
14. Resuspend the dried pellet in 100 μl of mass spectrometry grade water and acetonitrile (50:50 v/v).
15. Store samples at −80 °C.

3.4.2 Cartilage and Other Soft Tissues

1. Cut and dice cartilage, weight 100 mg of cartilage, and transfer to fast spin tubes.
2. Add 1 mL of 3:1 (v/v) methanol/water and two ceramic beads. Homogenize tissue using a tissue grinder for 2 h (*see* **Note** 7).
3. Transfer homogenized solution to a new microcentrifuge tube to isolate sample from ceramic beads and cartilage that did not break down.
4. Centrifuge at 16,100 × g at 4 °C for 10 min, transfer supernatant, and store pellet at −80 °C.
5. Add 100 μL of 80:20 methanol/water, vortex samples for 1 min, and incubate at −20 °C for 30 min.
6. Centrifuge at 16,100 × g for 10 min, transfer supernatant, and store pellet at −80 °C.
7. Dry down supernatant via vacuum concentrator (~3 h) (*see* **Note 9**).
8. Resuspend dried metabolites with 250 μL of 1:1 (v/v) acetonitrile/water and incubate for 1 h at −20 °C to precipitate macromolecules (*see* **Note 10**).
9. Remove samples from −20 °C, let thaw, and vortex samples for 1 min.
10. Centrifuge samples at 16,100 × g at 4 °C for 10 min, transfer supernatant, and store pellet at −80 °C.
11. Dry supernatant down via vacuum concentrator, resuspend dried metabolites with 100 μL of 1:1 acetonitrile/water, and store samples in −80 °C till LC-MS analysis.
12. Transfer 50 μL of sample to mass spectrometry caps (*see* **Note 8**).

3.4.3 Synovial Fluid

1. Thaw synovial fluid samples on ice for 10–20 min, aliquot 50 μL of SF to individual microcentrifuge tubes, and centrifuge at 500 × g at 4 °C for 5 min to remove cells and debris.
2. Transfer supernatant and add 1 mL of 80:20 methanol/water, vortex for 1 min, and incubate at −20 °C for 30 min. Store isolated pellet at −80 °C.
3. Centrifuge samples at 16,100 × g at 4 °C for 10 min, transfer supernatant, and store pellet at −80 °C.

4. Dry down supernatant via vacuum concentrator (~3 h) (*see* **Note 9**).
5. Resuspend dried metabolites with 250 μL of 1:1 acetonitrile/water, and incubate for 1 h at −20 °C to precipitate macromolecules (*see* **Note 10**).
6. Remove samples from −20 °C, let thaw, and vortex samples for 1 min.
7. Centrifuge samples at 16,100 × g at 4 °C for 10 min, transfer supernatant, and store pellet at −80 °C.
8. Dry supernatant down via vacuum concentrator, resuspend dried metabolites with 100 μL of 1:1 acetonitrile/water, and store samples in −80 °C till LC-MS analysis.
9. Transfer 50 μL of sample to mass spectrometry caps (*see* **Note 8**).

3.5 Liquid Chromatography–Mass Spectrometry (LC-MS)

1. Extracted metabolites were analyzed using HPLC-MS (Agilent 6538 Q-TOF mass spectrometer) in positive mode (resolution, ~20 ppm; accuracy, ~5 ppm; possible ionization adducts, H^+, Na^+) (*see* **Note 11**).
2. Peak intensity values for m/z values in the experimental sample set were identified and exported using Agilent MassHunter Qualitative Analysis software.

3.6 Metabolomics: Statistical Analysis (Fig. 3)

1. Following mass spectrometry analysis, XCMS is used to convert spectral data to excel format.
2. In excel, format file to include row ID, mass-to-charge ratios (m/z values), and individual samples. Information to exclude includes mzmax, maxint, and npeaks (*see* **Note 12**).

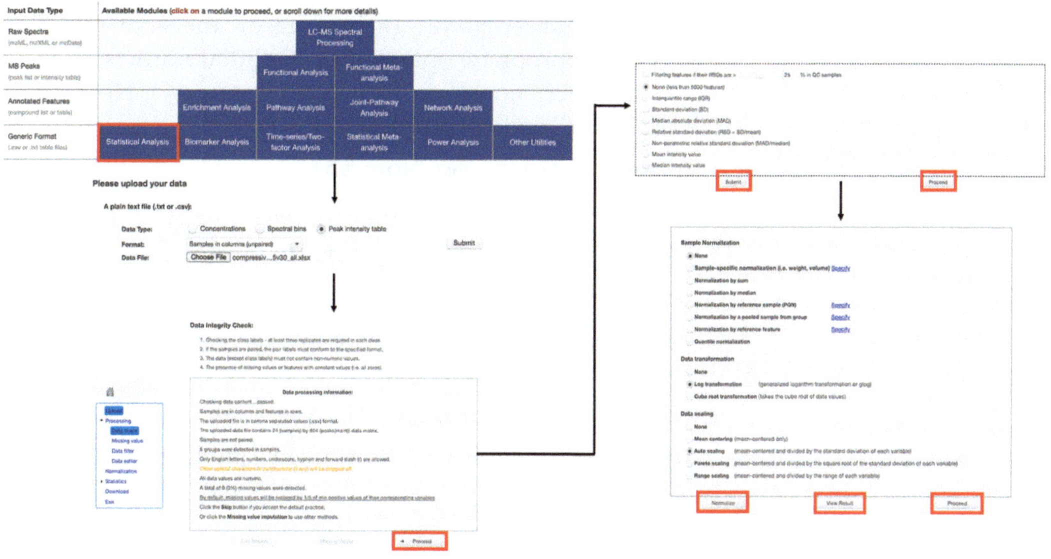

Fig. 3 Workflow to perform statistical analyses on metabolic data via MetaboAnalyst

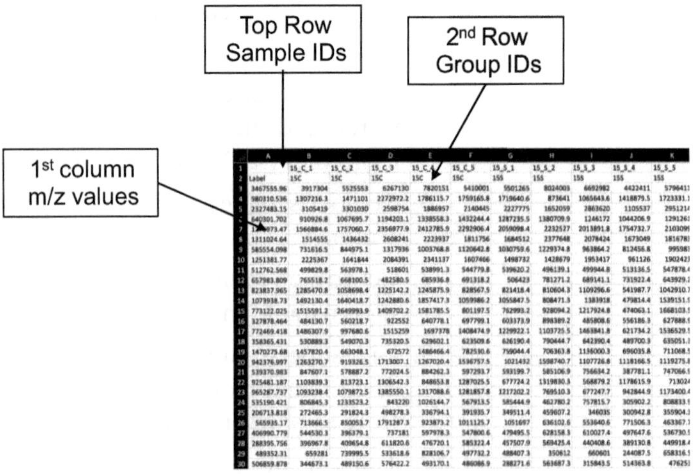

Fig. 4 Example input file for MetaboAnalyst. Required formatting for input to MetaboAnalyst

3. Sort individual samples within their cohort and assign group labels. Prior to submission to MetaboAnalyst, excel file must include m/z values, individual sample names, as well as group identifiers (Fig. 4).
4. Save excel file as a .csv or .txt file.
5. When uploading metabolite data to MetaboAnalyst, select "statistical analysis," select "peak intensity table" for data file type, select correct format, and upload file.
6. Steps when performing statistical analyses in MetaboAnalyst include data check, missing value, data filter, data editor, normalization, and statistics.
   (a) Data check + missing value: Select proceed.
   (b) Data filter + editor: Depending on the # of variables, select appropriate option. (i.e., if <5000 features are present, select "none (less than 5000 features)"). Next, select "submit" and "proceed."
   (c) Normalization: (*See* **Note 15**)
      (i) Sample normalization = none.
      (ii) Data transformation = log transform.
      (iii) Data scaling = auto scaling.
      (iv) Select normalize, view result, and proceed.
7. Following data filtering, editing, and normalization, view statistical results (*see* **Notes 13** and **14**).
8. To download results, select "download" and download zip file.

3.7 Metabolomics: Pathway Analysis (Volcano Plot, Fig. 5)

1. Open volcano.csv downloaded in the MetaboAnalyst zip file from your initial analysis and copy to a new excel sheet (*see* **Note 12**).
2. Custom sort the rows by the log2(FC) from high to low. Positive fold change values are now at the top of the excel file; these metabolites are considered "upregulated." Fold change values at the bottom of the excel file are metabolites that are "downregulated."
3. Copy the column with the m/z values, log2(FC), and raw.pval to a new sheet.
4. To identify upregulated metabolic pathways and associated metabolites, create a new column titled "p.value," and assign a *p*-value <0.05 to upregulated metabolites based on fold change values. Label the column with m/z values as "m.z."
5. Copy "m.z" column and the "p.value" column with the newly assigned *p*-values to a new excel file and save file as a .txt file.
6. Alternatively, to identify downregulated metabolic pathways and associated metabolites, sort rows by the log2(FC) from low to high. Create a new column titled "p.value," and assign a *p*-value <0.05 to downregulated metabolites. Similarly, copy and paste "m.z" and "p.value" columns to a new excel file and save as a .txt file.
7. When uploading .txt file to MetaboAnalyst, select "functional analysis."

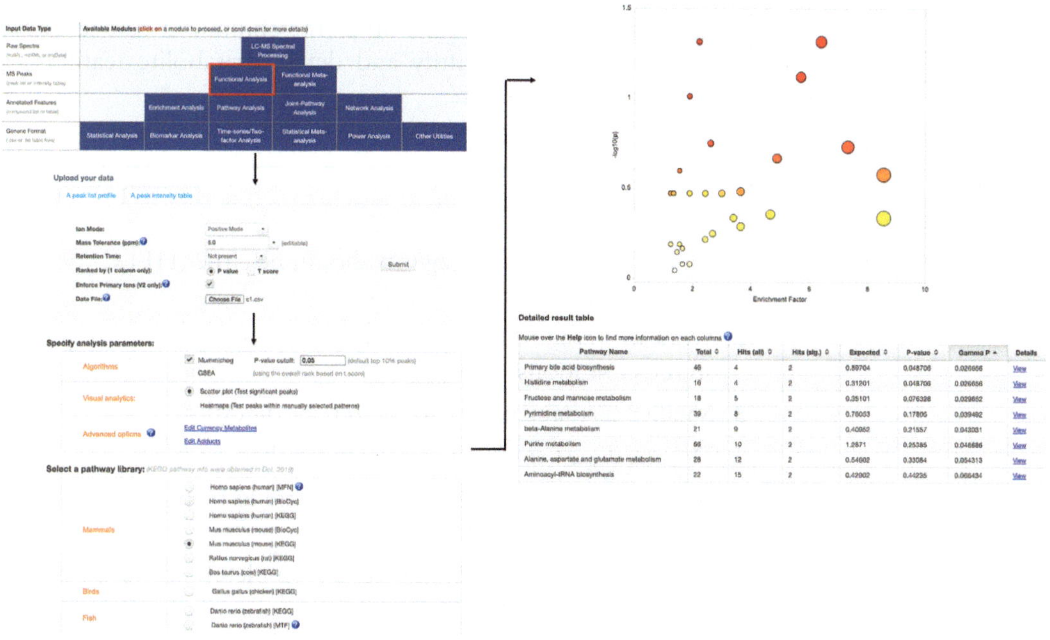

Fig. 5 Workflow to perform a functional pathway analysis via MetaboAnalyst

8. Within the "a peak table," tap select the appropriate parameters including ion mode, mass tolerance (ppm), retention time, rank (*p*-value or *t*-score), and enforcement of primary ions. Upload .txt file of interest and select "submit."
9. When specifying analysis parameters, algorithm options include Mummichog and GSEA. Input a *p*-value cutoff of 0.05 or less. Visual analytic options include scatter plot or heatmap.
10. Select the appropriate pathway library and select "submit" (*see* **Note 16**).
11. MetaboAnalyst will display an enrichment factor graph and pathways. Information provided include total metabolites in pathway, number of metabolites in pathway detected in dataset, significant metabolites in pathway detected in dataset, *p*-value, and gamma *P* values (*see* **Note 17**).
12. To download metabolic results, select "download."

3.8 Targeted Metabolite Analysis

1. Compile a list of metabolite features of interest (e.g., metabolites of glycolysis).
2. Spike isotopically labelled standards into samples to normalize metabolite concentrations across all samples and individual groups.
3. Set parameters for metabolites of interest on mass spectrometry instrument of choice.
4. Following LC-MS analysis, begin analyzing data via MetaboAnalyst as previously discussed.
5. Utilization of XCMS and other metabolomics databases are recommended to identify and detect metabolite features of interest within dataset.

4 Notes

1. All consumables and reagents should be either HPLC- or mass spectrometry-grade.
2. Work in sterile biosafety laminar flow hood for steps requiring sterile conditions.
3. Smaller cartilage pieces are better for digestion.
4. Do not digest the cartilage for more than 16 h.
5. All steps involving cell culture work are to be performed in the sterile biosafety laminar flow hood.
6. Make sure the sodium alginate is completely dissolved before proceeding with encapsulation.

7. When homogenizing solid and hydrogel samples, the goal is to produce a milky-looking solution. Once this stage is reached, transfer the top layer of solution to a new microcentrifuge tube. Cartilage won't break down completely; therefore, some chunks will remain at the end of this step.
8. When preparing for LC-MS analysis, utilize 50 μL for analysis, and save 50 μL at −80 °C n the case of error or needing to repeat a run.
9. When drying samples via vacuum concentration, drying time may vary, but we recommend checking on samples every 15–30 min to prevent overdrying of the sample.
10. When extracting metabolites, if after the first round of centrifugation and vacuum concentration large amount of protein is still present, consider leaving overnight.
11. To fully analyze and investigate metabolite features in samples, consider running samples in negative mode and in reverse mode.
12. We recommend to always save the original excel file outputted from XCMS prior to major editing.
13. Statistical analyses commonly performed to generate metabolomic profiles and to gain a better understanding of the metabolism at hand includes hierarchical clustering analysis (HCA), principal component analysis (PCA), partial least squares-discriminant analysis (PLS-DA), fold change, t-tests, volcano plot, and heatmap analysis. MetaboAnalyst will provide all statistical tests that are applicable to the data.
14. Depending on sample and objective of study, evaluate what statistical tests best fit the scope of interest. For example, consider using volcano plot analysis to determine metabolite features that are distinct and statistically significant to each individual cohort/experimental group.
15. When normalizing data via MetaboAnalyst, consider various options like pareto scaling and others. Additionally, log transformation is not always necessary.
16. When selecting a pathway library, we recommend utilizing KEGG first. To further investigate metabolic interactions, consider using BioCyc.
17. To view and search significant metabolites that were involved in a pathway and detected in dataset, click "view," and KEGG IDs will be listed. IDs in red are significant and can be found on KEGG.

References

1. Sophia Fox AJ, Bedi A, Rodeo SA (2009) The basic science of articular cartilage: structure, composition, and function. Sports Health 1(6):461–468. https://doi.org/10.1177/1941738109350438
2. Buckwalter JA, Mankin HJ (1998) Articular cartilage: tissue design and chondrocyte-matrix interactions. Instr Course Lect 47:477–486
3. Zheng L, Zhang Z, Sheng P et al (2021) The role of metabolism in chondrocyte dysfunction and the progression of osteoarthritis. Ageing Res Rev 66:101249. https://doi.org/10.1016/j.arr.2020.101249
4. Jewett MC, Hofmann G, Nielsen J (2006) Fungal metabolite analysis in genomics and phenomics. Curr Opin Biotechnol 17(2): 191–197. https://doi.org/10.1016/j.copbio.2006.02.001
5. Jutila AA, Zignego DL, Hwang BK et al (2014) Candidate mediators of chondrocyte mechanotransduction via targeted and untargeted metabolomic measurements. Arch Biochem Biophys 545:116–123. https://doi.org/10.1016/j.abb.2014.01.011
6. Zignego DL, Hilmer JK, June RK (2015) Mechanotransduction in primary human osteoarthritic chondrocytes is mediated by metabolism of energy, lipids, and amino acids. J Biomech 48(16):4253–4261. https://doi.org/10.1016/j.jbiomech.2015.10.038
7. Salinas D, Mumey BM, June RK (2019) Physiological dynamic compression regulates central energy metabolism in primary human chondrocytes. Biomech Model Mechanobiol 18(1):69–77. https://doi.org/10.1007/s10237-018-1068-x
8. Haudenschild DR, Carlson AK, Zignego DL et al (2019) Inhibition of early response genes prevents changes in global joint metabolomic profiles in mouse post-traumatic osteoarthritis. Osteoarthr Cartil 27(3):504–512. https://doi.org/10.1016/j.joca.2018.11.006
9. Hahn AK, Wallace CW, Welhaven HD et al (2021) The microbiome mediates epiphyseal bone loss and metabolomic changes after acute joint trauma in mice. Osteoarthr Cartil. https://doi.org/10.1016/j.joca.2021.01.012
10. Carlson AK, Rawle RA, Wallace CW et al (2019) Characterization of synovial fluid metabolomic phenotypes of cartilage morphological changes associated with osteoarthritis. Osteoarthr Cartil 27(8):1174–1184. https://doi.org/10.1016/j.joca.2019.04.007
11. Carlson AK, Rawle RA, Wallace CW et al (2019) Global metabolomic profiling of human synovial fluid for rheumatoid arthritis biomarkers. Clin Exp Rheumatol 37(3): 393–399

Chapter 12

Techniques for Visualization and Quantification of Primary Cilia in Chondrocytes

Huan Meng, Clare L. Thompson, Clarissa R. Coveney, Angus K. Wann, and Martin M. Knight

Abstract

Primary cilia regulate and coordinate a variety of cell signaling pathways important in chondrocyte physiology and cartilage development, health, and disease. Despite this, the chondrocyte primary cilium and its associated role in cartilage biology remains poorly understood. Key to elucidating primary cilia structure and function in chondrocytes is the ability to visualize this unique structure. Here we describe materials and methods for immunofluorescence labeling, microscopy, and measurement of chondrocyte primary cilia.

Key words Chondrocyte, Primary cilia, Immunofluorescence, Confocal microscopy, Super resolution, Quantification

1 Introduction

1.1 Cilia Structure and Function

The primary cilium is a solitary microtubule-based organelle that typically protrudes from the membrane of most eukaryotic cells. The axoneme of the primary cilium consists of a ring of nine-doublet microtubules in a "9 + 0" arrangement, lacking the central microtubule pair and dynein arms found in "9 + 2" motile cilia such as those of the airway rendering it immotile. The axoneme is surrounded by a specialized ciliary membrane and is anchored through a basal body originating from the mature mother centriole [1]. Thus, cilia are resorbed during cell division such that cilia prevalence is greater in quiescent, nondividing cells [2].

Cilia can be immunolabeled and visualized using antibodies targeting acetylated α-tubulin because of the enrichment of acetylation on the ciliary microtubules. The maintenance and function of primary cilia relies on the process of intraflagellar transport (IFT) in which multi-protein IFT complexes coupled to molecular motor proteins shuttle selected proteins along the axoneme [3, 4].

Martin J. Stoddart et al. (eds.), *Cartilage Tissue Engineering*,
Methods in Molecular Biology, vol. 2598, https://doi.org/10.1007/978-1-0716-2839-3_12,

Recent studies have revealed that primary cilia regulate and coordinate multiple signaling pathways including Hedgehog, Wnt, and growth factor signaling. In addition, primary cilia play a role in cellular mechanotransduction with enriched ciliary expression of integrins and mechanosensitive ions channels such as TRPV4, connexin 43, Piezo, and polycystin 1 and 2 observed within the axoneme [5]. However, the precise ciliary mechano-signaling mechanisms and how these may vary between different cell types remain unclear. Nevertheless, these cilia/IFT-dependent pathways are of fundamental importance in tissue development, homeostasis, and disease [6]. As a result genetic disruption of cilia expression results in a variety of ciliopathies while new studies continue to identify roles for primary cilia in other conditions including cancer, atherosclerosis, and osteoarthritis.

1.2 Primary Cilia in Cartilage and Chondrocytes

Primary cilia were first described in articular cartilage in pioneering work by Tony Poole and Brent Beaumont in New Zealand in 1985 [7]. However, at the time these organelles were widely viewed as vestigial remnants with no function in adult tissues. It was not until the widespread discovery of ciliopathies 20 years later that the scientific community started to take an interest in primary cilia.

In articular cartilage tissue, approximately 40–70% of chondrocytes express primary cilia with an axonemal length of 0.75–1.75 μm. However, work by McGlashan et al. [8] showed that cilia expression varies across the thickness of articular cartilage with lowest length and prevalence in the superficial zone. In addition, McGlashan reported that cilia expression was increased in osteoarthritic cartilage, although this may be partly confounded by a loss of superficial zone tissue [8]. It should be noted that difficulties in 3D visualization and quantification of cilia in situ within tissue may cause underestimates of cilia prevalence and length.

By contrast, it is more reliable to accurately measure cilia in isolated chondrocytes cultured in 2D monolayer where prevalence may reach 80–100% in serum starvation conditions that promote ciliogenesis. Furthermore, cilia length is typically longer than that reported in situ with average length of approximately 2 μm [9]. Chondrocytes cultured in monolayer provide a good model for measuring cilia expression as the cells are relatively quiescent and exhibit a high percentage of cilia on the basal surface of the cell which therefore lie flat against the substrate (Fig. 1).

Here we introduced methods for culturing chondrocytes, immunofluorescence labeling of primary cilia, microscopy imaging of primary cilia, and measurement of primary cilia length and prevalence.

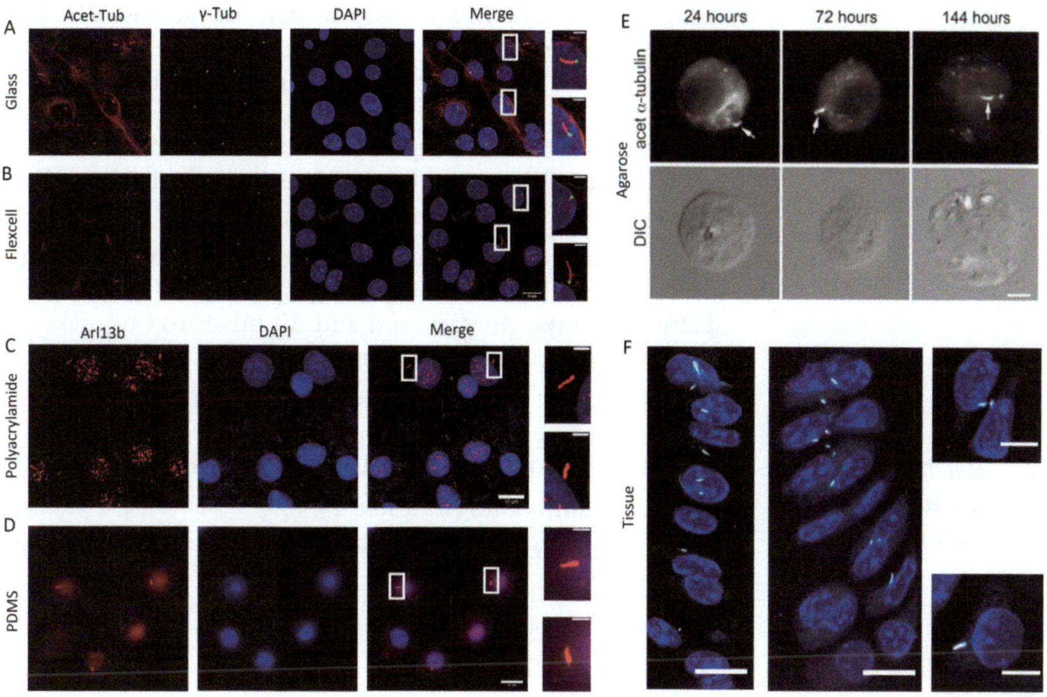

Fig. 1 Chondrocyte primary cilia labeled in monolayer on glass, Flexcell membrane, polyacrylamide and PDMS gel, and in 3D agarose culture and in situ with cartilage tissue. (**a**–**d**) Primary cilia expressed by bovine articular chondrocytes in 2D monolayer culture on (**a**) glass coverslips, (**b**) Flexcell membranes, (**c**) polyacrylamide gel, and (**d**) PDMS substrates. Cilia labeled with acetylated α-tubulin (red) and γ-tubulin (green) (**a** and **b**) or ARL13B (red) (**c** and **d**) with nuclei counterstained with DAPI (blue). Images obtained using confocal maximum projection (**a**–**c**) and epifluorescence (**d**). (**e**) Primary cilia (arrows) expressed by bovine chondrocytes cultured in 3D agarose gel for 24, 72, and 144 h and corresponding DIC images. Cilia labeled with antibodies against acetylated a-tubulin and visualized by widefield fluorescence. Images from work by Sue McGlashan and Tony Poole in collaboration with the lead author [19]. (**f**) Primary cilia in murine cartilage growth plate in situ. Cilia labeled with acetylated α-tubulin (cyan) and Arl13b (magenta) and imaged by confocal maximum projection. All scale bar represents 10 μm for original images and 2 μm for boxed images with cilia typically 1–3 μm in length

2 Materials

2.1 Materials for Chondrocyte Culture

1. Glass coverslips (13 mm in diameter).
2. 24-well plates.
3. Chondrocyte serum free culture medium: low-glucose (1000 mg/L) DMEM, 200 mM L-glutamine, 100× penicillin/streptomycin (10,000 units penicillin and 10 mg streptomycin/mL), 1 M HEPES buffer, L-ascorbic acid. To 500 mL of low glucose DMEM, add 5 mL of L-glutamine and 5 mL of Pen/Strep, 10 mL of HEPES, and 0.07 g of L-ascorbic acid. Mix well by stirring.

4. Chondrocyte 10% FBS culture media: low-glucose DMEM, fetal bovine serum (FBS), 200 mM L-glutamine, 100× Pen/-Strep (10,000 units penicillin and 10 mg streptomycin/mL), 1 M HEPES buffer, L-ascorbic acid. To 500 mL of low glucose DMEM, add 5 mL of L-glutamine and 5 mL of Pen/Strep, 10 mL of HEPES, and 0.07 g of L-ascorbic acid. Mix well by stirring. Remove 50 mL and add 50 mL of FBS.
5. Agarose gel solution: agarose, low gelling temperature (type VII, powder), 1× Earle's balanced salt solutions (EBBS). Weigh 1.2 g of agarose powder, and add 20 mL EBBS to make 6% agarose gel suspension (weight/volume).
6. Steel module of 5 mm in diameter and 5 mm in height.

2.2 Materials for Immunofluorescence of Primary Cilia

1. Primary antibodies (*see* Table 1, **Note 1** and Fig. 2).
2. Secondary antibodies (*see* Table 2).
3. Fresh defrosted 4% paraformaldehyde (PFA)/ice-cold methanol.
4. 1× phosphate-buffered saline (PBS).
5. 0.5% Triton-X/PBS.
6. 5% goat/donkey serum.
7. 0.1% bovine serum albumin (BSA) in PBS.
8. Mounting medium (e.g., Fluoromount G, *see* **Note 2**).
9. No. 10 disposable scalpel.
10. Nail varnish (clear).

Table 1
Typical antibodies used for immunofluorescence labeling of primary cilia

Target	Cilia feature	Cat no./species	Company	Dilution
Acetylated α-tubulin	Axoneme structure	T7451/mouse	Sigma Aldrich	1:2000
Arl13b	Ciliary membrane	17711-1-AP/ rabbit	Protein tech	1:2000
Glutamylated tubulin	Axoneme (possibly section within ciliary pocket)	GT355/mouse	AdipoGen	1:2000
Polycystin-2	Ciliary membrane protein	SC25749/ mouse	Santa Cruz	1:1000
IFT88	IFT protein	13967-1-AP/ rabbit	Protein tech	1:2000
γ-Tubulin	Basal body	sc-17787/ mouse	Santa Cruz	1:1000
Pericentrin	Basal body	Ab444/rabbit	Abcam	1:100

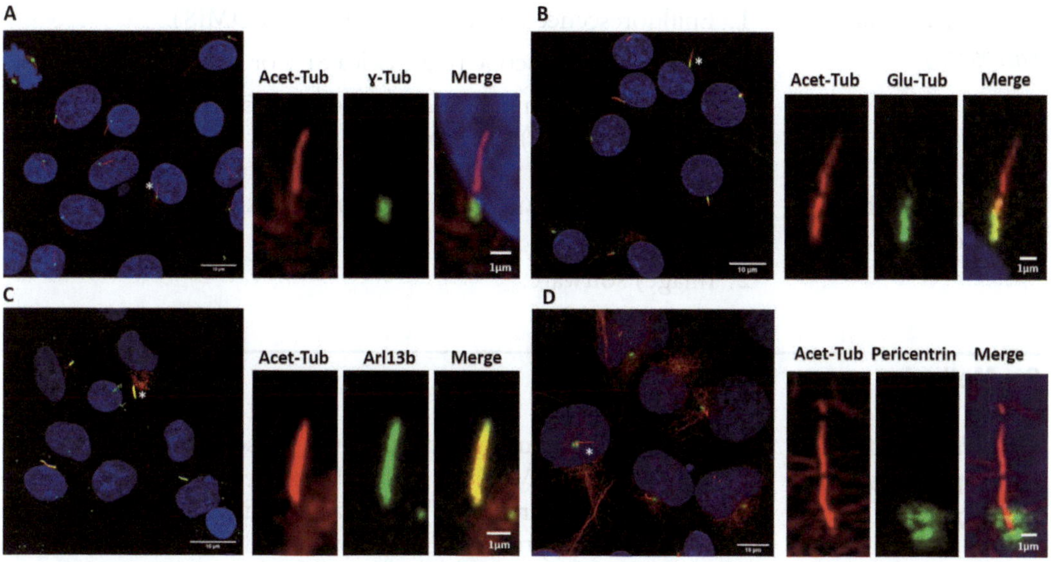

Fig. 2 Confocal imaging of chondrocyte primary cilia demonstrating the use of different immunolabeling of the axoneme and basal body. (**a**–**d**) Confocal maximum projection images showing primary cilia axoneme labeled with acetylated α-tubulin (red) and nuclei counter stained with DAPI (blue). The cilia are additionally labeled with (**a**) γ-tubulin, (**b**) Glu-tubulin, (**c**) Arl13b, and (**d**) pericentrin (all green). Cells are primary bovine chondrocytes cultured in 2D monolayer on glass coverslips. Scale bars represent 10 μm for fields of view images and 1 μm for magnified images. Typically, α-tubulin labels the entire axoneme; γ-tubulin labels the basal body; Glu-tubulin labels the basal region of the axoneme, possible associated with the ciliary pocket; Arl13b labels the entire ciliary membrane and pericentrin appears as diffuse staining around the basal body

Table 2
Excitation and emission of secondary antibodies

Fluorophore	Excitation (nm)	Emission (nm)
Alexa Fluor 488	488	500–550
Alexa Fluor 555	555	550–630
Alexa Fluor 633	633	640–750
DAPI	405	410–510

11. 10 × 10 mm glass coverslips.
12. Embedding medium (*see* **Note 3**).
13. Cryofilm type 3C (16UF) 2.5 cm (C-FUF304 Section-lab Co. Ltd).
14. Superfrost Plus Micro Slide. (VWR, 631-0108).
15. Spot stickers. (Sigma-Aldrich, Z740261).
16. Blocking buffer: 10% BSA, 5% goat serum in PBS.

2.3 Materials for Microscopy

1. Epifluorescence microscope (e.g., Leica DMi8).
2. Confocal microscope (e.g., Leica SP5 or Zeiss LSM710).
3. Structured Illumination Microscopy (Zeiss Elyra Super resolution microscope) (*see* **Note 4**).

2.4 Materials for Measurement of Cilia

1. Original maximum projection images exported from imaging software.
2. ImageJ software.

3 Methods

3.1 Chondrocyte Monolayer Culture for Imaging of Cilia

Primary cilia form during the G0 or G1 phase of the cell cycle and disassemble at G2/M phase [10]. Hence, serum-free media (also known as serum starvation) is typically used promote quiescence and induce cilia expression. Similarly cell-cell contact, as occurs in confluent monolayer cultures, induces cilia expression. Consequently, culture to confluence and 24 h serum starvation are widely used for promoting ciliogenesis.

Chondrocytes are readily cultured in monolayer on serum-coated glass coverslips which is ideal for visualization and quantification of primary cilia. However, there is inevitably some loss of chondrocytic phenotype with prolonged monolayer culture and changes in actin organization and primary cilia expression [11]. Seeding at high densities can help to limit this [12]. The protocol described below can be adapted for mechanobiology studies in which chondrocytes are exposed to tensile strain on flexible membranes using Flexcell™ [5] or other such loading devices. Similarly, the protocols can also be adapted for seeding cells into microfluidic organ-on-chip models such as the Emulate™ platform, in which cells are exposed to both tensile strain and fluid shear forces (Fig. 1).

1. Sterilize glass coverslips with autoclave (121 °C, 20 min), and then coat for 3 h with FBS at 37 °C with one coverslip in each well of a 24-well plate (*see* **Note 5**).
2. Seed chondrocytes at 2–4 × 10^4 cells/cm^2 onto the glass coverslips. Allow approximately 1–2 mL of DMEM+10% FBS culture media per well, and maintain cultures at 37 °C with 5% CO_2. Replace culture media with fresh media every 2–3 days. Primary articular chondrocyte monolayers should reach confluence in 8–9 days (*see* **Note 6**).
3. Once the chondrocyte monolayer has reached confluence, change to serum free media and culture for further 24 h to promote ciliogenesis prior to fixation/microscopy (*see* **Note 7**).

3.2 Chondrocyte Culture in 3D Agarose Gel

The chondrocyte-agarose model system is well characterized and shown to maintain chondrocytic phenotype in a rounded morphology and promote the synthesis of cartilage extracellular matrix [13]. The system is well suited for mechanobiology experiments due to the ability to apply physiological compressive strain to the isolated chondrocytes (*see* **Note 8**). Previous studies have characterized chondrocyte primary cilia expression in agarose cultures as well as the response to mechanical stimulation [14]. These protocols can be readily adapted for other 3D hydrogel model systems.

1. Autoclave 6% agarose gel solution at 121 °C for 20 min to melt the agarose and place into 80 °C oven for 20 min on rollers to avoid bubble formation. Then transfer to 37 °C incubator, and allow to cool so as not to cause loss of cell viability when mixed with chondrocytes (*see* **Note 9**).
2. Resuspend isolated chondrocytes in DMEM+20%FBS at a concentration of 8×10^6 cells/mL (i.e., twice the required final concentration of cells).
3. Add 20 mL cell suspension into 20 mL of molten agarose, and mix thoroughly on a roller to yield a final cell concentration of 4×10^6/mL in 3% (w/v) agarose. Do not shake otherwise bubbles will form in the gel which will not disperse due to the viscous nature (*see* **Note 10**).
4. Pour the chondrocyte/agarose suspension into sterile molds. We typically use molds to generate cylindrical 5 mm diameter × 5 mm high constructs (*see* **Note 11**). Place the mold in a large sterile petri dish in a 4 °C fridge for 15 min to cause gelling of the cell-agarose constructs.
5. When the constructs are solidified, transfer individual chondrocyte-agarose constructs into well plates and culture with 10%FBS DMEM media for further experiments.

3.3 Immunofluorescence of Primary Cilia in Monolayer Culture

1. Fix samples for 10 min with freshly defrosted 4% (PFA)/ice-cold methanol (*see* **Note 12**).
2. Wash 3× in PBS (5 min each).
3. Incubate at room temperature (RT) for 5 min with 300 μL 0.5% Triton-X/PBS to permeabilize the cells.
4. Incubate at RT for 30 min to 1 h with 300 μL 5% goat/donkey serum/PBS (*see* **Notes 13**).
5. Incubate at 4 °C overnight with 300 μL primary antibody solution made up in 0.1%BSA/PBS according to Table 1.
6. Wash 3× in 0.1%BSA/PBS (10 min each).
7. Incubate at RT for 1 h with 300 μL appropriate Alexa Fluor conjugated secondary antibody (1:1000) made up in 0.1% BSA/PBS.

8. Incubate at RT for 1 min with 300 μL 1 μg/mL DAPI (1: 5000) in 0.1%BSA/PBS.
9. Wash 3× in 0.1% BSA/PBS (5 min each).
10. Add 10 μL of mounting medium onto a glass slide, and transfer stained coverslips onto the slides (the side of the coverslip with the cells should be facing downwards in contact with the glass slide).
11. Mount the coverslips/other desired substrates and seal with nail varnish. Slides need to dry for 30–60 min at room temperature before imaging. Slides should be stored at 4 °C.

3.4 Immunofluorescence of Primary Cilia in 3D Cell-Seeded Gels

1. Fix cells for 20 min with freshly defrosted 4% paraformaldehyde (PFA).
2. Wash 3× in PBS (5 min each).
3. Using a scalpel, cut the agarose or other hydrogel scaffold into slices about 1 mm thick (*see* **Note 14**).
4. Incubate the gel slices at RT for 10 min with 300 μL 0.5% triton-X/PBS to permeabilize the cells.
5. Incubate at RT for 1 h with 300 μL 5% goat/donkey serum/PBS.
6. Incubate at 4 °C overnight with 300 μL primary antibody made up in 0.1%BSA/PBS according to Table 1.
7. Wash 3× in 0.1%BSA/PBS (20 min each).
8. Incubate at RT for 1 h with 300 μL appropriate Alexa Fluor conjugated secondary antibody (1:1000) made up in 0.1% BSA/PBS (*see* **Note 15**).
9. Incubate at RT for 1 min with 300 μL 1 μg/mL DAPI (1: 5000) in 0.1% BSA/PBS.
10. Wash 3× in 0.1% BSA/PBS (5 min each).
11. Add 10 μL of mounting medium onto glass slides and transfer stained gel slices onto slides.
12. Cover each gel slice with a 10 × 10 mm square coverslip and seal with nail varnish. Coverslips need to dry for 30–60 min at room temperature before imaging. Slides should be stored at 4 °C.

3.5 Immunofluorescence of Primary Cilia in Cartilage and Ossified Musculoskeletal Tissues

1. Collect tissue and fix immediately into ice-cold 4% freshly prepared or defrosted PFA. Incubate for 24 h at 4 °C.
2. Move samples into ice cold 10% sucrose solution (made in 1 × PBS), and incubate for a further 24 h at 4 °C. Repeat with 20% and 30% ice cold sucrose solution.
3. Embed samples on dry ice into Super Cryo Embedding Medium (C-EM001, Section-lab Co. Ltd) and store at −80 °C.

4. For cilia visualization collect sections of 10–14 μm using a precooled cryotome at −16 °C. Use Cryofilm type 3C (16UF) 2.5 cm C-FUF304 (Section-lab Co. Ltd). Important: Fix tape to slides (must be positively charged, e.g., Superfrost Plus, VWR, 631-0108) using touch spot stickers (Sigma Aldrich, Z740261). Make sure they are fixed.
5. Store sections at −80 °C until ready for immunofluorescence.
6. Hydrate sections for 5 min in 1× phosphate-buffered saline (PBS) (*see* **Note 16**).
7. Fix sections for 10 min with 4% freshly prepared or defrosted PFA, 0.2% Triton X-100 in PBS.
8. Incubate sections in 5% blocking buffer for 10 min.
9. Incubate sections with primary antibody diluted in blocking buffer for 45 min at room temperature.
10. Wash 3 × 10 min in PBS.
11. Incubate sections with Alexa-conjugated secondary antibodies for 1 h diluted in blocking buffer (1:500).
12. Wash 3 × 10 min in PBS.
13. Counterstain sections with 1:5000 DAPI diluted in PBS for 5 min.
14. Wash once with PBS for 5 min.
15. Gently tap slide to remove excess fluid and dry for a maximum of 5 min in the dark. Add 10 μL of mounting medium (*see* **Note 2**) and mount slide to maintain fluorescent signal. Seal the coverslip with nail varnish. Coverslips need to dry for 30–60 min at room temperature before imaging. Slides should be stored at 4 °C.

3.6 Epifluorescence Microscopy Imaging of Primary Cilia

1. Switch on and set up the epifluorescence microscope to epifluorescence mode and turn on the fluorescence lamp and camera connected to microscopy software according to the manufacturer's instructions.
2. Place the sample on the microscope stage. Samples typically consist of coverslips or tissue sections mounted on microscope slides. However, for inverted microscopes it is possible to visualize cells, tissue or 3D constructs on a coverslip placed on the stage or within a heated stage mount for live cell imaging.
3. Select an appropriate objective lens for imaging primary cilia, ideally with a high numerical aperture to ensure optimal resolution (*see* **Note 17**).
4. Locate desired field of view. This is typically done using the eyepieces, adjusting the focusing to get a clear view of the nuclei labeled with DAPI in order to select the cells (*see* **Note 18**).

5. Select the appropriate cilia channel (e.g., Arl13b or Acetylated α-tubulin as per Table 1) using the correct excitation and emission wavelengths. Use the eye pieces to visualize cilia adjusting the fine focus (*see* **Note 19** and Table 2).
6. Cilia prevalence is often best measured from the eyepiece in epifluorescence microscopy as this allows fine focal adjustments. However, cilia length measurement requires saved images. Hence, the light path should be switched from eyepieces to camera for image acquisition.
7. Set the appropriate camera setting for each channel to clearly show the cilia and other features (*see* **Note 20**).
8. Ensure appropriate pixel size to maximize image resolution based on the Nyquist criterion which states that the pixel size should be no larger than half the optical limit of resolution which is dependent on the objective numerical aperture, the emission wavelength (*see* **Note 21**). Typically, a pixel size of 0.10 × 0.10 μm is appropriate for epifluorescence microscopy.
9. Take images separately on different channels and merge. Ensure consistency across all images with any postacquisition adjustment of contrast or brightness, e.g., to reduce the background and enhance signal.
10. Export image files as .tiff files prior to cilia quantification.

3.7 Confocal Microscopy Imaging of Primary Cilia

1. Switch on microscope, control software, and fluorescence lamp and lasers according to manufacturer's instructions (e.g., Zeiss LSM710 Confocal and ZEN software).
2. Load samples onto the microscope stage, and choose ×63/ 1.4 NA oil immersion objective (*see* **Note 17**).
3. Locate field of views: (e.g., for Zeiss LSM710, choose "Locate" mode in ZEN software). On eyepieces, adjust the focusing to get a clear view of the nuclei on DAPI channel (UV). Then choose Arl13b/Acetylated α-tubulin channel and cilia should be visualized.
4. Set acquisition: (e.g., for Zeiss LSM710, choose "acquisition" mode in ZEN software). Set laser excitation and emission detection wavelengths according to target fluorescent labels (Table 2). Adjust laser strength taking care to avoid photobleaching or phototoxicity whilst maximizing signal to noise.
5. Set desired image frame/format size (nos pixels) and pixel size to give the optimum resolution without over sampling (*see* **Note 22**).
6. Select scan speed, gain, and related camera/detector parameters depending on the microscope system used.
7. Where it is possible to adjust the confocal pinhole, this should be set at the level to provide optimal confocality, although it may be useful to increase the pinhole in some instances.

8. Set the Z-step size for optimal z resolution or as desired (*see* **Note 23**).
9. Set the Z scan start and end positions: one way to achieve this in monolayer cultures is to adjust the focus to the top and bottom position ensuring all cilia are present within this range. However, it is important to be consistent in the z-range used across multiple fields of view or separate specimens.
10. Record the z-scan and repeat with separate fields of view and separate samples. The number of images required for quantification of cilia length and prevalence will depend on the degree of ciliation as discussed below (*see* **Note 24**).
11. Reconstruct the Z-series to create a 3D projection image from which prevalence and projected length can be measured. Typically a maximum intensity projection is used and the file exported as .tiff.

3.8 Super Resolution Structure Illumination Microscopy (sr-SIM) Imaging of Primary Cilia

The development of super resolution microscopy provides the opportunity to go beyond the resolution limit for optical and confocal microscopy of approximately 0.2 μm in the axial direction (*see* **Note 21**). Super resolution structured illumination microscope (sr-SIM) is based on the principle of Moire effect and provides approximately twice the resolution with a limit of resolution of approximately 0.1 μm. This is particularly useful for visualizing primary cilia structure including diameter, length, and protein distribution along the axoneme (Fig. 3). Other super resolution techniques such as STED, STORM, and PALM provide even

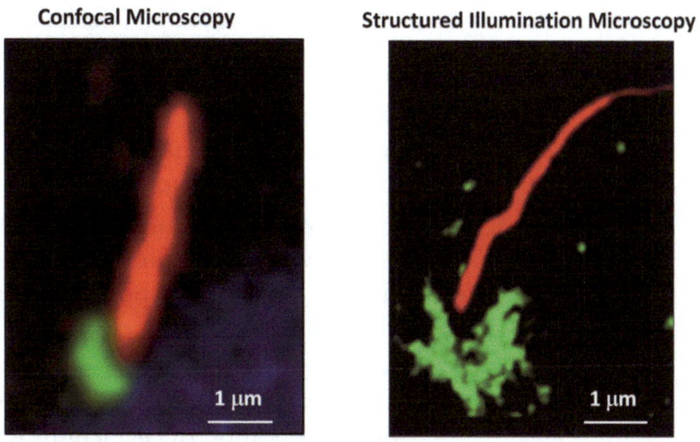

Fig. 3 Comparison between sr-SIM and confocal images of primary cilia. Immunofluorescence images of primary cilia obtained using confocal microscopy (left) and super resolution structured illumination microscopy (sr-SIM) (right). Cilia were labeled for acetylated α-tubulin (red) and pericentrin (green) and maximum intensity projection images created from z-stacks. Images show cilia expressed by primary bovine chondrocytes cultured on glass coverslips

greater resolution potentially down to single molecule location with a precision of 20–50 nm. This enables visualization of microtubule doublets and other ultrastructural details; however, the techniques are more complex and require significant optimization [15]. The following protocols are provided for sr-SIM imaging of cilia using the Zeiss Elyra super resolution microscope with Zen software.

1. Switch on microscope, control software, and fluorescence lamp and lasers according to manufacturer's instructions (e.g., Zeiss Elyra super resolution microscope and ZEN software).
2. Perform stage alignment according to the manufacturer's instructions.
3. Perform channel alignment using fluorescent beads with a 200 nm diameter according to the manufacturer's instructions.
4. Load samples onto the microscope stage, and choose ×63/1.4 NA oil immersion objective. Locate field of views: (e.g., for Zeiss Elyra super resolution microscope, choose "Locate" mode in ZEN software). On eyepieces, adjust the focusing to get a clear view of the nuclei on DAPI channel (UV). Then choose Arl13b/Acetylated α-tubulin channel and cilia should be visualized.
5. Set laser excitation and emission detection wavelengths according to target fluorescent labels (*see* Table 2). Adjust laser strength taking care to avoid photobleaching or phototoxicity while maximizing signal to noise.
6. Set desired image frame/format size (nos pixels), and pixel size to give the optimum resolution without over sampling. Frame size is by default 1280 × 1280 and can be zoomed in by decreasing the size. A frame size of 256 × 256 is highly recommended for SIM imaging of primary cilia.
7. Setting the grating to 5 rotations is recommended for better isotropic resolution enhancement. The resulting grid can be seen during continuous acquisition.
8. Set the Z-step size and Z scan start and end positions: one way to achieve this in monolayer cultures is to adjust the focus to the top and bottom position ensuring the cilium is present within this range (*see* **Note 23**).
9. Record the z-scan and repeat with separate cilia over multiple fields of view/samples. The number of cilia (1/image) required for quantification of cilia length will depend on the protein of interest and variability in expression.
10. Reconstruct the Z-series to create a 3D projection image from which projected length and protein distribution can be measured. Typically, a single z-stack through the center of the cilium is used and the file exported as .tiff.

3.9 Measurement of Cilia Prevalence

1. Select central field of each image, only cells with nuclear labeling with DAPI involved in the central square area will be taken in cilia measurement (*see* **Note 24**).
2. On DAPI channel, choose the nuclei within the central square area by adding region of interest (ROI) function of Image J, and label each nucleus with number (*see* **Note 25** and Fig. 4).
3. Count total cell number and ciliated cell number on acetylated α-tubulin/Arl13b and DAPI merged channel and record.
4. Repeat **steps 1–3** in at least 10 images each experimental condition to get at least 100 ciliated cells.
5. Calculate final cilia prevalence with total ciliated cells/total cells, and analyze statistics by chi-square and Fisher's exact test.

3.10 Measurement of Cilia Length

The chondrocyte monolayer model is suitable for cilia length measurement according to our previous studies [5] as chondrocyte cilia consistently lie flat on the coverslips on the basal or apical surface of the chondrocytes. However, vertical cilia do exist in chondrocytes. Measurement of vertical cilia is thought to be not accurate because the decreased Z-axis resolution caused by axial pixel spread in confocal microscopy. Therefore, vertical cilia are not accounted in cilia length measurement.

1. Identification of vertical cilia: check the Z-stacks of each cilia, and identify cilia observed in at least three sections (vertical height >1 μm) as vertical cilia. Exclude those vertical cilia when measuring cilia length (*see* **Note 26**).
2. Locate each ciliated cell using labels created in **step 2**, Subheading 3.9. Zoom image appropriately and manually draw segmented line from the base to the tip of each cilium (*see* **Note 27**).
3. Measure the length of drawn segmented line by measure function of ImageJ.
4. Repeat **step 2** until over 100 cilia from at least 10 representative images have been measured (*see* **Note 28**).
5. Add up total cilia length data, and analyze by nonparametric Mann-Whitney U-test (*see* **Note 29**).

3.11 Measurement of Ciliary Protein Distribution Along the Axoneme

1. Export confocal or super resolution microscopy images of primary cilia (note the resolution limit for confocal which may limit the usefulness of these data particularly for relatively short cilia <2 μm).
2. Zoom image appropriately and manually draw segmented line along the axoneme using line function of ImageJ. Axoneme length and the intensity of staining can be directly quantified from this line; however, for sr-SIM images, follow **steps 3–5** to provide higher quality data including axoneme diameter.

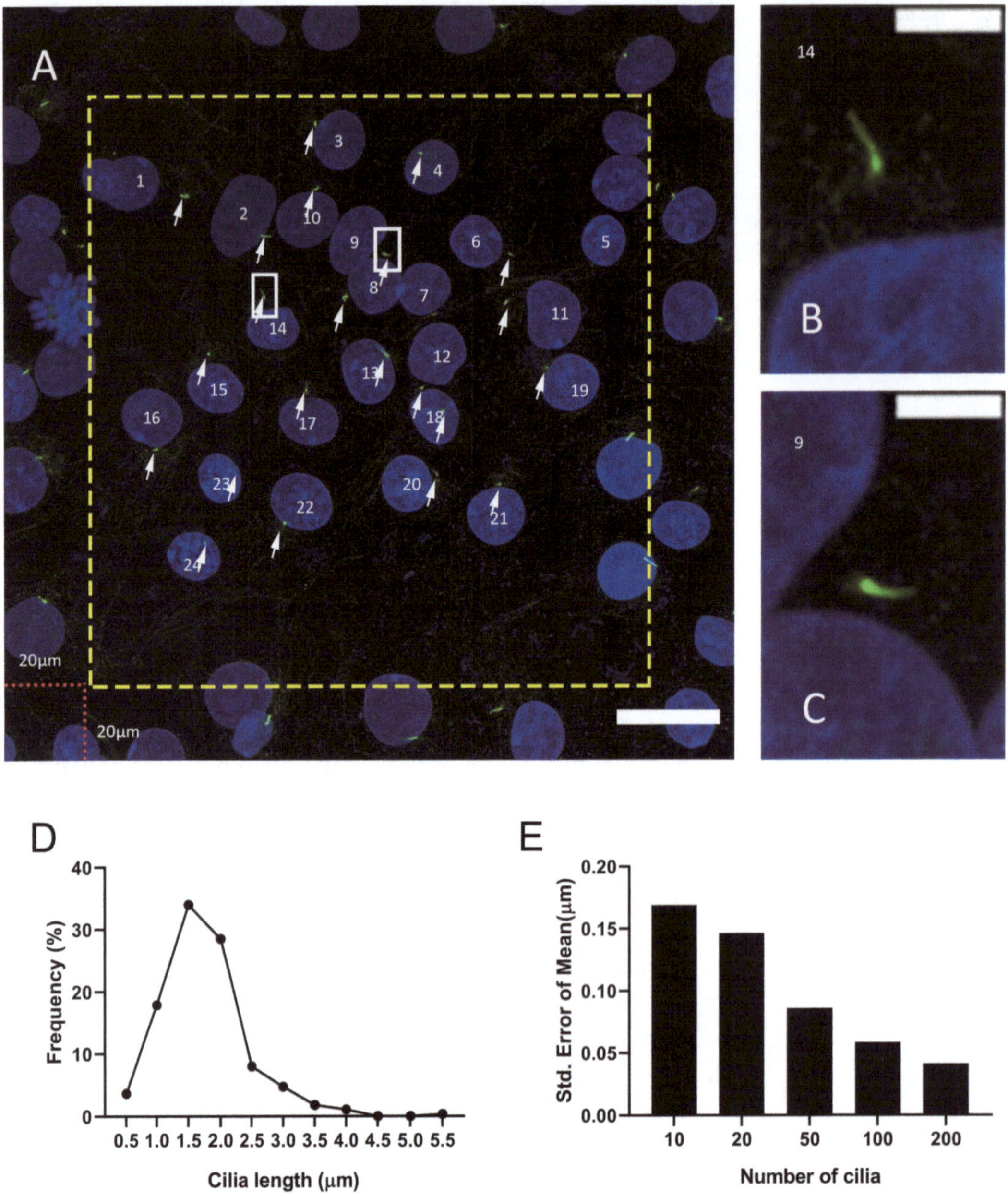

Fig. 4 Measurement of cilia prevalence and cilia length. (**a**–**c**) Confocal maximum projection image of primary bovine chondrocytes labeled with acetylated α-tubulin (green) and DAPI (blue). Cilia are highlighted with red arrows. Only cells in the central region are included for quantification of primary cilia prevalence. Scale bar represents 30 μm for field of view and 5 μm for magnified boxed images. (**d**) Frequency distribution of cilia axonemal length measured using the segmented line tool in ImageJ ($n = 200$ cilia) and (**e**) the corresponding precision of the sample mean length measurement represented by the standard error of mean (SEM). This demonstrated that a sample size of >100 cilia is typically required, producing a SEM of approximately +/− 0.05 μm

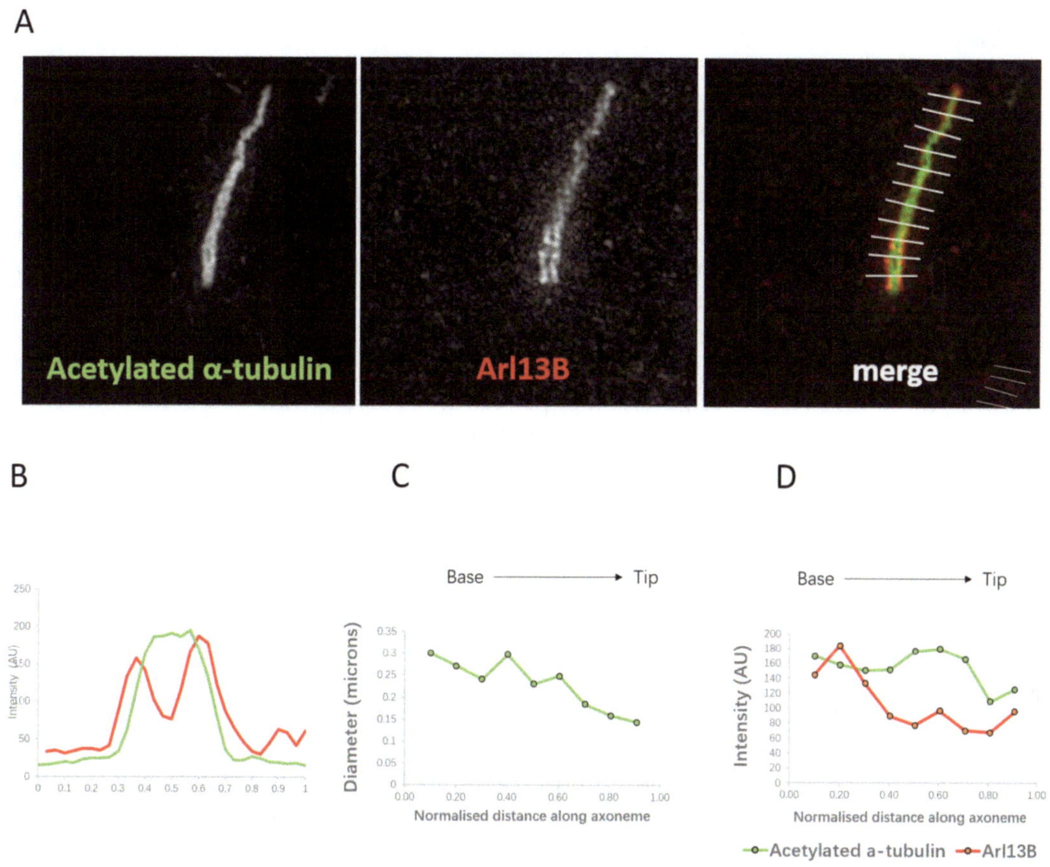

Fig. 5 Measurement of cilia diameter and protein distribution along cilia axoneme. (**a**) sr-SIM maximum projection image of primary bovine chondrocytes labeled with acetylated α-tubulin (green) and Arl13b (red). (**b**) Corresponding transverse intensity profile perpendicular to cilia axoneme demonstrating the arl13b labeling of the cilia membrane. The transverse intensity profile taken at regular spacing along the length of the axoneme were used to measure the variation in (**c**) cilia diameter and (**d**) mean labeling intensity from the base to the tip. The integrated intensity (data not shown) can also be used to provide an indirect estimate of relative total amounts of each protein along the axoneme. Data demonstrates a reduction in diameter of the axoneme based on acetylated α-tubulin (**c**) and a reduction in arl13b mean intensity (**d**) from base to tip

3. Create intensity profiles perpendicular to the axis of the cilium (segmented line from **step 2**). This is best done for ARL-13B and/or acetylated α-tubulin. Use the full width half maximum (FWHM) intensity threshold to define the boundary of the cilium and hence calculate the diameter at that position. Measure the mean and the integrated intensity for each channel within the boundary set by the FWHM. (It is also possible to use an absolute threshold or a different % threshold rather than the FWHM).
4. Repeat the intensity profiles at different positions along the axoneme from the base to the tip (*see* **Note 30**, Fig. 5).

4 Notes

1. A variety of different proteins are enriched on the axoneme of primary cilia which therefore serve as useful markers for immunofluorescence microscopy. Typically, the ciliary axoneme is labeled with antibodies to acetylated α-tubulin or the ciliary membrane protein, arl13b. However, researchers should be aware that some cilia may exist with reduced acetylation or altered arl13b expression such that cilia prevalence assessed by these two markers may differ slightly. As well as an axoneme marker, it is often useful to have a marker of the basal body such as γ-tubulin or pericentrin. This provides additional confidence in identifying individual cilia and enables analysis of spatial variation in protein expression from axoneme base to tip. The following antibodies (*see* Table 1) and protocols have already been optimized in bovine primary chondrocytes and can be provided as a guide but may need further optimization depending on cell source, culture conditions, and proteins of interest. It is normal in Arl13b antibody may have some nuclear staining.
2. The Fluoromount G (00-4958-02, Invitrogen) can be replaced by other water-based mounting medium. ProLong Antifade Mountants (Thermo Fisher) are also acceptable.
3. Super Cryo Embedding Medium (C-EM001, Section-lab Co. Ltd) is recommended.
4. There are a variety of light microscopy-based techniques that can be used for visualizing primary cilia ranging from standard epifluorescence to confocal or super-resolution microscopy techniques such as SIM, PALM, and STORM. Other new exciting microscopy techniques include correlative electron microscopy, high-throughput confocal imaging, expansion microscopy, and x-ray diffraction imaging [15, 16]. These techniques for imaging cilia, along with the widely used technique of transmission electron microscopy (TEM), are beyond the scope of this chapter. Instead, we provide protocols based on the following light microscopy imaging platforms.
5. For polyacrylamide gel surfaces, add 0.2 mg/mL sulfo-SANPAH and expose to UV for 10 min, remove Sulfo-SANPAH, and coat with 50 μg/mL collagen I for further 3 h. For PDMS surfaces, coat with 10 μg/mL fibronectin for 1 h. For Flexcell, coat with 50 μg/mL collagen I for 1 h before seeding cells.
6. The time for chondrocyte to reach confluence in monolayer is likely to depend on the cell source/species, initial seeding density, concentration and batch of FBS, previous passage, and culture substrate. Therefore, it is advisable to check

cultures daily using light microscopy to identify the time to confluence. Initial seeding density may also require optimization for different cell sources).

7. Serum starvation should not normally exceed 48 h; otherwise, cell viability and cell function may be compromised.
8. Under uniaxial compressive strain, the tangent modulus of 3% low gelling agarose is approximately 78.3 kPa ± 9.1 (mean ± standard deviation). Note that agarose is viscoelastic and hence the compressive stress reduces under a constant strain. For 3% agarose low gelling agarose, compressed to 20% strain, the percentage of relaxation at 5 min is 24%, and the corresponding relaxation modulus is 25.0 kPa. This viscoelastic behavior is likely to cause a reduction in peak cell deformation during cyclic compression and may also cause cell relaxation during periods of static compression as previously described [17].
9. Different concentrations of agarose gel may be used for chondrocyte-agarose cultures, typically in the range of 2–4% w/v. Increasing the agarose concentration will increase the mechanical properties of the resulting agarose constructs.
10. Low-gelling agarose (type VII) will set at temperatures under 30 °C, and hence, it is advisable to preheat agarose gel solution in the oven. Different agarose types are available, including ultralow gelling temperature agarose. However, researchers should be aware that the corresponding mechanical and structural properties are not the same. For example, the compressive modulus of 3% low gelling agarose is more similar to that of 4% ultralow gelling agarose.
11. Molds of different shapes and sizes may be used to create specific chondrocyte-agarose constructs. Alternatively, the cell-agarose suspension may be gelled as a thin layer (2–5 mm) at the base of a petri dish or multi-well plate. Researchers should be aware that limited nutrient delivery to cells within the center of thicker constructs may cause loss of viability [18].
12. PFA and methanol fixation can have different effects on antibody staining, and hence, it is therefore always advisable to perform an optimization to determine ideal fixation protocol and antibody concentrations. Indeed, in some cells methanol fixation is necessary to label the basal body.
13. Blocking of samples to avoid nonspecific binding is necessary to provide a clear background. 5% milk powder solution is also accepted for this purpose.
14. To cut identical slices of agarose gel, one tip is to fix two scalpel blades together separated by a glass slides.

15. Usually Alexa Fluor 488, 555, and 633 are suitable conjugated secondary antibodies for most epifluorescence and confocal microscopes.
16. Take care not to disturb the section. Always keep the sections as flat as possible. Do not let them dry out.
17. The imaging magnification increases with the objective lens. We recommend ×63/1.4 NA oil immersion objective for visualizing cilia. When loading samples, the side with coverslips should be faced down to oil immersion objective.
18. It is important to use an unbiased approach to select fields of view when recording cilia length and prevalence. Hence, it is sensible to select the view of view based on the DAPI channel prior to switching to the cilia channel.
19. Primary cilia may be expressed on either the apical or basal surface of cells in monolayer, and hence, it is useful to adjust the fine focus consistently when counting cilia prevalence using epifluorescence or to use a defined number of z sections when using confocal microscopy.
20. Be careful to avoid photobleaching from prolonged and/or high intensity excitation light exposure. To enable comparison between images, it is important to use the same setting such as the camera/detector exposure time, excitation intensity (e.g., confocal laser power), emission wavelength, and image format/pixel size. Optimal parameters of laser excitation and emission, laser strength, gain, and pinhole may vary from antibodies and cell types and procedure of imaging. Adjusting these properly to make images clear is necessary before z-stacks and scanning. Try to get best imaging quality by preview or pre-scan.
21. Abbe's limit of resolution ≈ half the wavelength of light. More specifically for confocal microscopy, the limit of resolution is $\frac{0.61\ \lambda}{\mathrm{NA}}$ in the lateral x-y axis and $\frac{2\ \lambda}{(n\ \mathrm{NA})^2}$ in the axial z axis, where λ is the wave length of emission light and NA is the objective numerical aperture. Typically, this yields lateral and axial limits of resolution of 0.2 and 0.5 μm, respectively.
22. Pixel size of 0.232 μm (image size 1024 × 1024 pixels equivalent to 238 × 238 μm) is usually used for confocal imaging output. For sr-SIM, a pixel size of 0.06 μm (image size: 1280 × 1280 pixels) is recommended.
23. Step size of 0.5 μm is appropriate for cilia confocal scanning; usually it will take 10 z-stack images for one field of view approximately. For sr-SIM, we recommend Z step size of 0.13 μm.
24. 10 or more field of images (no less than 100 cilia) are necessary for cilia quantification.

25. Exclude areas of 20 μm width around for cilia measurement, *see* Fig. 4 for details.
26. Assuming that the angle between the cilia and the horizontal plane is α, sin α = cilia height/cilia length, we exclude the cilia with an angle >30° from the horizontal plane. Since the average length of the cilia is about 2 μm, cilia height more than 1 μm will be considered $\alpha > 30°$. We set Z-stack size as 0.5 μm, so cilia present at least on three sections (height >1 μm) are considered as vertical cilia and excluded in length measurement.
27. When drawing the line, try to keep the segmented line inside the cilia and parallel to the direction of the cilia.
28. At least 100 cilia per experimental condition is essential for cilia length comparison due to the inherent variability (Fig. 4).
29. The length of cilia typically demonstrates a skewed frequency distribution with a few significantly longer cilia (Fig. 4); hence, nonparametric Mann-Whitney U-test may be needed for comparison of cilia length. However, for large sample sizes (>100), parametric tests such as Student t-test or ANOVA are usually sufficient.
30. In collaboration with Dr. Pavel Novak, we have developed an automated ImageJ plugin that performs this automatically to create ten equally space intensity profiles perpendicular to the axoneme.

Acknowledgements

The authors gratefully acknowledge the support of Dr. Pavel Novak for developing the automated ImageJ analysis of protein distribution along the length of the axoneme.
https://www.researchgate.net/profile/Pavel-Novak

References

1. Ishikawa H, Marshall WF (2011) Ciliogenesis: building the cell's antenna. Nat Rev Mol Cell Biol 12(4):222–234. https://doi.org/10.1038/nrm3085
2. Kim J, Jo H, Hong H et al (2015) Actin remodelling factors control ciliogenesis by regulating YAP/TAZ activity and vesicle trafficking. Nat Commun 6:6781. https://doi.org/10.1038/ncomms7781
3. Satir P, Pedersen LB, Christensen ST (2010) The primary cilium at a glance. J Cell Sci 123 (Pt 4):499–503. https://doi.org/10.1242/jcs.050377
4. Wren KN, Craft JM, Tritschler D et al (2013) A differential cargo-loading model of ciliary length regulation by IFT. Curr Biol 23(24): 2463–2471. https://doi.org/10.1016/j.cub.2013.10.044
5. Thompson CL, Chapple JP, Knight MM (2014) Primary cilia disassembly down-regulates mechanosensitive hedgehog signalling: a feedback mechanism controlling ADAMTS-5 expression in chondrocytes. Osteoarthr Cartil 22(3):490–498. https://doi.org/10.1016/j.joca.2013.12.016

6. Veland IR, Awan A, Pedersen LB et al (2009) Primary cilia and signaling pathways in mammalian development, health and disease. Nephron Physiol 111(3):p39–p53. https://doi.org/10.1159/000208212
7. Poole CA, Flint MH, Beaumont BW (1985) Analysis of the morphology and function of primary cilia in connective tissues: a cellular cybernetic probe? Cell Motil 5(3):175–193. https://doi.org/10.1002/cm.970050302
8. McGlashan SR, Cluett EC, Jensen CG et al (2008) Primary cilia in osteoarthritic chondrocytes: from chondrons to clusters. Dev Dyn 237(8):2013–2020. https://doi.org/10.1002/dvdy.21501
9. Wann AK, Knight MM (2012) Primary cilia elongation in response to interleukin-1 mediates the inflammatory response. Cell Mol Life Sci 69(17):2967–2977. https://doi.org/10.1007/s00018-012-0980-y
10. Plotnikova OV, Pugacheva EN, Golemis EA (2009) Primary cilia and the cell cycle. Methods Cell Biol 94:137–160. https://doi.org/10.1016/S0091-679X(08)94007-3
11. Thompson CL, Plant JC, Wann AK et al (2017) Chondrocyte expansion is associated with loss of primary cilia and disrupted hedgehog signalling. Eur Cells Mater 34:128–141. https://doi.org/10.22203/eCM.v034a09
12. Watt FM (1988) Effect of seeding density on stability of the differentiated phenotype of pig articular chondrocytes in culture. J Cell Sci 89 (Pt 3):373–378
13. Benya PD, Shaffer JD (1982) Dedifferentiated chondrocytes reexpress the differentiated collagen phenotype when cultured in agarose gels. Cell 30(1):215–224. https://doi.org/10.1016/0092-8674(82)90027-7
14. Wann AK, Zuo N, Haycraft CJ et al (2012) Primary cilia mediate mechanotransduction through control of ATP-induced Ca2+ signaling in compressed chondrocytes. FASEB J 26(4):1663–1671. https://doi.org/10.1096/fj.11-193649
15. Coltharp C, Xiao J (2012) Superresolution microscopy for microbiology. Cell Microbiol 14(12):1808–1818. https://doi.org/10.1111/cmi.12024
16. Sahabandu N, Kong D, Magidson V et al (2019) Expansion microscopy for the analysis of centrioles and cilia. J Microsc 276(3): 145–159. https://doi.org/10.1111/jmi.12841
17. Bader DL, Ohashi T, Knight MM et al (2002) Deformation properties of articular chondrocytes: a critique of three separate techniques. Biorheology 39:69–78
18. Sengers BG, Heywood HK, Lee DA et al (2005) Nutrient utilization by bovine articular chondrocytes: a combined experimental and theoretical approach. J Biomech Eng 127(5): 758–766. https://doi.org/10.1115/1.1993664
19. McGlashan SR, Knight MM, Chowdhury TT et al (2010) Mechanical loading modulates chondrocyte primary cilia incidence and length. Cell Biol Int 34(5):441–446. https://doi.org/10.1042/cbi20090094

Chapter 13

Co-culture of Human Articular Chondrocytes Seeded in Polyurethane Scaffolds and Human Mesenchymal Stromal Cells Encapsulated in Alginate Beads

Teresa Z. Brose, Yann D. Ladner, E. Johanna Kubosch, Martin J. Stoddart, and Angela R. Armiento

Abstract

Co-culturing is an essential method for unravelling the importance of cross talk and cellular interaction. This chapter describes the preparation of an indirect co-culture technique based on encapsulation of chondrocytes and mesenchymal stromal cells in polyurethane scaffolds and alginate beads, respectively. This way, both cell populations can communicate through paracrine effects in the absence of cell-cell contact. Due to the mechanical properties of polyurethane, this model can be employed in mechanobiology studies. The resulting engineered cultures can provide a more realistic environment, recreating the complex joints' microenvironment and physiology.

Key words Indirect co-culture, Tissue engineering, Cross talk, Alginate beads, Chondrocytes, Mesenchymal stromal cells

1 Introduction

Co-culture techniques are broadly applied to study interactions between cell populations. These methods are of great importance, as they can imitate a physiological and pathological microenvironment. The behavior of co-cultured cell populations is complex because it causes effects beyond the sum of their individual activities [1].

Co-culturing is a culture method composed of two or more distinct cell types sharing a medium. Co-cultures can be categorized into two major systems: direct and indirect co-cultures [2]. In direct co-cultures, the cells can establish cell-cell contacts, thus interacting through direct intercellular contact, as well as paracrine effects. Indirect co-cultures are defined by the physical separation of the two or more cell types under the same conditions but without

Martin J. Stoddart et al. (eds.), *Cartilage Tissue Engineering*,
Methods in Molecular Biology, vol. 2598, https://doi.org/10.1007/978-1-0716-2839-3_13,

direct cell-cell contact. In this way, cell communication is only possible through paracrine effects of soluble factors. Deciding which co-culture method is superior to another is still under constant investigation and depends on the desired experimental design. Parameters that affect the experiment's outcome, and therefore requiring optimization, are cell types/cell sources, cell ratios, type of cell culture medium and additional cytokines/growth factors, co-culture models, and potential biomaterials [3].

A composition of fibrin hydrogel and a polyurethane (PU) scaffold system has shown to be suitable for three-dimensional culture of chondrocytes and mesenchymal stromal cells (MSCs) [4, 5], which provides mechanical properties, and can withstand interfacial shear [6]. Through their resilience, PU scaffolds can be used within bioreactors simulating joint mechanical load, leading to an even more accurate in vitro recapitulation of the complex joint physiology [7]. Alginate, a polymer naturally occurring in brown algae, is widely used for cell immobilization [8] and as a biomaterial in cell culture, tissue engineering, drug delivery, and wound healing [9]. Alginate is not biologically active itself [10], is easy to handle, has low toxicity [9], and can be manipulated and adapted to the desired mechanical properties, such as permeability or stability [11].

Co-culture methods in cartilage tissue engineering, which harness chondrocytes and MSCs, along with its potential therapeutic usage, are dynamically researched and show promising results [12]. Generally, co-cultures of chondrocytes and MSCs offer several advantages to monocultures, such as a reduction of osteoarthritic catabolic markers [13], improvement of biomechanical properties of the newly formed matrix [14, 15], increased cell viability [14], and synergistic effects of co-culturing and growth factors [16]. Through paracrine effects, chondrocytes can induce a chondrogenic phenotype in human MSCs [17]. Simultaneously, MSCs can stimulate chondrocyte-driven cartilage formation through trophic effects [18].

This chapter describes the preparation of a co-culture, including human chondrocytes in PU scaffolds and human bone-marrow-derived MSCs in alginate beads. We mention the method to isolate human chondrocytes and elaborate on the seeding techniques in PU scaffolds and alginate beads.

2 Materials

All solutions are prepared at room temperature using MilliQ® water, and sterile filtered using a 0.22 μm filters, if not specified differently.

2.1 Cell Isolation, Culture, and Growth Media

1. Human femoral head (healthy or osteoarthritic, *see* **Note 1**).
2. Sterile scalpel blade (No. 20 or 21).
3. Sterile Petri dish (diameter: 10 cm).
4. Spinner flask.
5. Basic medium for human chondrocytes: to Dulbecco's modified Eagle's medium 4.5 mg/mL D-glucose (DMEM), add 10% FBS, 1 mmol/L of sodium pyruvate, 100 U/mL of penicillin, 100 μg/mL of streptomycin, 1% of nonessential amino acids, 100 mmol/L of HEPES buffer solution, and 0.29 mg/mL of glutamine. Stir the solution with a magnetic stirring bar and, after sterile filtering, store it at 4 °C.
6. Medium for human chondrocyte isolation: supplement each 100 mL of the basic medium for human chondrocytes with 0.2 g collagenase type II (256 U/mL). Stir the solution with a magnetic stirring bar, and, after sterile filtering, use immediately.
7. Cell strainer (40 μm).
8. Falcon tube (50 mL).
9. Tissue culture flasks (dimensions as needed).
10. Growth medium for human chondrocytes: add 1 ng/mL of TGF-β1 and 5 ng/mL of bFGF-2 to the basic medium for human chondrocytes (*see* **step 5**). Stir the solution with a magnetic stirring bar, and, after sterile filtering, use immediately.
11. Human bone marrow derived MSC (*see* **Note 2**).
12. Chondropermissive medium for co-culture of human chondrocytes and MSCs: to DMEM, add 1 mmol/L of sodium pyruvate, 100 U/mL of penicillin, 100 μg/mL of streptomycin, 1% of nonessential amino acids, 1% of insulin transferrin selenium-premix, 50 μg/mL of L-ascorbic acid 2-phosphate sesquimagnesium salt, and 100 nM of dexamethasone. Stir the solution with a magnetic stirring bar, and, after sterile filtering, use immediately.
13. 1× Trypsin-EDTA solution in PBS: Add 1 mL of the 0.5% trypsin/EDTA stock (concentration: 10×) to 9 mL of PBS.

2.2 PU-Scaffold Seeding

1. PU scaffolds: mix 9 g of presynthesized PU chips (*see* **Note 3**) in 63 mL of NN-dimethylformamide; dissolve it using a magnetic stirrer covered with Parafilm and incubate overnight at 50 °C. Add 39 mL of acetone dropwise, followed by a further 5 mL of MilliQ® water, added in 500 μL aliquots every 10 min. Mill 500 g of OmniPur sodium phosphate dibasic heptahydrate, and sieve the powder using four different sieves with various mesh widths (from 1.0 to 0.15 mm) to obtain particles

with a diameter of 0.15 mm to 0.3 mm. Add 1.36 times the weight of the PU solution from the pulverized sodium phosphate salt to the mixture and leave at 50 °C for 1 week for solvent evaporation. Wash the salt out with demineralized water for 1 week until there are no salt residues, and the pH is between 6.0 and 7.0. This is followed by the last wash step with 70% ethanol overnight and subsequent drying at room temperature for 1 week. Cut the PU scaffolds by waterjet, forming a cylindrical shape with a height of 4 mm and a diameter of 8 mm (*see* **Note 4**). Gas-sterilize PU scaffolds using ethylene oxide.

2. Petri dishes.
3. Sterile forceps.
4. DMEM: *see* **step 5** in Subheading 2.1 (*see* **Note 5**).
5. Desiccator.
6. 12-well tissue culture plates.
7. Sterile Eppendorf caps.
8. 0.9% (w/v) sodium chloride: dissolve 0.9 g of sodium chloride powder (NaCl) in 100 mL of MilliQ® water.
9. Fibrinogen: weigh the required fibrinogen from human plasma (50–70% protein, >80% of protein is clottable) under sterile conditions to obtain a concentration of 50 mg/mL. Dissolve the fibrinogen by layering it on top of the required volume of sterile 0.9% (w/v) sodium chloride. Place the tube in a 37 °C water bath for 4–6 h, with gentle agitation (*see* **Note 6**). Aliquots can be stored long-term at −20 °C.
10. Thrombin: prepare a solution of 0.1% (w/v) bovine serum albumin (BSA) in MilliQ® water. Dissolve the stock solution of thrombin (100 units) in 1 mL of the sterile 0.1% BSA solution to obtain a final concentration of 100 U/mL. Aliquots can be stored long-term at −20 °C (*see* **Note 7**).

2.3 Alginate Beads

1. Sterile Petri dish (diameter 10 cm).
2. Sterile 1 mL syringe.
3. Sterile 21G needle.
4. 0.9% (w/v) NaCl: *see* **step 8** in Subheading 2.2.
5. 1.2% (w/v) alginate solution: add 1.2 g of alginic acid sodium salt from brown algae to 100 mL of 0.9% sodium chloride until the salt is dissolved completely (*see* **Note 8**).
6. Polymerization solution: check the molar mass of the used dehydrated calcium chloride, calculate the required amount for a solution of 102 mM, and dissolve it in 1 L of MilliQ® water.

2.4 Harvest Cells from Alginate Beads

1. 0.9% (w/v) NaCl: *see* **step 8** in Subheading 2.2.
2. Sterile Falcon tube (15 mL).
3. 55 mM sodium citrate (dissolving buffer): Dissolve 1.617 g of trisodium citrate dihydrate with a formula weight of 294.10 g/mol in 100 mL of 0.9% NaCl (*see* **Note 9**). Stir the solution and then sterile filter using a 0.22 μm filter.
4. TRI-reagent.
5. Polyacryl Carrier.
6. Sterile Eppendorf tube (1.5 mL).

3 Methods

3.1 Isolation and Culture of Human Chondrocytes

This section describes the chondrocytes' isolation from human femoral heads, following the protocol of Francioli et al. [19]. However, the chondrocytes can also be isolated from other tissue sources. All procedures are performed under sterile conditions in a laminar flow hood and at room temperature, unless otherwise stated.

1. Shave the cartilage from the bone using a scalpel, and wash it once in PBS.
2. Mince the cartilage shavings into squares of 0.5 cm, transfer them into a spinner flask, add 10 mL of isolation medium per 1 g of tissue, and digest for 22 h at 37 °C, in a humidified atmosphere and 5% CO_2.
3. Pass the solution through a 40 μm cell strainer into a new 50 mL tube, and centrifuge for 5 min at 500 g. Remove the supernatant, wash the cells with growth medium, and repeat the centrifugation.
4. Plate the isolated chondrocytes in tissue culture flasks with a density of 10^4 cells/cm^2, add growth medium for human chondrocytes, and incubate them at 37 °C in a humidified atmosphere with 5% CO_2. Feed the cells twice a week by refreshing the growth media.

3.2 PU Scaffold Seeding

This section describes the seeding of human chondrocytes into PU scaffolds, but any other material where cells can be embedded can be used. All procedures are performed under sterile conditions in a laminar flow hood and at room temperature, unless otherwise stated. The seeding procedure is illustrated in Fig. 1.

1. Degas the PU scaffolds: add the scaffolds to DMEM in a Petri dish and squeeze them using sterile forceps to promote medium influx into the scaffolds. Close the Petri dish, place it into the desiccator, connect it to the vacuum pump, and degas the scaffolds for 2–3 h.

PU Scaffold Seeding

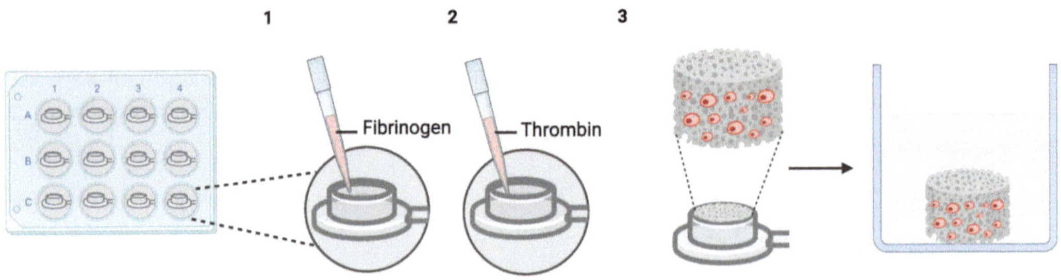

Fig. 1 PU scaffold seeding procedure. After trypsinizing the chondrocytes, the cells are seeded into PU scaffolds using the caps of 1.5/2 mL Eppendorf tubes as placement holders. The caps are placed into a 12-well plate, the fibrinogen-cell suspension [1] is mixed with thrombin [2], and the PU scaffold is added, soaking up the fibrinogen/thrombin/cell suspension [3]. (Created with Biorender.com)

2. Upon 80% confluency, detach the required number of cells (5×10^6 per each PU scaffold) by trypsinization (*see* **Note 10**) and centrifuge them at 500 g for 5 min at room temperature. Aspirate the supernatant.
3. Thaw the thrombin and dilute the solution to a final concentration of 2 U/mL using sterile MilliQ® water. Keep the thrombin on ice until use.
4. Squeeze the degassed scaffolds using sterile tweezers, and aspirate the released medium as much as possible.
5. Transfer a sterile cap of a 1.5 or 2 mL Eppendorf tube for every required scaffold into wells of a 12-well tissue culture plate.
6. Calculate the total amount of fibrinogen required (75 μL per scaffold), and use it to resuspend the cells carefully and slowly to avoid foam formation.
7. Add 75 μL of the fibrinogen-cell suspension into each Eppendorf cap.
8. Add 75 μL of thrombin solution to each cap, mix five times by pipetting the solution up and down. Add a PU scaffold into the lid, press it down using a pipette tip and forceps to force the PU scaffold to draw up the solution. Squeeze the PU scaffold three more times, invert the PU scaffold, and further squeeze the scaffold three more times.
9. Place the PU scaffolds in an incubator and incubate them at 37 °C, 5% CO_2, and 95% humidity, for 40–60 min (*see* **Note 11**).
10. Using sterile forceps, insert the scaffolds into a holder or vessel of choice. Add 3 mL of chondropermissive medium into each holder, and place them into the incubator.

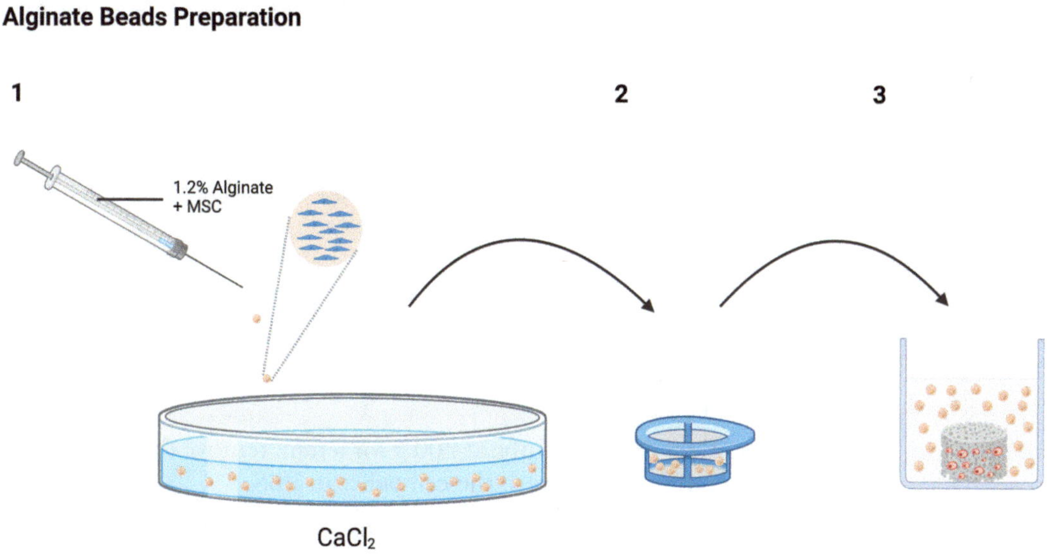

Fig. 2 Alginate bead preparation. The MSC-alginate solution is carefully dropped into the polymerization solution using a syringe with a 21G needle [1]. After polymerization, the beads are strained, washed [2], and then added into the culture medium of chondrocyte-laden scaffolds [3]. (Created with Biorender.com)

3.3 Alginate Bead Preparation

This section describes the encapsulation of MSC into alginate beads but any other cell type can be used. All procedures are performed under sterile conditions in a laminar flow hood and at room temperature, unless otherwise stated. A schematic representation of this procedure is shown in Fig. 2.

1. Detach the required number of MSCs by trypsinization (*see* **Note 12**), calculating 5×10^6 for each scaffold containing 5×10^6 chondrocytes (*see* **Note 13**).
2. Prepare a sterile Petri dish for each sample by adding 10 mL of the polymerization solution into the dish (*see* **Note 14**).
3. Centrifuge the required number of cells at 500 g for 5 min. Aspirate the supernatant and carefully add sterile 1.2% alginate solution until a final concentration of 5×10^6 cells/mL is reached (*see* **Note 15**).
4. Aspirate the solution into a sterile 1 mL syringe fitted with a 21G needle, and remove the air by inverting the syringe and flicking it gently with a finger (*see* **Note 16**).
5. Slowly release the alginate-cell solution dropwise into the polymerization solution. The beads polymerize instantly (*see* **Note 17**). Let the beads polymerize entirely for an additional 5 min (*see* **Note 18**).
6. Pour the bead-containing solution through a cell strainer, laid on top of a 50 mL tube, and discard the leftover solution.

7. Wash the beads one time in 0.9% NaCl and three times with PBS (*see* **Note 19**).
8. Add 5 × 10^6 cells (corresponding to 1 mL beads) into the culture medium of the chondrocyte-laden scaffolds.
9. Change the medium three times per week by carefully removing the medium (*see* **Note 20**) and then adding fresh medium.

3.4 Harvesting Cells from Alginate Beads

1. Wash beads once with 0.9% NaCl and twice with PBS.
2. Transfer beads into a centrifuge tube (15 mL) containing 5 mL of dissolving buffer.
3. Incubate the tube for 15 min at 37 °C or until the beads are dissolved.
4. Centrifuge the tube at 700 g for 5 min, remove the supernatant carefully, and wash the cell pellet in PBS.
5. For RNA isolation, remove the supernatant, break the pellet and resuspend it by adding 1 mL of TRI-Reagent and 5 μL polyacryl carrier (*see* **Note 21**).
6. Transfer the solution into a 1.5 mL tube, and store it at −80 °C until further use.

4 Notes

1. Isolate human articular chondrocytes following Francioli et al. [19] after informed consent by the patients and in accordance with the ethics committee.
2. Isolate human mesenchymal stromal cells from bone marrow aspirates as described previously by Armiento et al. [20], after informed consent and full approval of the local ethics committee (*see* Chapter 6).
3. It is recommended to use scaffolds made from the same batch of presynthesized PU chips for one experiment.
4. These dimensions can be modified as needed.
5. For this step, it does not need to be the complete basic medium.
6. As per the manufacturer's instructions, do not vortex the solution since it can lead to denaturation of the fibrinogen.
7. Aliquot the solutions in plastic tubes because thrombin solutions adsorb to glass. Do not refreeze the aliquots.
8. When sterile filtering the alginate solution, some positive pressure may be required.
9. If necessary, adjust the amount of trisodium citrate according to the formula weight.

10. Cells should be used at an early passage, not later than passage 5.
11. Even though the surface is dry already, leave the scaffolds in the incubator for the entire time.
12. MSCs should not be too confluent before seeding since the differentiation capacity decreases during culturing. Thus, use at an early passage.
13. When having a cell ratio of 1:1.
14. Alternatively, use a 30–100 mL sterile glass beaker with a stirring bar, add 40 mL of $CaCl_2$ solution, and set the stirring to low speed.
15. Carefully resuspend the cells in the alginate solution to avoid air bubbles. Keep in mind that changing the cell density in the alginate solution modifies the properties of the fabricated beads.
16. It is also possible to use a P1000 micropipette.
17. The drops should be released slowly, giving them time to polymerize, to avoid the beads sticking together.
18. If a stirring bar was used, remove the stirring bar with sterile forceps.
19. Use approximately 10 mL per washing step.
20. It is recommended to use a small tip to not aspirate the alginate beads. If the media is collected for further analysis, it is recommended to use low binding tips.
21. The preferred RNA isolation procedure can be used.

References

1. Goers L, Freemont P, Polizzi KM (2014) Co-culture systems and technologies: taking synthetic biology to the next level. J R Soc Interface 11(96):20140065. https://doi.org/10.1098/rsif.2014.0065
2. Kook Y-M, Jeong Y, Lee K et al (2017) Design of biomimetic cellular scaffolds for co-culture system and their application. J Tissue Eng 8: 2041731417724640. https://doi.org/10.1177/2041731417724640
3. Zhang Y, Guo W, Wang M et al (2018) Co-culture systems-based strategies for articular cartilage tissue engineering. J Cell Physiol 233(3):1940–1951. https://doi.org/10.1002/jcp.26020
4. Li Z, Kupcsik L, Yao SJ et al (2009) Chondrogenesis of human bone marrow mesenchymal stem cells in fibrin-polyurethane composites. Tissue Eng Part A 15(7):1729–1737. https://doi.org/10.1089/ten.tea.2008.0247
5. Lee CR, Grad S, Gorna K et al (2005) Fibrin-polyurethane composites for articular cartilage tissue engineering: a preliminary analysis. Tissue Eng 11(9–10):1562–1573. https://doi.org/10.1089/ten.2005.11.1562
6. Neumann AJ, Gardner OFW, Williams R et al (2015) Human articular cartilage progenitor cells are responsive to mechanical stimulation and adenoviral-mediated overexpression of bone-morphogenetic protein 2. PLoS One 10(8):e0136229–e0136229. https://doi.org/10.1371/journal.pone.0136229
7. Monaco G, El Haj AJ, Alini M et al (2021) Sodium hyaluronate supplemented culture medium combined with joint-simulating mechanical loading improves chondrogenic differentiation of human mesenchymal stem cells. Eur Cell Mater 41:616–632. https://doi.org/10.22203/eCM.v041a40

8. Smidsrød O, Skjåk-Braek G (1990) Alginate as immobilization matrix for cells. Trends Biotechnol 8(3):71–78. https://doi.org/10.1016/0167-7799(90)90139-o
9. Lee KY, Mooney DJ (2012) Alginate: properties and biomedical applications. Prog Polym Sci 37(1):106–126. https://doi.org/10.1016/j.progpolymsci.2011.06.003
10. Andersen T, Strand BL, Formo K et al (2012) Chapter 9 Alginates as biomaterials in tissue engineering. In: Carbohydrate chemistry, The Royal Society of Chemistry, vol 37, pp 227–258. https://doi.org/10.1039/9781849732765-00227
11. Pandolfi V, Pereira U, Dufresne M et al (2017) Alginate-based cell microencapsulation for tissue engineering and regenerative medicine. Curr Pharm Des 23(26):3833–3844. https://doi.org/10.2174/1381612823666170609084016
12. Brose TZ, Kubosch EJ, Schmal H et al (2021) Crosstalk between mesenchymal stromal cells and chondrocytes: the hidden therapeutic potential for cartilage regeneration. Stem Cell Rev Rep. https://doi.org/10.1007/s12015-021-10170-6
13. Diao HJ, Yeung CW, Yan CH et al (2013) Bidirectional and mutually beneficial interactions between human mesenchymal stem cells and osteoarthritic chondrocytes in micromass co-cultures. Regen Med 8(3):257–269. https://doi.org/10.2217/rme.13.22
14. Bian L, Zhai DY, Mauck RL et al (2011) Coculture of human mesenchymal stem cells and articular chondrocytes reduces hypertrophy and enhances functional properties of engineered cartilage. Tissue Eng A 17(7–8): 1137–1145. https://doi.org/10.1089/ten.tea.2010.0531
15. Yang Y-H, Lee AJ, Barabino GA (2012) Coculture-driven mesenchymal stem cell-differentiated articular chondrocyte-like cells support neocartilage development. Stem Cells Transl Med 1(11):843–854. https://doi.org/10.5966/sctm.2012-0083
16. Aung A, Gupta G, Majid G et al (2011) Osteoarthritic chondrocyte-secreted morphogens induce chondrogenic differentiation of human mesenchymal stem cells. Arthritis Rheum 63(1):148–158. https://doi.org/10.1002/art.30086
17. Kubosch EJ, Heidt E, Bernstein A et al (2016) The trans-well coculture of human synovial mesenchymal stem cells with chondrocytes leads to self-organization, chondrogenic differentiation, and secretion of TGFβ. Stem Cell Res Ther 7(1):64. https://doi.org/10.1186/s13287-016-0322-3
18. Kubosch EJ, Lang G, Furst D et al (2018) The potential for synovium-derived stem cells in cartilage repair. Curr Stem Cell Res Ther 13(3):174–184. https://doi.org/10.2174/1574888X12666171002111026
19. Francioli S, Cavallo C, Grigolo B et al (2011) Engineered cartilage maturation regulates cytokine production and interleukin-1β response. Clin Orthop Relat Res 469(10): 2773–2784. https://doi.org/10.1007/s11999-011-1826-x
20. Armiento AR, Ladner YD, Della Bella E, Stoddart MJ (2023) Isolation and in vitro chondrogenic differentiation of human bone marrow-derived mesenchymal stromal cells. In: Stoddart MJ, Della Bella E, Armiento AR (eds) Cartilage Tissue Engineering. Springer. https://doi.org/10.1007/978-1-0716-2839-3_6

Chapter 14

Isolation of Murine Articular Chondrocytes for Single-Cell RNA or Bulk RNA Sequencing Analysis

Jillian L. McCool, Nicholas R. Hum, Aimy Sebastian, and Gabriela G. Loots

Abstract

Single-cell RNA sequencing (scRNA-seq) is highly dependent on cellular composition of a tissue of interest. For soft tissues, isolation of individual cells from the extracellular matrix (ECM) while retaining viability and minimizing degradation within subpopulations is well established. In contrast, articular cartilage is comprised of sparsely positioned chondrocytes embedded within a dense ECM high in glycosaminoglycans, proteoglycans, and many fibrous proteins such as collagens, elastin, fibronectin, and laminins. This densely packed ECM makes it difficult to isolate viable chondrocytes for further single-cell analysis. This protocol highlights a successful technique optimized for isolating chondrocytes from the articulated joints of rodent animal models using a series of enzymatic digestions and chondrocyte enrichment using a double negative selection process through florescence-activated cell sorting (FACS).

Key words Single-cell RNA sequencing, scRNA-seq, Cartilage, Extracellular matrix (ECM)

1 Introduction

Orthopedic trauma and joint degenerative diseases such as osteoarthritis and rheumatoid arthritis contribute to major disability, worldwide, impacting more than 20% of the population. Affected individuals typically experience pain and impaired function of articulated joints, usually caused by significant physical and molecular alterations of the articular cartilage, synovium, tendons, ligaments, and sometimes muscle and subchondral bone [1, 2]. For the past 3 decades, scientists have pushed the limits of "omic" technologies to achieve high-quality genetic and molecular data on mammalian single cells towards cataloging every protein, gene expression, and epigenetic profile in various tissues, in health and disease states. Most recently the *Human Cell Atlas* (HCA) project has pushed these limits by aiming to understand the diversity in form and function of all ~37 trillion cells in the human body with the ultimate goal of aligning biological, clinical, and computational

Martin J. Stoddart et al. (eds.), *Cartilage Tissue Engineering*,
Methods in Molecular Biology, vol. 2598, https://doi.org/10.1007/978-1-0716-2839-3_14,

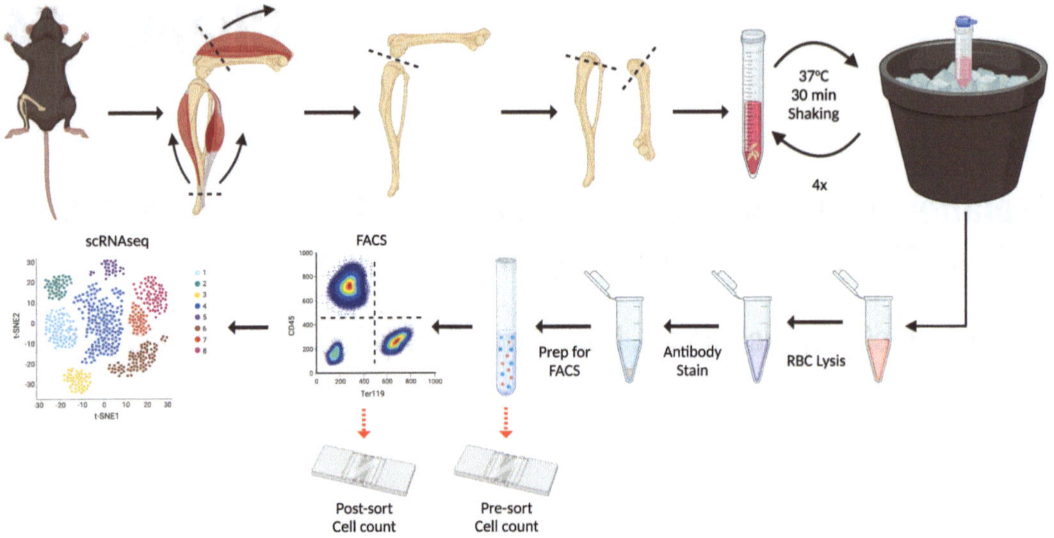

Fig. 1 Flowchart of chondrocyte isolation for scRNA-seq application

information to obtain new insights into how we can achieve cellular resilience, stop degeneration, and cure disease [3, 4].

However, despite the high burden of musculoskeletal disease, single-cell sequencing technology has provided the most insights in characterizing soft tissues (such as liver and lungs) [3] that require simpler cell dissociation approaches to yield a high percent of viable cells. In contrast, musculoskeletal tissues present significant technical challenges for releasing individual cells to generate high-quality scRNA-seq data due to the low-cellularity and dense, fibrous extracellular matrices encasing the cells. In humans, these challenges are further complicated by the lack of healthy tissue reference samples to allow us to define normal baselines. Animal models of disease are therefore essential for filling these gaps in knowledge and can assist the musculoskeletal community to build a cellular network representative of tissues and disease states that can parallel human disease pathology. Here we present a protocol optimized for isolating chondrocytes from the articulated joints of rodent animal models using a series of enzymatic digestions and chondrocyte enrichment through a double negative selection process by florescence-activated cell sorting (FACS) (Fig. 1).

2 Materials

Prepare all solutions using purifying molecular biology grade water and analytical grade reagents. Prepare all reagents fresh on the day of the experiment and, unless indicated otherwise, at room temperature (RT).

2.1 Appropriate Personal Protective Equipment (PPE)

1. Lab coat.
2. Gloves.
3. Eye protection.

2.2 Animal Euthanasia and Tissue Harvesting

1. CO_2 chamber at 1.72 L/min flow rate.
2. Medium tip sterile forceps.
3. 4.5″ sharp tip dissecting scissors.
4. Number (#)11 disposable sterile scalpel or scalpel blades.
5. Kimwipes and/or gauze pads.
6. Ice bucket.
7. 100 mm petri dishes.
8. Dulbecco's Minimal Essential Media (DMEM) supplemented with 10% Fetal Bovine Serum (FBS).
9. 5 and 10 mL serologic pipettes.
10. P1000 pipette and tips.

2.3 Articular Cartilage Digest

1. Sterile 15 and 50 mL conical tubes.
2. 70 μm Nylon cell strainer.
3. Digestion media: Dissolve 2 mg/mL of Collagenase 2 in Dulbecco's Minimal Essential Media with *no addition* of FBS.
4. Heated shaker at 37 °C.

2.4 Cell Preparation and Chondrocyte FACS Isolation

1. Refrigerated centrifuge with rotor for 50 mL conical tubes.
2. P20, P200, P1000 pipettes and tips.
3. Red blood cell lysing buffer Hybri-Max.
4. Phosphate buffered solution (PBS; diluted to 1× with sterile dH_2O).
5. Microcentrifuge.
6. 1.5 mL sterile Eppendorf tubes.
7. 15 mL Falcon Conical Tube and 50 mL Falcon Conical Tube.
8. DAPI (viability), APC-Cy7 anti-mouse CD45 Clone 30-F11 (immune cell marker), and APC anti-mouse Ter119 Clone Ter119 (erythroid cell marker).
9. 5 mL flow cytometry tubes (12 × 75 mm).
10. Trypan blue.
11. Thermo Fisher Countess™ (or similar instrument).
12. Hemocytometer.
13. BD FACS melody (or similar FACS instrument).

2.5 Library Preparation and Sequencing

1. Qiagen RNeasy Mini Kit.
2. Thermo Fisher Qubit Fluorometer.
3. Thermo Fisher Qubit RNA HS Assay Kit and tubes.
4. Illumina Stranded mRNA Prep, Ligation.
5. 10× Genomics Chromium Controller.
6. 10× Genomics Chromium Single Cell 3′ Library and Gel Bead Kit.
7. 10× Genomics Chromium Chip Single Cell.
8. 10× Genomics i7 Sample Indices.
9. Wide-bore pipette tips (*see* **Note 1**).
10. Glycerol.
11. Benchtop vortex.
12. PCR Tubes 0.2 mL 8-tube strips.
13. Thermal cycler.
14. Invitrogen Dynabeads™ MyOne™ SILANE Magnetic Beads.
15. 80% Ethanol.
16. Beckman Coulter SPRISelect Reagent Kit.
17. Qiagen Buffer EB.
18. Agilent Bioanalyzer High Sensitivity DNA Analysis Kit.
19. Thermo Fisher Qubit dsDNA HS kit & tubes.
20. 0.2 M NaOH.
21. Illumina NextSeq 500/550 High Output Kit v2.5 (75 cycles).
22. Illumina Nextseq 500 or comparable sequencer.

3 Methods

For the purposes of this chapter, murine knee joints (C57BL/6 strain) are used as the source material, but other tissue sources of chondrocytes (hips) are treated similarly but slight variations in the enzyme cocktail and duration of each step may be implemented. As a general note, cell viability decreases and transcriptional alterations increase as tissue processing time extends; therefore, it is recommended that the protocol is conducted continuously and as fast as possible to minimize the time between dissection and single-cell library preparation. Additionally, cells should be maintained on ice throughout the process when feasible.

3.1 Articular Cartilage Dissection from Long Bones

1. Euthanize the animals humanely under CO_2 at 1.72 L/min followed by cervical dislocation before dissection. All animal procedures should be performed according to approved institutional animal care protocols.

2. Skin manually the hindlimbs, and trim excess muscle from the leg by cutting the gastrocnemius at the achilleas tendon with a #11 disposable sterile scalpel and continuing up through the medial hamstring muscle until the hip joint is reached.
3. Remove the quadriceps femoris muscle by creating a single cut above the patella and peeling the muscle back towards the body using medium tip forceps.
4. Dislocate the hip joints by applying an upward force to the hip joint and femur separately, and trim the excess tensor fascia lata muscle to remove the entire hindlimb.
5. Remove excess muscle and fat from the intact hindlimb using a Kimwipe or gauze pad, while using 4.5″ dissection scissors to trim large pieces of the remaining muscle.
6. Separate the joints by lightly sawing through the connective tissue between the tibia and femur with a #11 disposable sterile scalpel, thus breaking the synovial capsule and exposing the articular cartilage of the knee joint.
7. Clean of muscle each epiphysis using a Kimwipe or gauze pad and remove connective tissue by trimming with 4.5″ dissection scissors carefully around the articular cartilage. Keep the separated bones on ice in a 100 mm petri dish submerged in 5 mL sterile DMEM+10% FBS until microdissection, ideally less than 30 min (min).
8. Once all bones are separated and cleaned of most connective and muscle tissues (Fig. 2a), use a #11 disposable sterile scalpel to remove ~1 mm of tissue from the articulated surface of each long bone (Fig. 2b) separating the articular cartilage from each long bone (Fig. 2c).

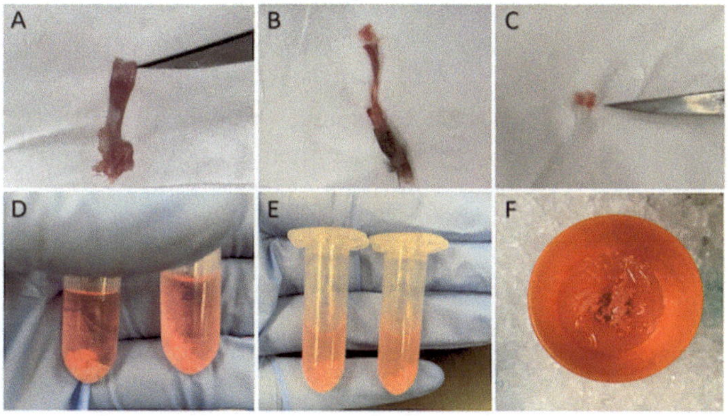

Fig. 2 Preparation of articular cartilage samples from mouse long bones

3.2 Articular Cartilage Dissociation and Digestion

1. Mince the separated articular cartilage (Fig. 2c) into smaller pieces using 4.5″ dissection scissors, and place them in a 15 mL conical tube (falcon with 5 mL of digestion media (Fig. 2d).
2. Digest the cartilage for 4 × 30 min increments on a heated shaker at 37 °C for a total digestion time of 2 h.
3. After each 30 min interval, digestion media should appear slightly cloudy (Fig. 2e). Filter each solution of digestion media through a 70 μm Nylon cell strainer into a 50 mL conical tube on ice, and then add equal amounts of DMEM *with* 10% FBS (Fig. 2f). Scrape the remaining cartilage back into the digestion tube for the next 30 min incubation.
4. Add 5 mL of fresh digestion media to digesting cartilage tissue before proceeding with the next 30 min incubation on a heated shaker at 37 °C.

3.3 Cell Preparation and Chondrocyte FACS Isolation

1. Pellet the digested cell suspension in 50 mL conical tubes by centrifugation at 4 °C for 10 min at 800 g. Pour off the supernatant and remove the remaining liquid using a P200 pipet for prevention of pellet disturbance.
2. Deplete the cell pellets of red blood cells by resuspending them in 1 mL of ACK red blood cell lysis buffer and incubating at RT for 5 min.
3. Bring the resuspended cells up to 2 mL total volume with PBS, and transfer to a fresh Eppendorf tube, and pellet at 600 g for 6 min at RT. Supernatant is removed with a P200 pipet.
4. Stain the cells for 15 min on ice in 100 μL per sample of antibody cocktail which includes: DAPI (1:100, viability), CD45 (APC, 1:100, immune cell marker), and Ter119 (APC-Cy7, 1:10, erythroid cell marker) diluted in PBS + 10% FBS.
5. Following incubation with antibody cocktail, add 1 mL of PBS and pellet the stained cells at 600 g for 6 min at RT. Remove the supernatant with a P200 pipet.
6. Resuspend the cells in 250 μL of PBS + 1% FBS solution, transfer them to flow cytometry tubes, and keep them on ice prior to FACS.
7. Count the cells and quantify the number of viable cells using trypan blue to stain for dead cell and the Thermo Fisher Countess™ instrument, per manufacturer guidelines.
8. Use FACS to isolate chondrocyte enriched populations ($CD45^-/Ter119^-$) cells into DMEM +10% FBS (Fig. 3). First gate the cell suspension on cellularity based on size (forward scatter) and granularity (side scatter) (Fig. 3a), then viability (Fig. 3b), and a finally chondrocyte-enriched population ($CD45^-/Ter119^-$) (Fig. 3c). This cell preparation

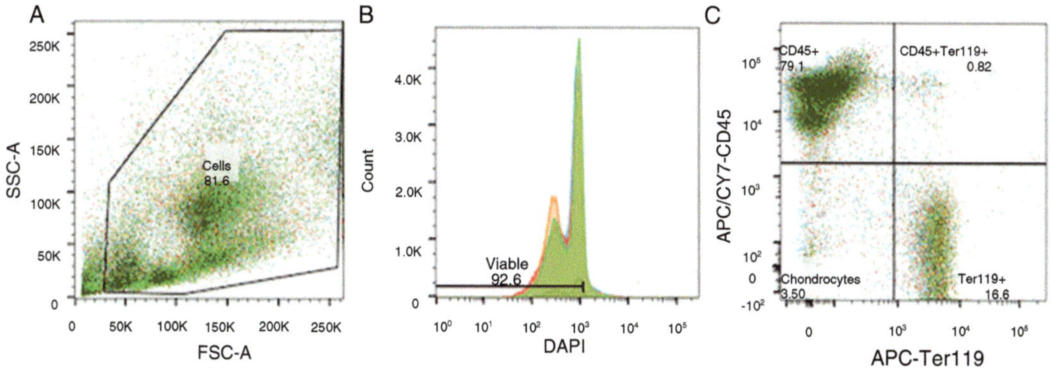

Fig. 3 Viability and FACS gating of chondrocyte population

typically yields ~80% cells and ~90% of those cells are viable with ~3.5% of those being nonerythroid or nonimmune cells of interest.

9. Following FACS, pellet the cells at 600 × *g* for 6 min at RT. A small pellet is expected at this stage due to low cell quantity. Carefully remove the supernatant with a P200 pipet to not to disturb the pellet; then resuspend in 40 μL of PBS + 1% FBS using a wide bore tip.
10. Determine final cell yield using a hemocytometer.

3.4 Bulk Library Preparation

Bulk RNA sequencing preparation can be performed on enriched chondrocyte populations. This protocol recommends using Qiagen RNeasy columns for RNA isolation followed by Illumina Stranded mRNA library preparation kits however other comparable RNA isolation and library preparation should also work with this cell suspension.

1. Centrifuge cell suspension at 600 g for 6 min, then resuspend in Qiagen Buffer RLT lysis buffer supplemented with 10 μL β-mercaptoethanol/1 mL RLT.
2. Mix by pipetting to ensure thorough lysis of cells.
3. Perform column isolation using Qiagen RNeasy spin columns per manufacturer's recommendations then elute into 30 μL of RNase-free water.
4. Quantify RNA content using Qubit RNA HS Assay and Thermo Fisher Qubit system.
5. Generate RNA library using Illumina Stranded mRNA Prep and ligation kit per manufacturer's recommendations.
6. Perform quantitative analysis of sequencing library with Qubit dsDNA HS kit.
7. Perform qualitative analysis of sequencing library with Agilent Bioanalyzer High Sensitivity DNA Chips to determine purity and average amplicon size (Fig. 4). Average library fragment size ~250 bp.

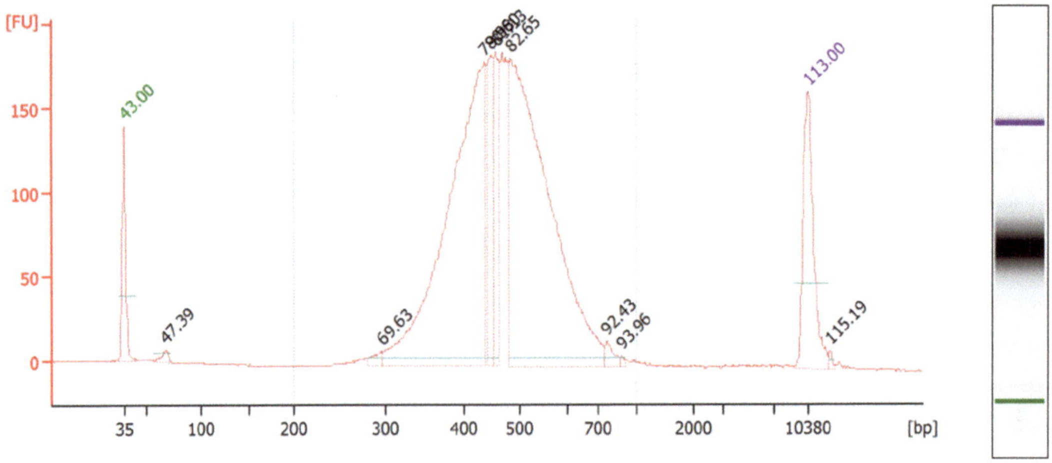

Fig. 4 Representative bioanalyzer trace from completed single-cell sequencing library preparation

3.5 Single-Cell Library Preparation and Sequencing

While the preparation of cells can be adapted to other technologies and approaches, the following section describes the implementation of 10× Genomics' Drop-seq based approach for scRNA-seq then sequenced on an Illumina NextSeq 500 instrument [5]. Manufacturer's protocols are closely adhered to for library preparation and sequencing, and it is recommended to consult reagent protocols for updated details. An abridged version of the protocol is outlined, and critical steps are detailed below.

1. Load the 10× Genomics Chip (Chip) into loading holder and fill unused wells with diluted glycerol (50% glycerol, 50% water by volume) prior to loading reaction reagents.
2. Dilute the cell suspension to a 2×10^5–1×10^6 cells/mL range for loading onto the Chip (*see* **Note 1**).
3. Prepare the Master Mix on ice containing reverse transcriptase and template switch oligonucleotide fragments according to manufacturer's specifications in DNA LoBind strip cap tubes.
4. Determine the volume of cell suspension to add to Master Mix based on target cell quantity to be sequenced and cell concentration. Add water first to bring volume up to 80 μL final Master Mix reaction, then finally add cell suspension. Mix gently by pipetting gently 10 times with a wide bore tip.
5. Gently load 75 μL of Master Mix into 10× Genomics Chip and accompanying gel beads and partitioning oil into appropriate wells.
6. Cover chip with 10× supplied rubber gasket, load Chip into Chromium instrument, then initiate single-cell program (~8.5 min).

7. Carefully transfer gel emulsions to a new LoBind strip cap tube then transfer to a thermocycler to perform RT reaction and addition of cellular barcodes and unique molecular identifier barcode (UMI) (*see* **Note 2**).
8. Dissociate the gel emulsion, and isolate DNA using magnetic separation with Dynabeads.
9. Generate cDNA libraries and perform DNA magnetic separation with SPRIselect (*see* **Notes 2** and **3**).
10. Fragment the cDNA into ~500 bp pieces for sequencing then ligate adapters onto the ends of the fragments to allow for addition of sequencing index via PCR (*see* **Note 4**).
11. Add sample indices and amplify the cDNA library via a PCR reaction with 12 amplification cycles. Perform size selection and cleanup using SPRIselect magnetic beads (*see* **Note 2**).
12. Perform quantitative analysis of sequencing library with Qubit dsDNA HS kit. Typical yields for 2500 cell = 500–1000 ng of DNA.
13. Perform qualitative analysis of sequencing library with Agilent Bioanalyzer High Sensitivity Chips to determine purity and average amplicon size (Fig. 4). Average library fragment size ~500 bp.

3.6 Sequencing

Suggested depth for bulk sequencing samples is 20 million reads per samples, and for single-cell sequencing, a depth of 40,000–50,000 reads/cell is suggested to identify chondrocyte subpopulations.

1. Pool sequencing libraries that will be sequenced together by equimolar addition of each library into a new tube. The following equation can be used to calculate molarity:

$$\frac{\text{DNA concentration (ng/}\mu\text{L)}}{(660 \times \text{Average library size (bp))}} \times 10^6 = \text{Molarity (nM)}.$$

2. Dilute pooled library to 5 nM, then denature using fresh 0.2 M NaOH.
3. Dilute denatured libraries to a final concentration of 1.8 pM without additional PhIX controls (*see* **Note 5**)
4. Load sequence on an Illumina NextSeq 500 using the 75 bp high output kits with the following sequencing run lengths:
   (a) Bulk, single-end; single index, index 1, 10 bp; read 1, 76 bp
   (b) Single cell, paired-end; single index sequencing, index 1, 8 bp; read 1, 28 bp; read 2, 56 bp; index 2, 0.

4 Notes

1. Wide bore tips reduce shear force and provide a wider opening for maintaining cell viability and gel emulsion integrity. Wide bore tips are used for all steps containing cells and gel-emulsion/beads to ensure integrity of single-cell reactions. Extra care should be taken to pipette slowly and gently handle all cells and gel-emulsion/beads.
2. **Safe stopping point:** 4 °C overnight or −20 °C for up to a week.
3. cDNA can be quantified using DNA HS Qubit dsDNA HS kit to ensure cDNA generation. Typical yields for this preparation: 2500 cells = ~200–300 ng cDNA.
4. SPRIselect magnetic beads are used for reaction cleanup, size selection, and DNA isolation throughout this step.
5. The complexity of the RNA libraries is sufficient to sequence without the addition of the PhIX control library.

References

1. Cieza A, Causey K, Kamenov K et al (2021) Global estimates of the need for rehabilitation based on the Global Burden of Disease study 2019: a systematic analysis for the Global Burden of Disease Study 2019. Lancet 396(10267): 2006–2017. https://doi.org/10.1016/S0140-6736(20)32340-0
2. McMahon SB, Dargan P, Lanas A et al (2021) The burden of musculoskeletal pain and the role of topical non-steroidal anti-inflammatory drugs (NSAIDs) in its treatment. Ten underpinning statements from a global pain faculty. Curr Med Res Opin 37(2):287–292. https://doi.org/10.1080/03007995.2020.1847718
3. Mereu E, Lafzi A, Moutinho C et al (2020) Benchmarking single-cell RNA-sequencing protocols for cell atlas projects. Nat Biotechnol 38(6):747–755. https://doi.org/10.1038/s41587-020-0469-4
4. Hon CC, Shin JW, Carninci P et al (2018) The human cell atlas: technical approaches and challenges. Brief Funct Genomics 17(4):283–294. https://doi.org/10.1093/bfgp/elx029
5. Sebastian A, McCool JL, Hum NR et al (2021) Single-cell RNA-Seq reveals transcriptomic heterogeneity and post-traumatic osteoarthritis-associated early molecular changes in mouse articular chondrocytes. Cell 10(6). https://doi.org/10.3390/cells10061462

Chapter 15

Modulation of MicroRNA Expression During In Vitro Chondrogenesis

Austin Bell-Hensley, Hongjun Zheng, and Audrey McAlinden

Abstract

Since their discovery in 1993, microRNAs (miRNAs) are now recognized as important epigenetic regulators of many mammalian cellular processes including proliferation, apoptosis, metabolism, and differentiation. These small non-coding RNAs function by interacting with specific regions in the 3′-untranslated region of mRNAs, thereby resulting in mRNA degradation or suppression of translation. Since miRNAs have the ability to target many mRNAs within a given cell type, a number of cellular pathways and networks may be regulated as a result. To study the function of miRNAs, a number of methods can be used to modulate their activity in cells such as synthetic mimics or antagomirs for short-term assays or viral-based approaches for longer-term experiments such as cell differentiation assays. In this chapter, we provide our methodology to constitutively overexpress a desired miRNA during in vitro chondrogenesis of human cartilage progenitor cells (CPCs). Specifically, we describe how we obtain CPCs from human articular cartilage specimens, how we generate and titrate lentivirus engineered to overexpress a precursor miRNA, how we transduce CPCs with lentivirus and differentiate them toward the chondrocyte lineage, and how we extract RNA and measure expression levels of the miRNA of interest during in vitro chondrogenesis. We also provide some data from our laboratory demonstrating that we can achieve and maintain miRNA overexpression for up to 14 days in cartilage pellet cultures. We predict that these lentiviral-based approaches will also be useful to study how miRNA modulation of progenitor cells affects cell differentiation and extracellular matrix production within three-dimensional biomaterial scaffolds.

Key words MicroRNA, miRNA, Lentivirus, Differentiation, Chondrogenesis, Progenitor cell, Chondrocyte, Cartilage

1 Introduction

Completion of the human genome project informed us that only approximately 1–2% of our DNA contains genes that are transcribed to protein-coding mRNAs [1]. The remainder of the genome contains DNA that does not provide instructions for generating protein. We know from various research endeavors,

---

Austin Bell-Hensley and Hongjun Zheng contributed equally with all other contributors.

Martin J. Stoddart et al. (eds.), *Cartilage Tissue Engineering*,
Methods in Molecular Biology, vol. 2598, https://doi.org/10.1007/978-1-0716-2839-3_15,

including the Encyclopedia of DNA Elements (ENCODE) project [2, 3], that a lot of non-coding DNA is not "junk," but actually play important functional roles. Such non-coding DNA sequences include gene promoters, enhancers, and silencers, as well as structural chromosomal elements including telomeres and satellite DNA. Other regions of DNA are actively transcribed to form non-coding RNAs that have critical functions in regulating gene and protein expression. These include transfer RNAs (tRNAs), ribosomal RNAs (rRNAs), long non-coding RNAs, snoRNAs, and microRNAs. This latter class of ncRNAs are the subject of this chapter with respect to methodologies related to modulating their expression during chondrogenesis.

MicroRNAs (miRNAs) are short (~17–24 nucleotides; nt) non-coding RNAs that function as epigenetic regulators by suppressing expression of their target mRNAs [4]. Genes encoding miRNAs are commonly found in intergenic regions of the genome or within introns of protein-coding genes. They are first transcribed to form large primary stem-loop precursors (pri-miRNAs) either as single entities or as clusters. These pri-miRNAs are processed in the nucleus by a Drosha-containing complex to form shorter (~100–150 nt) precursor miRNAs (pre-miRNAs) that are exported into the cytosol. A Dicer-containing complex then processes these pre-miRNAs to short miRNA duplexes containing a 5p and 3p strand [5]. One of these strands represents the functional miRNA strand that will enter the RNA-induced silencing complex (RISC) and interact, via its seed sequence (position 2–8 of the mature miRNA strand), with a complementary site in the 3′UTR of its target mRNA. The end result is either degradation of the mRNA or suppression of mRNA translation [6, 7]. The reader is referred to Fig. 1 of our recent review article demonstrating the various steps leading to generation of mature miRNAs in the cell [8]. Although miRNAs account for only 1–5% of the human genome [9], it has been estimated that at least 60% of protein-coding genes are regulated by miRNAs [10].

Short-interfering RNAs (siRNAs) target one mRNA, and suppression of this mRNA is robust due to 100% complementary binding with the siRNA. However, miRNA suppression of target mRNAs is modest in comparison given the imperfect pairing between the miRNA and target mRNA (except for the seed sequence interaction). As such, one miRNA may have the potential to target and modestly suppress tens to hundreds of mRNAs thereby affecting many cellular pathways and networks [11]. This feature, together with their small size, renders these RNAs as attractive therapeutic agents. A search on clinicaltrials.gov website reveal some miRNA targeting therapies currently under phase I/II clinical trials as potential treatments for fibrosis or wound healing, for example. According to the current miRBase website (http://mirbase.org), over 1900 mature miRNAs have been identified in human and over 1200 in mice. Also of note is the high degree of

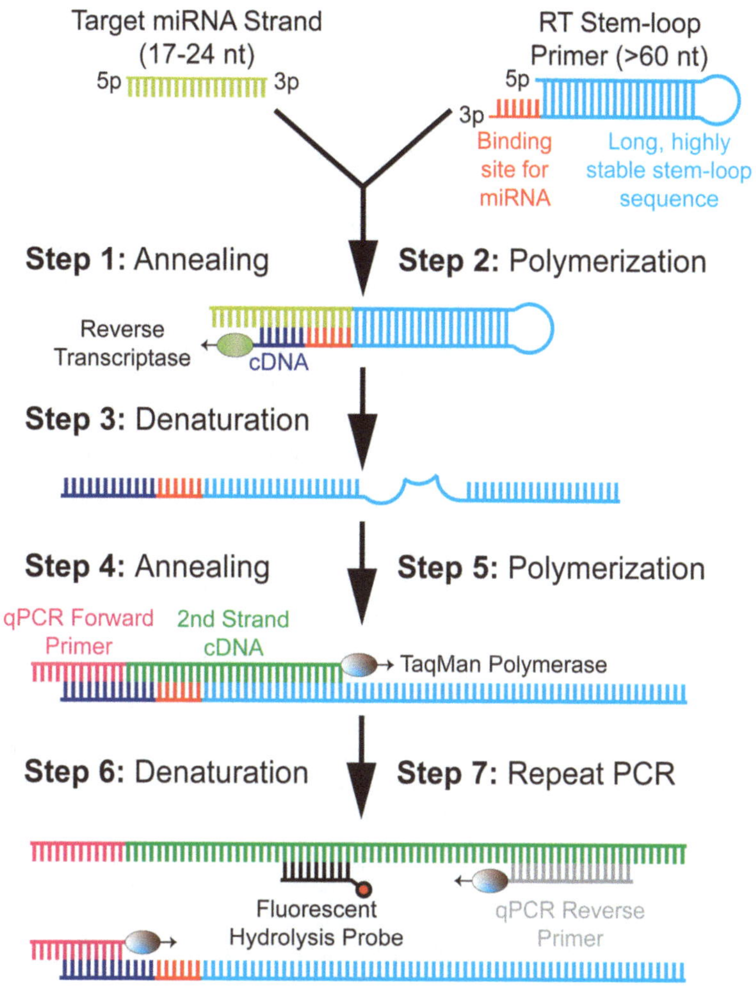

Fig. 1 Methodology to detect and quantify levels of small, mature miRNAs using the Custom TaqMan™ Small RNA Assays (Applied Bioscience). This diagram shows the stem-loop primer that is specifically designed to anneal to the mature miRNA strand of interest during reverse transcription (**steps 1** and **2**). The remaining steps represent the qPCR stage involving a specific forward and reverse primer pair and TaqMan™ probe to amplify a region of cDNA corresponding to the mature miRNA sequence of interest

conservation between human and mouse mature miRNA sequences, thereby permitting preclinical studies in mouse models to study the effects of miRNA modulation on tissue development and disease processes.

The published literature on miRNAs is vast and informs us that they are important regulators of many cellular processes and are often dysregulated in disease. There are hundreds of reports (summarized in various review articles) describing the effects of miRNAs on chondrocyte function and cartilage tissue formation. Among these reports, specific miRNAs have been shown to regulate chondrogenesis as well as proliferation, metabolism, apoptosis, growth factor signaling, and inflammatory responses in chondrocytes [12–17]. Many studies also report on the effects of miRNA modulation in suppressing cartilage degradation in rodent models of osteoarthritis (OA) thereby suggesting that therapeutic targeting of miRNAs could be a promising strategy to treat OA [18–24]. With respect to cartilage development, we previously determined miRNA expression patterns in chondrocytes at distinct stages of differentiation in the human developing growth plate of long bones [25]. We are still exploring the functional roles of some of the miRNA candidates identified from that microarray study, two of which (miR-138 and miR-181a/b cluster) are mentioned in this chapter since we are currently modulating their expression to investigate their effects on chondrogenesis in vitro.

Common methods used to modulate expression/activity of miRNAs in vitro or in vivo involve utilizing short synthetic double-stranded "mimics" or single-stranded "antagomirs." As the names suggest, commercially available mimics have the same sequence as the endogenous double-stranded miRNA duplex of interest, while the antagomir is designed to bind, via complementary base pairing, to a specific endogenous functional miRNA strand in the cell. In doing so, the endogenous miRNA can no longer bind to its target mRNAs. In general, mimics or antagomirs can be transiently transfected into cells in vitro, and the effects of miRNA overexpression/activation (mimics) or suppression of miRNA activity (antagomirs) on specific cellular processes can be determined.

Other methods to overexpress or inhibit a miRNA of interest include plasmid or viral approaches. To suppress miRNA activity, an expression plasmid can be engineered to express a specific short hairpin RNA (shRNA). This shRNA would then be processed in the cell to generate the antisense (or antagomir) strand of the miRNA of interest. Lentiviral reagents (i.e., miRZip™ anti-miRNA expression lentivectors) are commercially available to generate shRNAs that will inhibit the function of specific miRNAs (System Biosciences). To determine if miRNA suppression has been achieved, it is recommended to analyze mRNA and/or protein expression levels of a known miRNA target (expression of the miRNA target should increase if the endogenous miRNA has been suppressed by the antagomir).

For miRNA overexpression, cDNA encoding the precursor miRNA (pre-miRNA) of interest should be cloned into an expression plasmid (for transient transfections) or into a transfer plasmid for lentivirus generation. The expressed pre-miRNA will be processed in the cytoplasm by the Dicer-containing complex to generate the short functional mature miRNA strand. Successful overexpression of the miRNA can be determined by carrying out specific reverse transcription-quantitative PCR (RT-qPCR) assays (as described in this chapter). An additional assay to confirm increased expression of the miRNA may involve measuring mRNA and/or protein levels of a known mRNA target of the miRNA in question (levels of the target mRNA/protein will be expected to decrease if the miRNA has been overexpressed).

While short synthetic mimics/antagomirs or expression plasmids are useful for studying the function of miRNAs in vitro, these approaches involve transient transfection of cells and, therefore, are appropriate for short-term functional assays (i.e., up to 3–4 days). For longer-term assays, such as cell differentiation/tissue engineering protocols, a lentiviral-based system is attractive. Lentiviruses can infect nearly all mammalian cell types, including hard-to-transfect primary cells and non-dividing cells, and they can be used to generate stable cell lines due to integration of the transgene into the host genome [26]. Also, one report comparing miRNA overexpression via synthetic mimics or lentivirus in vitro suggested that lentivirus was superior given the off-target effects caused by transfection with mimics [27]. While lentiviruses to express your gene of interest are commercially available, we generate our own lentiviral reagents to overexpress a miRNA of interest.

We have previously published on our lentiviral-based approach to conditionally overexpress miR-483 (via doxycycline-mediated induction) during in vitro chondrogenesis of human bone marrow-derived mesenchymal stromal cells (MSCs) [28]. In this chapter, we provide our methodology to constitutively overexpress a desired miRNA during in vitro chondrogenesis of primary human cartilage progenitor cells (CPCs). Note that the same lentiviral-based approaches can be applied to MSCs as well. Specifically, we describe (1) how we obtain and culture human CPCs; (2) how we generate batches of lentivirus particles; (3) how we transduce primary CPCs with lentivirus, (4) in vitro chondrogenesis of lentiviral-transduced CPCs, and (5) how we isolate total RNA from chondrogenic pellet cultures and utilize a specific RT-qPCR approach to determine overexpression levels of the miRNA of interest.

2 Materials

All reagents are prepared using sterilized, deionized water unless specified otherwise. All cell culture medium and associated reagents are handled and aliquoted under sterile conditions within tissue culture hoods.

2.1 Key Equipment and Tools

- Agilent 2100 Bioanalyzer (Agilent Technologies).
- NanoDrop™ 2000 Spectrophotometer (Thermo Scientific).
- Spinner Flask (Bellco) and Spinner Plate (Dura-Mag).
- Pellet Pestle Motor and Pellet Pestle Mixers (Kontes).
- Polyethersulfone (PES) Syringe Filters, 0.22 μm and 0.45 μm (Millipore Sigma).

2.2 Reagents, Cell Culture Media, and Kits

2.2.1 Isolation and Culture of Human Cartilage Progenitor Cells (CPCs)

- *Cartilage digestion medium*: 0.025% collagenase (Roche) and 0.025% pronase (Roche) in DMEM/F12 (Gibco) with 10% FBS.
- *Cartilage Progenitor Cell (CPC) Growth Medium*: 10% FBS (R&D Biosystems), 2.5 μg/mL amphotericin B (Corning), 1% penicillin/streptomycin (Gibco) in DMEM/F12 (Gibco).
- *Fibronectin-coating solution*: 10 μg/mL bovine fibronectin (Sigma), 1 mM $CaCl_2$ (Sigma), and 1 mM $MgCl_2$ (Sigma) in 0.1 M PBS (Gibco).
- *Fibronectin-blocking solution*: 1% bovine serum albumin (Sigma), 1 mM $CaCl_2$ (Sigma), and 1 mM $MgCl_2$ (Sigma) in 0.1 M PBS (Gibco).
- DPBS (Gibco).
- 0.25% Trypsin in 1X EDTA (Gibco).

2.2.2 Chondrogenic Induction Media

- 1% ITS+ premix (Corning or R&D Systems).
- 1 mM sodium pyruvate (Corning).
- 2 mM L-glutamine (Corning).
- 1% penicillin-streptomycin (Gibco).
- 40 μg/mL L-proline (Sigma).
- 100 μM ascorbate-2-phosphate (Sigma).
- 100 nM dexamethasone (Sigma).
- High glucose (4.5 g/L D-Glucose) DMEM (Gibco).
- 10 ng/mL TGF-β3 (R&D Systems) (*see* **Note 1**).

2.2.3 Lentiviral-Induced Overexpression of microRNAs

- DNeasy Blood & Tissue Kit (Qiagen).
- HotStar HiFidelity Polymerase Kit (Qiagen).
- QIAquick Gel Extraction Kit (Qiagen).
- Human embryonic kidney (HEK)-293T cells (ATCC).
- *Human embryonic kidney (HEK)-293T Cell Growth Medium*: 10% FBS (R&D Biosystems), 2 mM L-glutamine (Corning), 1% penicillin/streptomycin (Gibco) in high glucose (4.5 g/L D-Glucose) DMEM (Gibco).
- Gibson Assembly Master Mix (New England Biolabs).
- pLemiR-NS (lentiviral transfer plasmid) (*see* **Note 2**).
- psPAX2 (lentiviral packaging plasmid) (Addgene) (https://www.addgene.org/12260/).
- pMD2.G (lentiviral envelope plasmid) (Addgene) (https://www.addgene.org/12259/).
- 100 μg/mL polyethyleneimine (PEI) (Sigma) in Opti-MEM (Gibco).
- Lenti-X™ qRT-PCR Titration Kit (Takara Bio USA).
- Protamine sulfate (Sigma).

2.2.4 RNA Isolation and Detection of miRNA Expression

- Total RNA Purification Kit from Norgen Biotek.
- Custom TaqMan™ Small RNA Assay from Applied Biosystems™ (*see* **Note 3**).
- TaqMan™ MicroRNA Reverse Transcription Kit from Applied Biosystems.

3 Methods

3.1 Isolation of Cartilage Progenitor Cells from Human Articular Cartilage

We utilize human osteoarthritic knee joint specimens following total knee arthroscopy to obtain batches of primary articular chondrocytes or cartilage progenitor cells (CPCs) (*see* **Note 4**). Prior to processing the tissue specimens, appropriate equipment (i.e., tweezers, scalpel blade holder, gauze pads, spinner flasks) should be autoclaved ahead of time. Please refer to our previously published chapter for additional information on storage of human knee joint tissue specimens to maximize tissue quality for the purpose of RNA isolation directly from whole tissue [29]. Here we describe our procedure for isolation of a population of cartilage progenitor cells from articular cartilage using a fibronectin selection step. Fibronectin adhesion assays have been used to isolate cells with high colony-forming efficiency for nearly three decades [30]. This assay has been adapted to identify and isolate cartilage progenitor cells (CPCs) from articular cartilage tissue [31–34]. While additional techniques such as FACS and clonal expansion have been used to further characterize these cells, we simply pool the entire cell isolate for use in chondrogenesis assays. Note that these cells also have the potential to differentiate toward the osteogenic and adipogenic lineages when cultured in the appropriate induction medium.

1. In advance, prepare fibronectin-coated tissue culture plates.
2. Apply fibronectin coating solution (*see* Subheading 2.2.1) to a sterile culture flask (T75/T150) and incubate overnight at 37 °C.
3. Add fibronectin blocking solution (*see* Subheading 2.2.1) to the fibronectin-coated tissue culture flask for 30 min at room temperature.
4. Remove the fibronectin blocking solution and store the coated flask at 4 °C. (These coated flasks can be stored at 4 °C for approximately 1 month).
5. Prepare 50 mL of Digestion Media (*see* Subheading 2.2.1). Filter through a 0.22 μm PES filter and place in a 37 °C incubator. Pre-warm DPBS as well.

6. In a laminar cell culture flow hood, prepare a sterile dissection workspace.
7. While holding the human tissue specimen in place with sterile tweezers, remove blood using a vacuum hose, warmed DPBS, and the gauze pads.
8. A scalpel blade (size #10) is then used to remove layers of full-thickness articular cartilage from the knee joint specimen. Place the harvested cartilage pieces into DPBS.
9. Dice the cartilage into small (~1 cm^3) pieces in DPBS using the scalpel.
10. Rinse the cartilage pieces multiple times in DPBS.
11. Transfer the cartilage pieces into the spinner flask using the tweezers and add 50 mL of pre-warmed Digestion Media.
12. Place the spinner flask on a spinner plate inside a 37 °C incubator.
13. Activate the spinner plate to rotate at 200 rpm for 16 h.
14. Pour the media containing digested cartilage tissue through a cell strainer (70 μm) into a sterile conical tube. This will remove undigested tissue and debris.
15. Centrifuge the conical tube at 1500 rpm for 5 min at 25 °C. Aspirate the supernatant and resuspend the cells in warmed DPBS.
16. Repeat **steps 14–15** until the DPBS is clear after centrifuging.
17. Centrifuge the conical tube at 1500 rpm at 25 °C for 5 min.
18. Aspirate the supernatant and resuspend the cells in warmed Cartilage Progenitor Cell (CPC) Growth Media (*see* Subheading 2.2.1).
19. Plate the cells in a fibronectin-coated tissue culture flask.
20. Incubate the cells on the fibronectin-coated tissue culture flask for 20 min at 37 °C.
21. After 20 min, aspirate the CPC Growth Media (and any non-adherent cells floating in the media). Add fresh CPC Growth Media to the culture flask (e.g., 10 mL to a T75 or 20 mL to a T150 flask) and incubate at 37 °C refreshing the CPC Growth Media every 3–4 days (*see* **Note 5**).
22. After fibronectin selection, expand the cells using the CPC Growth Media in standard non fibronectin-coated tissue culture flasks. Passage when the cells reach 90% confluence using 0.25% Trypsin in 1X EDTA.
23. CPCs are usually passaged up to four times prior to freezing in liquid nitrogen. For chondrogenesis assays, aim to use CPCs between P4 and P6 (*see* **Note 6**).

3.2 Generation of Lentivirus Particles to Overexpress microRNAs

The procedure described in this chapter involves utilizing a second-generation lentivirus system (*see* **Note 7**). See Subheading 2.2.3 for the transfer plasmid, envelope plasmid and packaging plasmid we use. Note that you may choose to utilize a "safer" third-generation lentivirus system for your purposes.

Among the miRNAs we have overexpressed using this lentivirus system include miR-138 and the miR-181a/b-1 cluster. To achieve this, we first clone a cDNA fragment encoding the precursor form of the miRNA of interest into the pLemIR-NS transfer plasmid (*see* **Note 2**). We isolate genomic DNA from human chondrocytes (using the DNeasy Blood and Tissue Kit), and use specific primers to amplify the desired region of the pre-miRNA sequence. The purified cDNA amplicon is then cloned into the pLemiR transfer plasmid. Using the Gibson Assembly Master Mix (New England Biolabs), we remove the non-silencing (NS) cDNA fragment and replace with the cDNA amplicon encoding the pre-miRNA following digestion of the plasmid with MluI and NotI. Sanger sequencing is used to confirm the integrity of the cloned amplicon. Note that we have already shown successful overexpression of the mature functional miRNA strands of miR-138 and miR-181a/b using this lentiviral-based procedure to determine the downstream effects on osteogenesis [35, 36]. The following step-by-step protocol below describes how we generate and titrate lentivirus particles for cell transduction.

3.2.1 Lentivirus Packaging and Titering

1. Expand HEK-293T cells in Cell Growth Medium (*see* Subheading 2.2.3 for the media components).
2. Seed HEK-293T cells in 10 cm sterile cell culture dishes (~8×10^6 cells per dish) with 5 mL of HEK-293T Cell Growth Media and incubate overnight at 37 °C.
3. Prepare the lentivirus transfection mixture: For 10 cm culture dish, 6 μg transfer plasmid (pLemiR-NS or pLemiR-miRNA), 4.5 μg packaging plasmid (psPAX2), 1.5 μg envelope plasmid (pMD2.G), 0.1 mg/mL polyethyleneimine in Opti-MEM. Approximately 650 μL of this mixture will be adequate per 10 cm dish of HEK-293T cells.
4. Vortex the lentivirus transfection mixture and incubate at room temperature for 20 min.
5. Add the transfection mixture in a dropwise fashion in a circular pattern to each 10 cm dish containing the HEK-293T cells, and gently swirl the mixture to coat the entire surface of the plate.
6. Incubate the cells with the lentivirus transfection mixture at 37 °C for 16–18 h.
7. Remove the media and transfection mixture. Add 5 mL of fresh HEK-293T Cell Growth Media.
8. Incubate the cells at 37 °C or 36–48 h.

9. Withdraw the 5 mL of HEK culture media (which now contains lentivirus) with a syringe.
10. Filter the media through a 0.45 μm polyethersulfone (PES) filter into a 50 mL conical tube (*see* **Note 8**).
11. Add 1.67 mL of Lenti-X concentrator to the 50 mL conical tube for every 5 mL of lentivirus suspension.
12. Incubate the 50 mL conical tube at 4 °C for at least 30 min.
13. Centrifuge the 50 mL conical tube at 4 °C, 1500 × *g* for 45 min. Remove the supernatant.
14. Concentrate the virus 100× relative to the volume of the lentivirus suspension in **step 13** using ice-cold DPBS.
15. Transfer the lentivirus concentrate from the 50 mL conical tube to a 1.5 mL micro-centrifuge tube.
16. Proceed directly to lentivirus titering or store the lentivirus concentrate at −80 °C.
17. For lentivirus titering, make a 10× dilution of a small aliquot of lentivirus concentrate in DPBS in a fresh 1.5 mL micro-centrifuge tube (recommended: 15 μL of lentivirus concentrate in 135 μL of DPBS).
18. Measure the titer of the virus using the Lenti-X™ qRT-PCR Titration Kit (Takara Bio USA) according to the manufacturer's instructions.

3.3 Lentiviral-Induced Modulation of microRNAs During Chondrogenesis

The following subsections provide information on the steps required for (1) transduction of human CPCs with lentiviral particles, (2) chondrogenic induction of these transduced CPCs using the well-established pellet culture assay system [37], (3) isolation of total RNA from chondrogenic pellets, and (4) determining the levels of miRNA overexpression achieved in the cells of the chondrogenic cultures.

3.3.1 Lentiviral Transduction of Human CPCs

1. Seed CPCs at an appropriate density according to Table 1 and incubate overnight at 37 °C (*see* **Note 9**).
2. Determine how much media will be necessary for cell transduction according to Table 2.
3. Generate Transduction Media by dissolving protamine sulfate (0.1 mg/mL) in CPC Growth Media (*see* Subheading 2.2.1). It will take at least an hour for the protamine sulfate to fully dissolve.
4. Filter the protamine sulfate dilution through a 0.22 μm PES syringe filter.
5. Add lentivirus to the Transduction Media protamine sulfate dilution according to the desired multiplicity of infection (MOI), the titer of the lentivirus calculated in Subheading 3.2.1, and the number of cells seeded (*see* Table 1).

Table 1
Cell seeding density for lentiviral transduction of progenitor cells

Culture dish	Number of cells to seed
96-well plate	1.0×10^4
24-well plate	1.6×10^4
12-well plate	3.2×10^4
6-well plate	8.0×10^4
10 cm dish	4.6×10^5
T75	6.0×10^5

Table 2
Volume of media required for lentiviral transduction of progenitor cells

Culture dish	Transduction media amount [mL]
96-well plate	0.04
24-well plate	0.1
12-well plate	0.2
6-well plate	0.5
10 cm dish	3
T75	4.5

6. Invert the lentivirus-containing Transduction Media three to four times to mix.
7. Aspirate the media from the CPCs seeded in **step 1**, and then add the appropriate amount to each dish according to Table 2.
8. Incubate CPCs at 37 °C for about 24 h to permit lentivirus uptake by the cells.
9. Replace the transduction media with fresh CPC expansion media according to Table 3.
10. Incubate the lentiviral-transduced CPCs at 37 °C for an additional 24–48 h (*see* **Note 10**).

3.3.2 Chondrogenic Induction of Lentiviral-Transduced CPCs

1. Combine cell subcultures transduced with the same lentivirus and count the cells (*see* **Note 11**).
2. Centrifuge the cells at 1500 rpm at 25 °C for 5 min.
3. Resuspend the cells in warmed chondrogenic induction media from Subheading 2.2.2 at a concentration of $2.5–3 \times 10^5$ cells/mL.
4. Pipette 1 mL of well-mixed cell suspension into a fresh 15 mL conical tube.

Table 3
Expansion media volume for lentiviral transduction of progenitor cells

Culture dish	Expansion media amount [mL]
96-well plate	0.2
24-well plate	0.5
12-well plate	1
6-well plate	2
10 cm dish	10
T75	13

5. Repeat **step 15** using a fresh 15 mL conical tube each time until all cell suspension has been transferred.
6. Centrifuge the conical tubes at 400 × *g* at 25 °C for 5 min to form pellet cultures.
7. Slightly loosen the cap of each 15 mL conical tube to allow gas to flow in and out.
8. Incubate all conical tubes, which now contain pellet cultures with chondrogenic induction media, at 37 °C with 5% CO_2 (*see* **Note 12**).
9. Refresh chondrogenic induction media three times per week (*see* **Note 1**).

3.3.3 Isolation of RNA from Lentiviral-Transduced Pellet Cultures

1. Transfer chondrogenic pellet cultures from 15 mL conical tubes into 1.5 mL micro-centrifuge tubes using a 1 mL pipette tip (*see* **Note 13**). Remove excess media from the pellets.
2. Add 50 μL of Buffer RL provided by the Norgen Biotek Total RNA Purification Kit to each micro-centrifuge tube.
3. Flash freeze the pellets and Buffer RL by submerging them for 5 s in liquid nitrogen.
4. Use the Pellet Pestle Motor with a Pellet Pestle Mixer to break up the chondrogenic pellets (*see* **Note 14**).
5. Repeat **steps 3** and **4** until the pellets have dissolved in Buffer RL.
6. Add 300 μL of Buffer RL to the micro-centrifuge tube to bring the final volume to ~350 μL.
7. Repeat **steps 3–6** for each micro-centrifuge tube containing chondrogenic pellets using a fresh Pellet Pestle Mixer for each micro-centrifuge tube.
8. Add 200 μL of 100% ethanol to each micro-centrifuge tube containing dissolved pellets.

9. Mix the contents of each micro-centrifuge tube by pipetting up and down ten times or vortexing for 15 s.
10. Isolate total RNA (including small RNAs) using the Total RNA Purification Kit (Norgen Biotek) according to the manufacturer's instructions starting at **step 2**.
11. Determine RNA concentration and quality using a NanoDrop™ 2000 Spectrophotometer or an Agilent 2100 Bioanalyzer (*see* **Note 15**). Store RNA at −80 °C until further use.

3.3.4 Determining miRNA Expression Levels in Lentiviral-Transduced Cells

1. Obtain Custom TaqMan™ Small RNA Assays (Applied Bioscience) that contain reagents designed to specifically determine expression levels of the mature miRNA strand of interest. Each assay kit provides one vial of stem-loop primers designed to anneal to the mature miRNA strand during the reverse transcription step. This kit also provides one vial containing a mixture of the forward and reverse PCR primers and the fluorescently labeled TaqMan™ probe for the qPCR step. Figure 1 shows how these stem-loop primers work and provides a diagrammatic step-by-step description of the reverse transcription step and the TaqMan™ real-time qPCR steps to measure expression levels of your mature miRNA strand of interest.
2. Dilute the RNA samples (*see* Subheading 3.3.3) to a concentration of 5 ng/μL. A total of 25 ng of RNA will be used for the reverse transcription step. Store the diluted RNA samples on ice until use.
3. Reverse-transcribe the RNA using the TaqMan™ MicroRNA Reverse Transcription Kit (Applied Biosystems) according to the manufacturer's instructions. The instructions will refer to an RT Primer which corresponds to the stem-loop primer from the Custom TaqMan™ Small RNA Assay (Applied Bioscience).
4. With the cDNA synthesized in **step 3**, TaqMan™ real-time PCR is carried out according to the manufacturer's instructions. Here, the PCR primer pair set and the fluorescently labeled TaqMan™ probe from the Custom TaqMan™ Small RNA Assay kit (Applied Bioscience) mentioned in **step 1** will be used.
5. Analyze results using the $2^{-\Delta\Delta Ct}$ method. For our purposes, we are interested in determining the level of miRNA overexpression achieved following lentiviral transduction during chondrogenesis. Following normalization of miRNA expression to a reference gene (e.g., RNU44), we then calculate the fold change increase in expression of the miRNA of interest in the experimental group (e.g., lentivirus overexpressing miR-138) compared to the control group (lentivirus overexpressing the non-silencing, NS, RNA). Figure 2 shows data from our laboratory on the fold change increase in expression of miR-138,

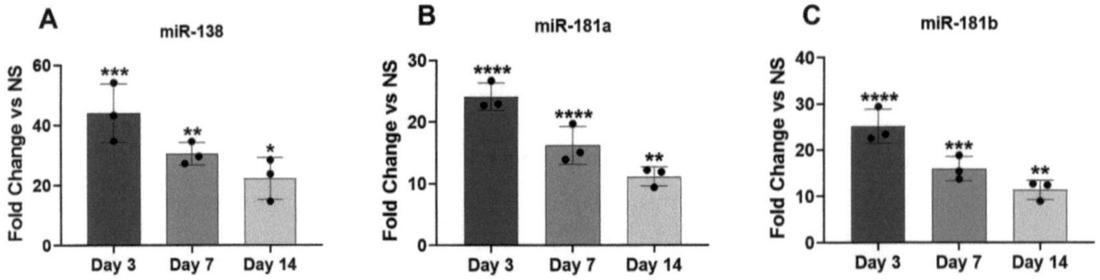

Fig. 2 Fold change increase in expression levels of miR-138, miR-181a, and miR-181b following chondrogenic induction of lentiviral-transduced CPCs in pellet culture. Custom TaqMan™ Small RNA Assays (Applied Bioscience) were used to determine expression levels of the mature 5p strands of miR-138, miR-181a, and miR-181b in RNA samples generated from CPC pellet cultures at day 3, 7, or 14 of chondrogenic induction. CPCs in these pellet cultures had been transduced with either lentivirus particles expressing precursor-miR-138 or precursor-miR-181a/b-1 cluster of the non-silencing (NS) control RNA. Data show the fold change increase in levels of each mature miRNA strand when compared to NS control group at the different time points during the chondrogenesis assay. Note that lentiviral-induced overexpression of miR-138 or miR-181a/b is achieved in pellet cultures up to 14 days of chondrogenic induction. Data are expressed ± SD and statistical significance was measured using one-way ANOVA. $n = 3$; * $p < 0.05$; ** $p < 0.01$; *** $p < 0.001$; ****$p < 0.0001$

miR-181a, and miR-181b found in cells from CPC pellet cultures at day 3, 7, and 14 of chondrogenic induction. These data confirm successful overexpression of the miRNAs during in vitro chondrogenesis of CPCs up to 2 weeks following transduction with the appropriate lentiviral particles (*see* **Notes 16** and **17**).

4 Notes

1. The chondrogenic induction medium mixture, devoid of TGF-β3, can be prepared in bulk and stored at 4 °C. For chondrogenesis assays, TGF-β3 is freshly added to an appropriate amount of induction medium needed for your pellet cultures. The same procedure is followed every 3 days when fresh medium is applied to the chondrogenic pellets.
2. pLemiR-NS is a gift from Jerry Crabtree (Addgene plasmid # 32809; http://n2t.net/addgene:32809; RRID: Addgene 32809) [38]. This is a replication incompetent HIV-1-based plasmid containing cDNA encoding a precursor miRNA (pre-miRNA)-like sequence. This is based on the human pre-miR-30a sequence, but the stem part of the hairpin loop has been replaced with a nonspecific miRNA sequence. This is used as the lentiviral control group whereby a small non-silencing (NS) RNA will be overexpressed. Data obtained from this control group can be directly compared to the experimental group overexpressing the pre-miRNA of interest [35, 36,

38]. Note that this plasmid also contains cDNA encoding RFP to enable fluorescent-based approaches to detect cell uptake of the lentivirus.

3. We have overexpressed precursor forms of miR-138 and the miR-181a/b cluster using lentiviral technology described in this chapter. The TaqMan™ Small RNA Assays (Applied Biosystems) we use to detect expression of mature 5p strands of these miRNAs are miR-138: Assay ID 002284; miR-181a: Assay ID 000480; and miR-181b: Assay ID 001098. Note that we use RNU44 as the normalizing gene to quantify levels of miRNA overexpression. RNU44: Assay ID 001094.
4. In collaboration with Orthopedic Surgeons in our Department, we regularly obtain osteoarthritic knee joint specimens within an hour following total knee replacement surgery. These specimens are from de-identified patients. Included in the information we receive is the sex, age, BMI, smoker status, and diabetes status of the patient. On obtaining specimens, we process them as soon as possible and harvest regions of full-thickness articular cartilage. We avoid cartilage regions that are clearly damaged and degraded. Note that it has been reported that osteoarthritic tissue likely contains more CPCs than non-diseased articular cartilage [39–41].
5. In some cases, the cell density will be low following fibronectin selection, and it may take up to 2 weeks until the first passage.
6. The ability of human CPCs or human bone marrow-derived MSCs to differentiate decreases with increasing passage number. This is why we usually avoid using cells higher than P6. Also, for CPCs, we often check the ratio of *COL2A1* to *COL1A1* gene expression by qPCR. Compared to primary human articular chondrocytes that produce high levels of *COL2A1*, CPCs should generate higher levels of *COL1A1*.
7. We ensure that we follow NIH biosafety and recombinant DNA policies when handling lentivirus in the laboratory. Synthesis of lentiviral reagents and subsequent transduction of lentiviral particles into primary cell lines is always carried out in a BSL2+ biosafety cabinet tissue culture hood. We also utilize a tissue culture incubator that is restricted for culturing lentiviral-transduced cells. When handling lentivirus, researchers are double-gloved and wear a disposable lab coat, face mask, and eye protection. All liquid waste generated from handling lentivirus is decontaminated with bleach while solid waste is immersed in a bleach solution for a minimum of 1 h before disposal. The biosafety cabinet is decontaminated with Envirocide surface disinfectant immediately following use.

8. A PES filter is recommended because it has very low binding affinity for the virus particles. If the viral medium is not centrifuged to remove cell debris, it is important to use a larger filter pore size (e.g., 0.45 μm) to avoid clogging the filter.
9. The same cell seeding densities are required for lentiviral transduction of human bone marrow-derived MSCs, another common cell type used by us and other laboratories to study in vitro chondrogenesis.
10. Given that RFP is also expressed in addition to the pre-miRNA of interest, lentivirus-transduced CPCs can be visualized by fluorescence microscopy. The presence of red fluorescent-labeled cells gives an indication that the transduction has been successful. An additional approach to confirm successful transduction (in the event your transfer plasmid does not contain a fluorescent marker) would be to isolate RNA from the transduced cells (prior to establishing chondrogenic cultures) and carry out specific RT-qPCRs to determine expression levels of the miRNA of interest (*see* Subheading 3.3.4).
11. If desired, one can specifically select for cells that have been transduced with lentivirus. In this case, one could perform FACS analysis to isolate only RFP-positive cells (i.e., cells that have been transduced with the lentivirus). In our case, we did not perform a selection step prior to setting up the chondrogenesis assays.
12. Chondrogenesis assays are commonly performed in hypoxia incubators (2–5% oxygen) for optimal conditions, although chondrocyte differentiation can still occur in a 20% oxygen environment. If we are studying chondrocyte hypertrophy or osteogenesis, then we culture in incubators containing 20% oxygen.
13. Multiple pellets from the same lentiviral condition should be pooled into a single tube to obtain sufficient RNA for analysis. For qPCR, harvest two to three pellets for time points up to 14 day of chondrogenic induction and four to five for endpoints up to 28 days.
14. Flash frozen pellets and Buffer RL will gradually melt. Once this mixture has melted, repeat flash freezing steps. Frozen pellets are more brittle and easier to break up using a pestle.
15. The RNA 260/280 ratio obtained from the NanoDrop™ 2000 Spectrophotometer Nanodrop should be ~1.9 for qPCR analysis. The RNA Integrity Number (RIN) as determined from the Agilent 2100 Bioanalyzer should ideally be ~8.0 for use in RNA-Seq or microarray projects. Lower RINs may be acceptable if the goal is to determine small RNA expression levels by RNA-Seq or microarray.

16. We have also detected increased levels of these miRNAs in lentiviral-transduced cartilage pellets following 28 days of chondrogenic induction (results not shown).

17. There are a number of approaches to determine if modulation of your gene/miRNA of interest has affected the process of in vitro chondrogenesis. Common methods include extraction of RNA or protein from chondrogenic pellets and measuring levels of chondrocyte markers such as SOX9, aggrecan, type II collagen, and others by qPCR or western blotting. Isolated RNA may also be used for microarray or RNA-Seq analysis to provide more information on how global gene expression and cellular pathways have been affected. Paraffin or frozen sections of chondrogenic pellets may be obtained and stained with safranin O/fast green or toluidine blue to assess the proteoglycan distribution as well as chondrocyte phenotype within the cartilage pellets. A more quantitative approach to assess levels of proteoglycans within the cartilage pellets would be to apply the dimethylmethylene blue (DMMB) assay. Some cartilage pellet tissue sections may also be stained with a calcein/ethidium homodimer mixture (to determine the presence of live/dead cells via fluorescence microscopy) or immunostained with specific antibodies to further assess the protein composition of the extracellular matrix.

Acknowledgments

This work was supported by NIH grants R01AR075730 and R21AR077203 (to AM). We thank Dr. Ryan Nunley and Dr. Regis O'Keefe in the Department of Orthopedic Surgery at Washington University for providing us with human osteoarthritic articular specimens that we use as our source of CPCs. We also thank our Research Assistant, Jin Liu, for assistance in cell culture and in carrying out miRNA RT-qPCR assays.

References

1. Lander ES (2011) Initial impact of the sequencing of the human genome. Nature 470(7333):187–197. https://doi.org/10.1038/nature09792
2. Rosenbloom KR, Dreszer TR, Pheasant M et al (2010) ENCODE whole-genome data in the UCSC Genome Browser. Nucleic Acids Res 38 (Database issue):D620–D625. https://doi.org/10.1093/nar/gkp961
3. Consortium EP, Moore JE, Purcaro MJ et al (2020) Expanded encyclopaedias of DNA elements in the human and mouse genomes. Nature 583(7818):699–710. https://doi.org/10.1038/s41586-020-2493-4
4. Bartel DP (2004) MicroRNAs: genomics, biogenesis, mechanism, and function. Cell 116(2): 281–297. https://doi.org/10.1016/s0092-8674(04)00045-5
5. Ha M, Kim VN (2014) Regulation of microRNA biogenesis. Nat Rev Mol Cell Biol 15(8): 509–524
6. Eulalio A, Huntzinger E, Izaurralde E (2008) Getting to the root of miRNA-mediated gene

silencing. Cell 132(1):9–14. https://doi.org/10.1016/j.cell.2007.12.024

7. Djuranovic S, Nahvi A, Green R (2012) miRNA-mediated gene silencing by translational repression followed by mRNA deadenylation and decay. Science 336(6078):237–240. https://doi.org/10.1126/science.1215691
8. Hensley AP, McAlinden A (2021) The role of microRNAs in bone development. Bone 143:115760. https://doi.org/10.1016/j.bone.2020.115760
9. Berezikov E, Guryev V, van de Belt J et al (2005) Phylogenetic shadowing and computational identification of human microRNA genes. Cell 120(1):21–24. https://doi.org/10.1016/j.cell.2004.12.031
10. Friedman RC, Farh KK, Burge CB et al (2009) Most mammalian mRNAs are conserved targets of microRNAs. Genome Res 19(1):92–105. https://doi.org/10.1101/gr.082701.108
11. Bartel DP (2009) MicroRNAs: target recognition and regulatory functions. Cell 136(2):215–233. https://doi.org/10.1016/j.cell.2009.01.002
12. Gibson G, Asahara H (2013) microRNAs and cartilage. J Orthop Res 31(9):1333–1344. https://doi.org/10.1002/jor.22397
13. Asahara H (2016) Current status and strategy of microRNA research for cartilage development and osteoarthritis pathogenesis. J Bone Metab 23(3):121–127. https://doi.org/10.11005/jbm.2016.23.3.121
14. Razmara E, Bitaraf A, Yousefi H et al (2019) Non-coding RNAs in cartilage development: an updated review. Int J Mol Sci 20(18). https://doi.org/10.3390/ijms20184475
15. McAlinden A, Im GI (2018) MicroRNAs in orthopaedic research: disease associations, potential therapeutic applications, and perspectives. J Orthop Res 36(1):33–51. https://doi.org/10.1002/jor.23822
16. Swingler TE, Niu L, Smith P et al (2019) The function of microRNAs in cartilage and osteoarthritis. Clin Exp Rheumatol 37 Suppl 120(5):40–47
17. Shvedova M, Kobayashi T (2020) MicroRNAs in cartilage development and dysplasia. Bone 140:115564. https://doi.org/10.1016/j.bone.2020.115564
18. Sondag GR, Haqqi TM (2016) The role of MicroRNAs and their targets in osteoarthritis. Curr Rheumatol Rep 18(8):56. https://doi.org/10.1007/s11926-016-0604-x
19. Oliviero A, Della Porta G, Peretti GM et al (2019) MicroRNA in osteoarthritis: physiopathology, diagnosis and therapeutic challenge. Br Med Bull 130(1):137–147. https://doi.org/10.1093/bmb/ldz015
20. Miyaki S, Asahara H (2012) Macro view of microRNA function in osteoarthritis. Nat Rev Rheumatol 8(9):543–552. https://doi.org/10.1038/nrrheum.2012.128
21. Fathollahi A, Aslani S, Jamshidi A et al (2019) Epigenetics in osteoarthritis: novel spotlight. J Cell Physiol 234(8):12309–12324. https://doi.org/10.1002/jcp.28020
22. Si HB, Zeng Y, Liu SY et al (2017) Intra-articular injection of microRNA-140 (miRNA-140) alleviates osteoarthritis (OA) progression by modulating extracellular matrix (ECM) homeostasis in rats. Osteoarthr Cartil. https://doi.org/10.1016/j.joca.2017.06.002
23. Wang H, Zhang H, Sun Q et al (2017) Intra-articular delivery of antago-miR-483-5p inhibits osteoarthritis by modulating matrilin 3 and tissue inhibitor of metalloproteinase 2. Mol Ther 25(3):715–727. https://doi.org/10.1016/j.ymthe.2016.12.020
24. Zhang X, Wang C, Zhao J et al (2017) miR-146a facilitates osteoarthritis by regulating cartilage homeostasis via targeting Camk2d and Ppp3r2. Cell Death Dis 8(4):e2734. https://doi.org/10.1038/cddis.2017.146
25. McAlinden A, Varghese N, Wirthlin L et al (2013) Differentially expressed microRNAs in chondrocytes from distinct regions of developing human cartilage. PLoS One 8(9):e75012. https://doi.org/10.1371/journal.pone.0075012
26. Durand S, Cimarelli A (2011) The inside out of lentiviral vectors. Viruses 3(2):132–159. https://doi.org/10.3390/v3020132
27. Jin HY, Gonzalez-Martin A, Miletic AV et al (2015) Transfection of microRNA mimics should be used with caution. Front Genet 6:340. https://doi.org/10.3389/fgene.2015.00340
28. Anderson BA, McAlinden A (2017) miR-483 targets SMAD4 to suppress chondrogenic differentiation of human mesenchymal stem cells. J Orthop Res 35(11):2369–2377. https://doi.org/10.1002/jor.23552
29. Zheng H, McAlinden A (2021) RNA isolation from articular cartilage tissue. Methods Mol Biol 2245:121–133. https://doi.org/10.1007/978-1-0716-1119-7_9
30. Jones PH, Watt FM (1993) Separation of human epidermal stem cells from transit amplifying cells on the basis of differences in integrin function and expression. Cell 73(4):713–724. https://doi.org/10.1016/0092-8674(93)90251-k

31. Dowthwaite GP, Bishop JC, Redman SN et al (2004) The surface of articular cartilage contains a progenitor cell population. J Cell Sci 117(Pt 6):889–897. https://doi.org/10.1242/jcs.00912
32. Khan IM, Bishop JC, Gilbert S et al (2009) Clonal chondroprogenitors maintain telomerase activity and Sox9 expression during extended monolayer culture and retain chondrogenic potential. Osteoarthr Cartil 17(4): 518–528. https://doi.org/10.1016/j.joca.2008.08.002
33. Williams R, Khan IM, Richardson K et al (2010) Identification and clonal characterisation of a progenitor cell sub-population in normal human articular cartilage. PLoS One 5(10):e13246. https://doi.org/10.1371/journal.pone.0013246
34. Anderson DE, Markway BD, Weekes KJ et al (2018) Physioxia promotes the articular chondrocyte-like phenotype in human chondroprogenitor-derived self-organized tissue. Tissue Eng Part A 24(3–4):264–274. https://doi.org/10.1089/ten.TEA.2016.0510
35. Zheng H, Ramnaraign D, Anderson BA et al (2018) MicroRNA-138 inhibits osteogenic differentiation and mineralization of human dedifferentiated chondrocytes by regulating RhoC and the actin cytoskeleton. JBMR Plus:1–15. https://doi.org/10.1002/jbm4.10071
36. Zheng H, Liu J, Tycksen E et al (2019) MicroRNA-181a/b-1 over-expression enhances osteogenesis by modulating PTEN/PI3K/AKT signaling and mitochondrial metabolism. Bone 123:92–102. https://doi.org/10.1016/j.bone.2019.03.020
37. Johnstone B, Hering TM, Caplan AI et al (1998) In vitro chondrogenesis of bone marrow-derived mesenchymal progenitor cells. Exp Cell Res 238(1):265–272. https://doi.org/10.1006/excr.1997.3858
38. Yoo AS, Sun AX, Li L et al (2011) MicroRNA-mediated conversion of human fibroblasts to neurons. Nature 476(7359):228–231. https://doi.org/10.1038/nature10323
39. Fellows CR, Williams R, Davies IR et al (2017) Characterisation of a divergent progenitor cell sub-populations in human osteoarthritic cartilage: the role of telomere erosion and replicative senescence. Sci Rep 7:41421. https://doi.org/10.1038/srep41421
40. Koelling S, Kruegel J, Irmer M et al (2009) Migratory chondrogenic progenitor cells from repair tissue during the later stages of human osteoarthritis. Cell Stem Cell 4(4):324–335. https://doi.org/10.1016/j.stem.2009.01.015
41. Alsalameh S, Amin R, Gemba T et al (2004) Identification of mesenchymal progenitor cells in normal and osteoarthritic human articular cartilage. Arthritis Rheum 50(5):1522–1532. https://doi.org/10.1002/art.20269

Chapter 16

Purification and Isolation of Proteins from Hyaline Cartilage

Makenna J. Hardy, Xinzhu Pu, and Julia Thom Oxford

Abstract

Proteins from hyaline or articular cartilage can be isolated and purified using a series of chemical extraction steps and various identification techniques including mass spectrometry and immunoblotting. The isolation and purification of proteins from cartilage will facilitate the study of specific proteins and multimeric complexes of cartilage proteins to better understand their functions in normal healthy cartilage as well as pathological conditions of cartilage. Cartilage tissue engineering efforts rely on the comprehensive understanding of the composition of cartilage and the function of each of the protein constituents.

Key words Cartilage, Protein, Collagen, Proteoglycan, Purification

1 Introduction

Damaged cartilage has poor regenerative potential; however, tissue engineering approaches hold promise to restore function in the case of damage, improving the quality of life for patients. There are a number of questions that still remain to be investigated before tissue engineering approaches can be translated into new restorative methods. Cartilage tissue is composed of hundreds of proteins that carry out unique functions. The major components are collagens, highly sulfated anionic proteoglycans, hyaluronic acid, and hundreds of other proteins found in lower abundance [1–9]. Type II collagen is the most abundant collagen in cartilage [10] with types IX and XI forming heterotypic fibrils with type II [4, 11, 12] and other collagen types present including types III, VI, XII, and XIV [13]. Method development for the isolation of proteins from cartilage has been a focus of laboratories since 1963. A combination of homogenization and extraction methods provides the best approach for a comprehensive analysis of protein composition of cartilage. Analytical methods that take advantage of unique characteristics of cartilage extracellular matrix molecules include pepsin resistance and conversely the collagenase sensitivity of the

Martin J. Stoddart et al. (eds.), *Cartilage Tissue Engineering*,
Methods in Molecular Biology, vol. 2598, https://doi.org/10.1007/978-1-0716-2839-3_16,

collagen triple helix. Differential salt precipitation has been used to separate distinct collagen types based on their solubility properties. Highly effective protein identification techniques include mass spectrometry [14–25] and antibody-based detection methods such as immunoblotting.

2 Materials

Prepare all solutions using ultrapure water (prepared by purifying deionized water, to attain a sensitivity of 18 MΩ-cm at 25 °C) and analytical grade reagents. Prepare and store all reagents at 4 °C (unless otherwise indicated). Follow all waste disposal regulations when disposing waste materials (*see* **Note 1**).

2.1 Homogenization of Cartilage

1. Cartilage: Hyaline cartilage can be obtained from the ends of long bones of fetal bovine hind and forelimbs using a scalpel while maintaining samples at 4 °C.
2. Phosphate-buffered saline (PBS).
3. Homogenization buffer: 0.15 M NaCl, 50 mM Tris–HCl, pH 7.4 with protease inhibitors 1 mM AEBSF, 0.8 μM aprotinin, 20 μM leupeptin, 40 μM bestain, 15 μM pepstatin A, 14 μM E-64, and 10 mM EDTA.
4. Homogenizer (such as a Kinematica Polytron, Brinkmann Instruments, Westbury NY).
5. Ice.
6. Centrifuge.

2.2 Sequential Protein Extraction

1. Low salt extraction buffer: 0.15 M NaCl, 50 mM Tris–HCl, pH 7.4 containing protease inhibitors as described for homogenization buffer in 2.1.
2. High salt extraction buffer: 1 M NaCl, 50 mM Tris–HCl, pH 7.4 containing protease inhibitors as described for homogenization buffer and low salt extraction buffer.
3. Guanidine–HCl extraction solution: 4 M Guanidine–HCl, pH 7.4.
4. Deionized water.

2.3 Pepsin Digestion

1. Pepsin solution: 0.5 M acetic acid containing 1 μg/mL pepsin.

2.4 Differential Salt Precipitation

1. Dialysis tubing.
2. 0.5 M acetic acid, 0.9 M NaCl.
3. 0.5 M acetic acid, 1.2 M NaCl.
4. 0.5 M acetic acid, 2 M NaCl.

2.5 Collagenase Digestion

1. 50 mM Tris–HCl, pH 7.4 containing 20 mM $CaCl_2$, protease inhibitors, 100 U/mL bacterial collagenase (CLSPA, Worthington Biochemical, Lakewood, NJ).
2. Heat block or incubator capable of maintaining 37 °C.
3. Centrifuge capable of 100,000 × g at 4 °C.

2.6 Sodium Dodecyl Sulfate Polyacrylamide Gel Electrophoresis

1. Nonreducing sample buffer (5×): 10% w/v SDS, 50% v/v glycerol, 312.5 mM Tris–HCl pH 6.8, 0.125% w/v bromophenol blue.
2. Reducing sample buffer (5×): 10% w/v SDS, 50% v/v glycerol, 312.5 mM Tris–HCl pH 6.8, 0.125% w/v bromophenol blue, 0.5 M dithiothreitol (DTT).
3. Native polyacrylamide gel electrophoresis sample buffer (5×): 50% glycerol, 312.5 mM Tris–HCl, pH 6.8, 0.125% w/v bromophenol blue.
4. Coomassie Blue R-250.
5. Stacking gel: 3.5% polyacrylamide.
6. Running gel: concentration varies depending upon application: 5% polyacrylamide used for separation of collagens and 9% polyacrylamide used for separation of smaller proteins.

2.7 In-Gel Trypsin Digestion for Mass Spectrometry Analysis

1. Gel washing solution: 50 mM ammonium bicarbonate, prepared freshly.
2. Coomassie blue destaining solution: 40% acetonitrile in 50 mM ammonium bicarbonate at room temperature, prepared freshly (*see* **Note 2**).
3. Reduction solution: 10 mM DTT in 50 mM ammonium bicarbonate, prepared freshly.
4. Alkylation solution: 55 mM iodoacetamide in 50 mM ammonium bicarbonate, prepared freshly.
5. Trypsin solution (sequencing grade trypsin): 20 μL/mL in 100 mM ammonium bicarbonate.
6. Formic acid, 5% v/v.

2.8 Immunoblot Identification of Proteins

1. PVDF membrane.
2. Bovine serum albumin blocking solution: 3% bovine serum albumin in TBST (Tris-buffered saline with 0.1% Tween 20).
3. Primary antibodies of interest.
4. Secondary antibodies corresponding to source of primary antibodies.
5. Enhanced chemiluminescence (ECL) detection reagent.

3 Methods

Carry out all procedures at 4 °C unless otherwise indicated. The following methods are for 0.1 g cartilage as starting material. Methods can be scaled up to accommodate larger amounts of cartilage as starting material (*see* **Note 1**).

3.1 Homogenization of Cartilage

1. Harvest hyaline cartilage from the ends of long bones of fetal bovine hind and forelimbs using scalpel. Keep tissue at 4 °C during tissue collection. Wash cartilage with PBS and store at –20 °C.
2. Mince 0.1 g of cartilage with scalpel and transfer to a 15 mL conical screw cap tube containing 1 mL of low salt extraction buffer with protease inhibitors.
3. Homogenize cartilage in low salt extraction buffer with protease inhibitors on ice until the cartilage is no longer visible, approximately 10 s.
4. Incubate at 4 °C for 16 h with agitation or stirring.

3.2 Sequential Protein Extraction

1. Centrifuge at 100,000 × g at 4 °C for 1 h to separate supernatant containing extracted proteins from the insoluble pellet. Save the 1 mL of supernatant for analysis (label sup. 0.15 M).
2. Add 1 mL of high salt extraction buffer to pellet and homogenize for 10 s on ice.
3. Incubate at 4 °C for 16 h with agitation or stirring.
4. Centrifuge at 100,000 × g at 4 °C for 1 h to separate supernatant containing extracted proteins from the insoluble pellet. Save the 1 mL of supernatant for analysis (label sup. 1.0 M).
5. Add 1 mL of guanidine–HCl extraction buffer to pellet and homogenize for 10 s on ice.
6. Incubate at 4 °C for 16 h with agitation or stirring.
7. Centrifuge at 100,000 × g at 4 °C for 1 h to separate supernatant containing extracted proteins from the insoluble pellet. Save the 1 mL of supernatant for analysis (label sup. GuHCl_1).
8. Add 1 mL of guanidine–HCl extraction buffer to pellet and homogenize for 10 s on ice.
9. Incubate at 4 °C for 16 h with agitation or stirring.
10. Centrifuge at 100,000 × g at 4 °C for 1 h to separate supernatant containing extracted proteins from the insoluble pellet. Save the 1 mL of supernatant for analysis (label sup. GuHCl_2).

11. Wash pellet in deionized water. Use homogenizer to mix while on ice. Centrifuge at 100,000 × g for 1 h at 4 °C to collect insoluble material. Discard supernatant. Repeat wash step two additional times, discarding supernatant each time. Store washed pellet at −20 °C.

3.3 Pepsin Treatment

1. Resuspend washed pellet in acetic acid containing pepsin. Incubate for 24 h at 4 °C.
2. Centrifuge at 100,000 × g for 1 h at 4 °C. Save supernatant for differential salt precipitation.

3.4 Collagenase Treatment

1. Resuspend washed pellet containing insoluble material after guanidine–HCl extraction in 50 mM Tris–HCl, pH 7.4 containing 20 mM $CaCl_2$, and protease inhibitors at 4 °C.
2. Add 100 U/ mL of bacterial collagenase.
3. Incubate at 37 °C for 24 h.
4. Centrifuge at 100,000 × g at 4 °C for 1 h.

3.5 Differential Salt Precipitation

1. Transfer the supernatant from pepsin treatment of Gu–HCl pellet to dialysis tubing.
2. Dialyze against acetic acid containing 0.9 M NaCl. Collect precipitate for analysis. Precipitate will contain type II collagen. Save supernatant for additional dialysis.
3. Transfer supernatant to new dialysis tubing and dialyze against acetic acid containing 1.2 M NaCl. Collect precipitate for analysis. Precipitate will contain type XI collagen. Save supernatant for additional dialysis.
4. Transfer supernatant to new dialysis tubing and dialyze against acetic acid containing 2.0 M NaCl. Precipitate will contain type IX collagen fragments. Collect precipitate for analysis.

3.6 In-Gel Trypsin Digestion

1. Above fractions can be analyzed for protein identity by mass spectrometry and immunoblot or other antibody-based identification techniques.
2. Add one-fifth volume of 5× sample buffer to supernatant fractions generated by above procedure.
3. Perform separation of proteins by discontinuous gel electrophoresis.
4. After electrophoresis, visualize proteins by staining with Coomassie Blue R-250.
5. Excise bands from gels using new razor blade.
6. Destain gel band with destaining solution at room temperature for 15 min.
7. Remove the supernatant.

8. Repeat the destaining step until the gel pieces are completely destained.
9. Remove supernatant and dry gel band under vacuum.
10. Reduce and alkylate protein within the gel band using 50 mM ammonium bicarbonate containing 10 mM DTT at 56 °C for 45 min.
11. Remove supernatant and replace with 50 mM ammonium bicarbonate containing 55 mM iodoacetamide for 30 min in the dark.
12. Remove supernatant and wash gel band two times with 50 mM ammonium bicarbonate at room temperature for 15 min each time.
13. Dry gel band under vacuum.
14. Add trypsin solution to dried gel band and incubate at 37 °C for 20 h.
15. Remove supernatant containing tryptic peptides and transfer to new tube.
16. Dry down peptides under vacuum.
17. Rehydrate peptides in 5% v/v formic acid.
18. Load sample on capillary reverse phase column and elute using a gradient of 5–60% acetonitrile in 0.1% v/v formic acid over 110 min, followed by 60–80% acetonitrile in 0.1% v/v formic acid over 2 min.
19. Peptides eluted from the column will be analyzed by mass spectrometry.
20. Match peptide spectrum to reference database using a proper search engine for positive identification of proteins contained within gel bands (*see* **Notes 3** and **4**).

3.7 Immunoblot

1. Separate proteins by SDS-polyacrylamide gel electrophoresis.
2. Transfer to a PVDF membrane.
3. Block membrane with 3% BSA solution for 1 h at room temperature.
4. Incubate with primary antibody diluted in 3% BSA blocking solution overnight at 4 °C (*see* **Notes 5** and **6**).
5. Wash with TBST three times for 5 min (*see* **Note 6**).
6. Incubate membrane with secondary antibody diluted in 3% BSA blocking solution for 1 h at room temperature (*see* **Notes 5** and **6**).
7. Visualize positively identified protein bands using chemiluminescence reagent.

4 Notes

1. Keratin contamination is a common problem in MS-based shotgun proteomics. The main source of keratin contamination is dust that contains dead skin cells from humans. Efforts should be taken to minimize dust exposure of samples, supplies, reagents, and workspace during reagent preparation and sample handling.
2. 50 mM ammonium bicarbonate in 50% acetonitrile is commonly used to destain Coomassie Blue-stained gel pieces. We found that 50 mM ammonium bicarbonate in 40% acetonitrile is a more efficient destaining buffer.
3. A computer algorithm, commonly called database search engine, is usually used to identify the peptide and protein sequence from ms/ms spectra in the dataset. There are many different search engines available. Each search engine has its unique searching algorithm. It has been shown that searching the same datasets using multiple search engines and then combining the search results usually improves the analysis and gives better protein coverage [26]. We recommend using multiple search engines and combining the search results when possible.
4. Proline and lysine residues in collagens are extensively hydroxylated. Thus, database search parameters should include proline and lysine hydroxylation (mass shift +15.995 Da) for accurate collagen identification.
5. Commercially available primary and secondary antibodies have suggested concentrations. A typical dilution range for a primary antibody is 1:1000 to 1:5000. A typical dilution for a secondary antibody is 1:1000. It is recommended to test a series of dilutions in order to optimize concentration for a specific antibody in an experiment.
6. During washing and incubation steps, it is recommended to rock or shake in order to gently agitate the blot.

Acknowledgments

This work was supported by NIH grants P20GM109095, R01AR47985, K02AR48672, and R15HD059949 to JTO and the Biomolecular Research Center at Boise State University.

References

1. Mow VC, Ratcliffe A, Poole AR (1992) Cartilage and diarthrodial joints as paradigms for hierarchical materials and structures. Biomaterials 13(2):67–97. https://doi.org/10.1016/0142-9612(92)90001-5
2. Heinegard D (2009) Fell-Muir lecture: proteoglycans and more--from molecules to biology. Int J Exp Pathol 90(6):575–586. https://doi.org/10.1111/j.1365-2613.2009.00695.x
3. Schaefer L, Schaefer RM (2010) Proteoglycans: from structural compounds to signaling molecules. Cell Tissue Res 339(1):237–246. https://doi.org/10.1007/s00441-009-0821-y
4. Buckwalter JA, Mankin HJ (1998) Articular cartilage repair and transplantation. Arthritis Rheum 41(8):1331–1342. https://doi.org/10.1002/1529-0131(199808)41:8<1331::AID-ART2>3.0.CO;2-J
5. Roughley PJ, Mort JS (2014) The role of aggrecan in normal and osteoarthritic cartilage. J Exp Orthop 1(1):8. https://doi.org/10.1186/s40634-014-0008-7
6. Kiani C, Chen L, Wu YJ et al (2002) Structure and function of aggrecan. Cell Res 12(1):19–32. https://doi.org/10.1038/sj.cr.7290106
7. Schumacher BL, Block JA, Schmid TM et al (1994) A novel proteoglycan synthesized and secreted by chondrocytes of the superficial zone of articular cartilage. Arch Biochem Biophys 311(1):144–152. https://doi.org/10.1006/abbi.1994.1219
8. Muller C, Khabut A, Dudhia J et al (2014) Quantitative proteomics at different depths in human articular cartilage reveals unique patterns of protein distribution. Matrix Biol 40:34–45. https://doi.org/10.1016/j.matbio.2014.08.013
9. Lorenzo P, Bayliss MT, Heinegard D (1998) A novel cartilage protein (CILP) present in the mid-zone of human articular cartilage increases with age. J Biol Chem 273(36):23463–23468. https://doi.org/10.1074/jbc.273.36.23463
10. Martel-Pelletier J, Boileau C, Pelletier JP et al (2008) Cartilage in normal and osteoarthritis conditions. Best Pract Res Clin Rheumatol 22(2):351–384. https://doi.org/10.1016/j.berh.2008.02.001
11. Mendler M, Eich-Bender SG, Vaughan L et al (1989) Cartilage contains mixed fibrils of collagen types II, IX, and XI. J Cell Biol 108(1):191–197. https://doi.org/10.1083/jcb.108.1.191
12. Wu JJ, Woods PE, Eyre DR (1992) Identification of cross-linking sites in bovine cartilage type IX collagen reveals an antiparallel type II-type IX molecular relationship and type IX to type IX bonding. J Biol Chem 267(32):23007–23014
13. Reginato AM, Olsen BR (2002) The role of structural genes in the pathogenesis of osteoarthritic disorders. Arthritis Res 4(6):337–345. https://doi.org/10.1186/ar595
14. Hansen KC, Kiemele L, Maller O et al (2009) An in-solution ultrasonication-assisted digestion method for improved extracellular matrix proteome coverage. Mol Cell Proteomics 8(7):1648–1657. https://doi.org/10.1074/mcp.M900039-MCP200
15. Didangelos A, Yin X, Mandal K et al (2010) Proteomics characterization of extracellular space components in the human aorta. Mol Cell Proteomics 9(9):2048–2062. https://doi.org/10.1074/mcp.M110.001693
16. Angel LA (2011) Study of metal ion labeling of the conformational and charge states of lysozyme by ion mobility mass spectrometry. Eur J Mass Spectrom (Chichester) 17(3):207–215. https://doi.org/10.1255/ejms.1133
17. Rob T, Wilson DJ (2012) Time-resolved mass spectrometry for monitoring millisecond timescale solution-phase processes. Eur J Mass Spectrom (Chichester) 18(2):205–214. https://doi.org/10.1255/ejms.1176
18. Wilson R, Bateman JF (2008) A robust method for proteomic characterization of mouse cartilage using solubility-based sequential fractionation and two-dimensional gel electrophoresis. Matrix Biol 27(8):709–712. https://doi.org/10.1016/j.matbio.2008.07.007
19. Wilson R, Belluoccio D, Bateman JF (2008) Proteomic analysis of cartilage proteins. Methods 45(1):22–31. https://doi.org/10.1016/j.ymeth.2008.01.008
20. Guo Y, Xiao P, Lei S et al (2008) How is mRNA expression predictive for protein expression? A correlation study on human circulating monocytes. Acta Biochim Biophys Sin Shanghai 40(5):426–436. https://doi.org/10.1111/j.1745-7270.2008.00418.x
21. Wu Y, Li Q, Chen XZ (2007) Detecting protein-protein interactions by far western blotting. Nat Protoc 2(12):3278–3284. https://doi.org/10.1038/nprot.2007.459
22. Önnerfjord P, Khabut A, Reinholt FP et al (2012) Quantitative proteomic analysis of eight cartilaginous tissues reveals characteristic differences as well as similarities between subgroups. J Biol Chem 287(23):18913–18924. https://doi.org/10.1074/jbc.M111.298968

23. Hermansson M, Sawaji Y, Bolton M et al (2004) Proteomic analysis of articular cartilage shows increased type II collagen synthesis in osteoarthritis and expression of inhibin betaA (activin a), a regulatory molecule for chondrocytes. J Biol Chem 279(42):43514–43521. https://doi.org/10.1074/jbc.M407041200
24. Zhen EY, Brittain IJ, Laska DA et al (2008) Characterization of metalloprotease cleavage products of human articular cartilage. Arthritis Rheum 58(8):2420–2431. https://doi.org/10.1002/art.23654
25. Polacek M, Bruun JA, Johansen O et al (2010) Differences in the secretome of cartilage explants and cultured chondrocytes unveiled by SILAC technology. J Orthop Res 28(8):1040–1049. https://doi.org/10.1002/jor.21067
26. Shteynberg D, Nesvizhskii AI, Moritz RL et al (2013) Combining results of multiple search engines in proteomics. Mol Cell Proteomics 12(9):2383–2393. https://doi.org/10.1074/mcp.R113.027797

Chapter 17

Histological Preparation and Evaluation of Cartilage Specimens

Salim E. Darwiche, Milena Tegelkamp, Katja Nuss, and Brigitte von Rechenberg

Abstract

In this chapter, an introduction is given into histological techniques to research related to hyaline cartilage and subchondral bone. Emphasis is placed on the importance to investigate cartilage and bone as a unit, which includes the transition zone of the calcified cartilage and tidemark. Reasons for the appropriate selection of histological methods are presented such as when to use (decalcified) specimens for routine paraffin embedding including immunohistology, cryosections of cartilage alone, or non-decalcified specimens for embedding in polymethylmethacrylate with or without additional biomaterials. Appropriate staining methods are also outlined. Apart from detailed laboratory protocols for different embedding and staining methods including open communication about difficulties related to the various techniques, also practical instructions for state-of-the-art evaluation methods and their strengths and weaknesses are given. Sample figures for scoring methods are included.

Key words Histology techniques, Articular cartilage, Subchondral bone, Histological evaluation

1 Introduction

Scientific literature dealing with hyaline cartilage is abundant, and even in the modern Internet, guides for how to prepare histological specimens including special staining methodologies are available. For the latter, the classic methods like staining sections with alcian blue, toluidine blue, safranin O, picrosirius red, and Masson-Goldner can be easily downloaded with exact laboratory protocols [1]. Therefore, the simple question arises, whether another chapter about the histological preparation of cartilage specimens is really needed, when also in the meantime international scoring systems for cartilage can be downloaded from the Internet (using a search engine with keywords such as "ICRS+scoring+system").

Martin J. Stoddart et al. (eds.), *Cartilage Tissue Engineering*,
Methods in Molecular Biology, vol. 2598, https://doi.org/10.1007/978-1-0716-2839-3_17,

Mostly, the literature focuses on cartilage alone and does not highlight the importance of preparing specimens together with the calcified cartilage zone and the subchondral bone as *a unit*. Even if this is respected, quite often specimens are decalcified, stained with routine H&E, and, thus, lose an important aspect of the transition zone between bone and cartilage. The latter may be useful for routine pathologies, where diagnostics of cellular pathologies such as malignancies are more important. But if histology is used for scientific investigations in cartilage degeneration, repair, or regeneration or even studying the efficacy of cartilage replacements with biomaterials, routine paraffin histology and hematoxylin-eosin (H&E) staining is not sufficient, and preparation of specimens with more sophisticated methods is warranted. This is also, where the situation becomes more complicated, last but not least also depending on the animal species used for investigations.

Interestingly, cartilage research is often focused only on cartilage alone with the overall assumption found in many classic textbooks that the cartilage and subchondral bone are strictly separated units divided by the tidemark and calcified cartilage zone (also called transition zone). This assumption is still preserved and considered valid, although fluid exchange has been demonstrated, and cells originating from the subchondral bone can cross the tidemark into the deep cartilage zone and are regularly detected in the transition zone (Fig. 1). Another assumption is that cells found in the calcified cartilage zone are considered as hypertrophic chondrocytes (especially in cases of osteoarthritis (OA)) [2, 3] like in growth plates, although their morphology is quite different, and the only similarity is their expression of collagen type X. The latter is

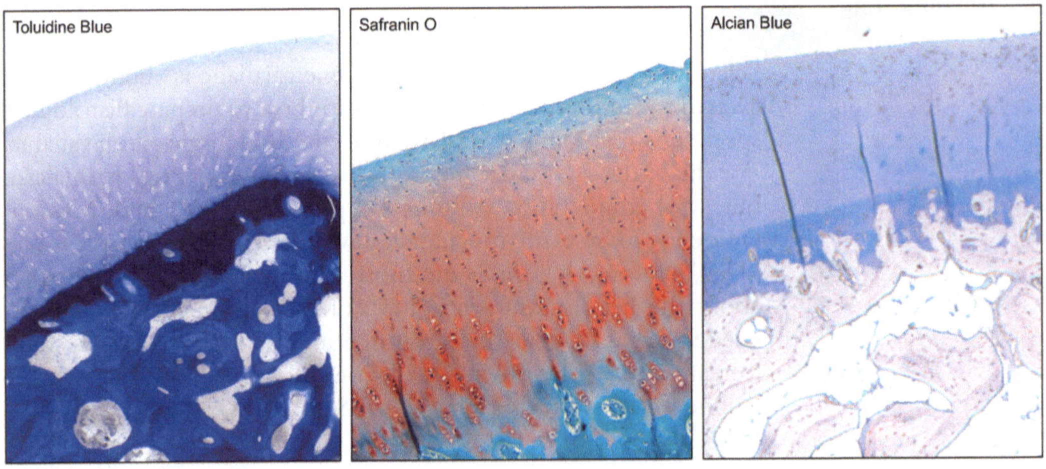

Fig. 1 Osteochondral samples stained, from left to right, with toluidine blue, safranin O, and alcian blue. The slides are all oriented to show articular cartilage at the top and subchondral bone at the bottom, with well-preserved transition zones. Note the penetration of cells through the transition zone and their elongated form arranged in columns in the middle picture (safranin O)

assumed to be specific for hypertrophic chondrocytes. However, it may also just be the link of any chondrocyte attaching to calcified tissue, a theory which still needs proof. Nevertheless, this sample illustrates that histology of cartilage specimens should clearly include a detailed histological approach to the transition zone and that decalcified tissue samples stained with H&E are not enough evidence for serious research studies of hyaline cartilage in joints.

This transition zone presents already the first complication for good cartilage histology. The composition of cartilage and bone is drastically different considering their calcium and water content. While water content of bone is between 10 and 20%, it is up to 80% in cartilage. In hyaline cartilage the water is bound by osmotic pressure to the macromolecules (mainly proteoglycan) providing the cushion for exudation and imbibition during normal mechanical joint loading. Hydroxyapatite as the main calcium component up to 65% in the bone and also in the calcified cartilage zone is completely missing in all other layers of hyaline cartilage [4]. If the transition zone has to be preserved as part of the research interest or if biomaterials are involved that cannot be cut with routine paraffin or cryosections, the solution is using non-decalcified specimens embedded in polymethylmethacrylate (PMMA). There are several commercial products available (e.g., Kulzer Technovit) which are ready to use, with some materials even allowing immunohistology. However, materials that allow immunohistology are not hard enough to work on calcified sheep or goat bones; they mainly work only on small laboratory rodents (mice, rats, rabbits). If bone marrow is the target, they may also work in human specimens. In sheep and goats only the classic, hard setting PMMA works, especially if the very dense subchondral bone is involved. The classic PMMA can easily be prepared in the laboratory, which will be less cost intensive than commercial products.

The PMMA technique will allow ground sections to be prepared using diamond-coated blades in bone saws (EXAKT, Norderstedt, Germany) and thin sections with a microtome (Leica RM2255 microtome). For ground sections, a 300–400 μm slice is cut and then ground and polished to sections between 20 and 100 μm, depending on the scope. This technique will preserve the transition zone very well, allowing correct interpretations of cracks in the calcified cartilage and cell penetration in the transition zone, including the tidemark, to be made. The disadvantage of this technique, however, is that the cellular situation of the cartilage is not sufficiently detailed in ground sections. Therefore, thin sections (ca 6 μm) have to be prepared in addition. They allow cellular evaluation but have the disadvantage that artificial cracks occur in the transition zone since hydroxyapatite can only be broken, but not cut. The combination of both, ground and thin sections, gives the best results complementing each other for structure preservation and also cellular details.

PMMA sections do not only preserve the structure of the transition zone better but also the cell morphology. In decalcified cartilage sections stained with H&E, the chondrocytes always are shrunk with cells seeming almost pyknotic obscuring details of the real state of matrix chondrocytes. In contrast, cellular details are beautifully preserved in PMMA sections with adequate preparation of thin sections. This allows also to distinguish apoptotic cells (e.g., hypertrophic chondrocytes) from still viable cells, for instance, in the transition zone. Cutting thin cartilage sections embedded in PMMA needs experience, especially since the difference of water content makes the cartilage part swollen but not the bone part resulting in folds during cutting and mounting on glass slides.

PMMA sections are also indicated when biomaterials are involved, especially, if these are not resorbable or are resorbed only slowly and show a strong structure themselves. In these cases, paraffin or cryosections are usually not possible. Depending on the materials, however, the preparation of histology using xylene for defatting the specimens may dissolve those materials (e.g., polymers like PGA, PLLA, and gels), and in the resultant sections only empty spaces may be visible. Skipping the xylene step does not work in tissues including bone, since the morphology in the sections will be severely disturbed due to the missing step of defatting. In this case, ground sections are very valuable, since the structure will be preserved and still allow interpretation (empty spaces = resorbed material). This is also true for immature tissue in cartilage repair studies, where it is difficult to preserve contact between filling and surrounding tissue including adjacent original cartilage. In this type of studies, side integration is very important to evaluate aspects such as ruptures without artifacts. Again, remodeling or assessing the possibly drastically altered tidemark and calcified cartilage zone in such cases is also possible with PMMA.

Staining of specimens will depend on preparation of the specimens. If paraffin sections are used on decalcified specimens or cartilage alone, H&E sections can be used for simple assessment of morphology. However, to receive information about the biochemical status of cartilage, there are more suitable staining methods available, such as toluidine blue, alcian blue, and safranin O [1]. All three dyes stain cartilage for proteoglycan, chondrocytes, and type II collagen such that the varying staining intensity indicates the proteoglycan (PG). Since loss of PG is a strong primary and early indicator of OA, these stains, or respectively loss of stain, give important information about the degradation process. Alcian blue and toluidine blue similarly stain proteoglycan, hyaluronic acid components including other macromolecules of connective tissue including collagen. Alcian blue is often used with alizarin red counterstaining in cartilage and bone specimens. Both, toluidine and alcian blue, can be used for surface staining in ground sections [5], whereas this was not possible for safranin O, at least in our laboratory. The advantage of toluidine blue is that it is very simple and fast and also distinguishes clearly between new woven and old lamellar bone, which allows interpretation of bone remodeling in connection with the calcified cartilage zone and subchondral bone.

The only real disadvantage of PMMA sections is that immunohistology is not possible in classic PMMA sections unless specified Kulzer Technovits are used in small laboratory animals, like rats, mice, and rabbits or human bone marrow sections. In these animals, antibodies are normally available. For specimens of sheep and goats, many commercially available antibodies are not applicable, since homology (ca 85–89%) is not sufficient for valuable cross-reactions. Even if antibodies would cross-react, very often results are not good enough if hard subchondral or cortical bone is involved and EDTA decalcification was used. The long period of decalcification results in swelling of tissue. Therefore, staining quality is poor and interpretation is impossible due to background staining. Additionally, after successful decalcification the tissue hardens again during embedding in paraffin and "jumps" out of the block during cutting or the morphology is just not adequate anymore.

So, in summary, before a technique of processing is chosen over another, the original research question, biomaterials, technical possibilities, and animal species involved have to be weighed against each other. In general, if basic molecular mechanisms are investigated, paraffin or cryosections most likely would be useful, whereas if more applied cartilage restitution, like defect filling, osteochondral grafts or biomaterials are in the focus, PMMA sections involving the subchondral bone area should be the method of choice. In any case, samples are often too small to apply several methods with one specimen, and one has to focus on the most important research question. Although the 3R principles (reduce, refine, and replace) for ethics in animal experiments need to be taken seriously, in these cases separate animal groups allocated to answer questions separately may be considered, such as specimens of one group can be embedded in PMMA including the subchondral bone, whereas those of the other group only the cartilage may be harvested without or only a small rim of the subchondral bone for cryo- or paraffin sections and immunohistology.

Research in hyaline cartilage is challenging and that includes histological preparation and evaluation of specimens. As shown in this chapter, there is no "one method solves it all" method as commonly used with paraffin sections and H&E staining, but rather a specific approach tailored to the original research question. Planning an experiment needs to include careful planning of which method, evaluation, and how many specimens per group are required to acquire meaningful data. Evaluation of cartilage sections should not be limited to one, but may involve several score systems either on different or even the same sections. The main message also comprises that without including subchondral bone in the evaluation, research with hyaline cartilage has serious shortcomings and is not complete. As many laboratories may not have all methods, especially embedding in PMMA, available and lack experience thereof, it may be advisable to collaborate with other laboratories, where all of these methods belong to the common routine.

2 Materials

1. PMMA solution is prepared by mixing a 1 : 0.005 : 0.1 volume solution of methylmethacrylate, Perkadox 16 (an initiator also known as Di(4-tert-butylcyclohexyl) peroxydicarbonate), and dibutylphthalate (a plasticizer, also known as DBP), respectively. The freshly prepared solution can be stored in the fridge for up to 5 days prior to use.
2. 4% buffered formalin (Merck, Darmstadt, Germany).
3. PMMA-GS white opal slides (3 mm thick) (Maagtechnic, Dübendorf, Switzerland).
4. Cementit™ speed glue (Merz+Benteli AG, Niederwangen, Switzerland).
5. Chrome alum-gelatin (preparation of chrome alum-gelatin (2.75 g gelatin) in 500 mL deionized (DI) water and 20 mL chrome alum solution (1 g chrome III, potassium sulfate-dodecahydrate in 25 mL DI water). Addition of few thymol crystals as fungicide.
6. 2-Methoxyethyl acetate (Merck, Darmstadt, Germany).
7. 100% ethanol (Thommen-Furler AG, Rüti bei Büren, Switzerland).
8. Xylene (Thommen-Furler AG, Rüti bei Büren, Switzerland).
9. A toluidine blue working solution is prepared by mixing one part of 1% toluidine blue stock solution (powdered toluidine blue O reconstituted in DI water) and 9 parts of a 0.1 M sodium phosphate buffer solution (pH 8.0). Before use, the working solution must be filtered and warmed to ca. 60 °C.
10. 0.7% formic acid (Merck, Darmstadt, Germany).
11. The standard solution of alcian blue can be prepared from its powder form (alcian blue 8GX) and diluted in acetic acid and DI water, with a final pH of ca. 2.5.
12. Periodic acid (Merck, Darmstadt Germany).
13. Schiff's reagent (Merck, Darmstadt, Germany).
14. 0.5% sodium pyrosulfate (Merck, Darmstadt, Germany).

3 Methods

Most techniques described below are standard histological processing procedures applicable to multiple tissue types. However, in order to assess the cartilage/bone tissue as a unit and to preserve the osteochondral interface as best as possible, particular care must be taken at key steps in the process. These will be highlighted throughout the methodology described below.

As mentioned in the introduction depending on the research question and the sample contents, e.g., if the sample contains any biomaterials, osteochondral tissue may be processed for polymethylmethacrylate (PMMA) embedding to produce non-decalcified sections or paraffin embedding to produce decalcified sections. Cryosectioning may also be used; however this is only possible with non-calcified tissue or decalcified tissue. As most methodological differences lie between PMMA processing and paraffin processing, the methodology for these two techniques covering sample fixation, dehydration, decalcification (when applicable), embedding, cutting, and staining with toluidine blue, safranin O, and alcian blue (Fig. 1) will be described below.

3.1 Processing Non-decalcified Samples Embedded in Polymethylmethacrylate

3.1.1 Sample Preparation and Fixation

As methylmethacrylate processing does not require decalcification of the bone tissue, the cartilage, underlying subchondral bone, and deeper bone layers can all be included in the block. This brings an advantage to evaluating the bone/cartilage unit, including deeper parts of the bone which may include subchondral bone cysts, for example. This technique is also useful when osteochondral samples contain metal, ceramic, or polymeric biomaterials which would have a significantly lower water content compared to surrounding tissues and therefore would not be retained in a paraffin embedded decalcified sample.

1. Trim the fresh osteochondral sample, taking care to include enough bone tissue to assess the entirety of a subchondral bone cyst (*see* **Note 1**). If processing a sample with a distinct cartilage defect area, or even a repaired cartilage defect area, care must be taken to also include enough healthy surrounding cartilage tissue (ca. 10 mm, if possible). This is done to allow for a comprehensive evaluation of cartilage repair, looking not only at the defect filling but also the interface with surrounding cartilage (to assess lateral integration) as well as the effect on healthy surrounding cartilage. The samples should be kept hydrated as much as possible during these manipulations, by wrapping them in saline-soaked gauze.
2. Once the sample is trimmed, place in 10–20 times the volume of the sample of a 4% solution of buffered formalin (*see* **Note 2**), preferably as soon as possible after original sample

harvest. The fixation duration of samples will vary depending on the size. Smaller osteochondral samples (ca. 1 cm^3) may take 7 days, while larger samples may take up to 3 weeks, the latter requiring a weekly exchange of the fixation solution.

3.1.2 Dehydration, PMMA Embedding, and Sectioning

1. After fixation, formalin-fixed samples are then washed with tap water (time depending on sample size, usually 3 × 20 min as a minimum).
2. Samples are then placed in 50% ethanol to start the dehydration process (*see* **Note 3**).
3. The samples are then moved through an ascending ethanol series, starting with a solution of 50% ethanol (3 × 30 min) onto 70% ethanol (exchange 2×/day), then 80% ethanol (exchange 2×/day), 90% ethanol (exchange 1×/day), 96% ethanol (exchange 1×/day), and finally 100% ethanol (exchange 4× over 2 days). The duration of each ethanol step depends on the size of the samples and may be longer for larger samples.
4. Once dehydrated, the samples can then be defatted in xylene, for at least 4 days if under vacuum in a glass desiccator (with one xylene solution exchange). This is a minimum duration for xylene infiltration, as this step will take longer with larger samples.
5. Blot the samples to remove excess xylene and place in the PMMA solution in a desiccator at 4 °C for 4–7 days (max. 14 days) to allow infiltration of the liquid PMMA solution.
6. After infiltration, the samples are placed in molds with PMMA solution and, if small enough to fit the standard Teflon molds, are left to polymerize under a chemical hood for 3–10 days. Larger samples are placed in PMMA solution in larger molds (e.g., glass jars) and allowed to polymerize slowly at 4 °C, in order to avoid bubble formation (*see* **Note 4**).
7. A diamond saw is then used to produce thick sections of ca. 300–700 microns. Thick sections can be used to perform surface stains and evaluate osteochondral tissue structure and composition, but as mentioned before do not show enough detail to evaluate cellular components (*see* **Note 5**).
8. After cutting, thick sections are then glued onto preferably PMMA-GS white opal slides (3 mm thick) using an adhesive (e.g., Cementit™ speed glue), taking care to avoid air bubbles between the slide and the section.
9. Sections are then ground down to 200–300 microns, primarily to obtain an even surface, and then polished. These slides are then ready for surface staining.

10. Thin sections of ca. 6 microns can also be prepared using a microtome. Unlike ground sections, thin sections are useful to evaluate cells and other detailed structures, at the expense of the integrity of the osteochondral interface, which may be disrupted during processing.
11. Thin sections are placed on glass slides coated with chrome alum-gelatin to ensure the adhesion of the section through the staining process. Note that the recommended glass slides to be coated are simple microscopy glass slides with a frosted edge and not pre-treated slides which are recommended for paraffin sections such as SuperFrost Plus™ slides (Fisherbrand™).
12. Prior to staining thin sections, a deplasticization step is needed to remove the polymethylmethacrylate and rehydrate the samples. The thin sections are placed in three sequential baths of 2-methoxyethyl acetate (30 min each) followed by two baths of xylene (5 min each) then through a descending ethanol series to rehydrate the sample (one bath of absolute ethanol followed by washes with absolute ethanol, 96% ethanol, 70% ethanol, then finally deionized water). Once in water, the section is ready to be stained. This procedure must be performed immediately prior to staining.

3.1.3 Staining Non-decalcified PMMA Sections

3.1.3.1 Toluidine Blue

Toluidine blue is a simple yet versatile stain which can differentiate important structures in an osteochondral sample by showing, among other things:

- Calcified tissues in various shades of blue depending on their maturity.
- Cell nuclei in dark blue.
- Glycosaminoglycan-rich matrix in purple.

The following protocol can be used to perform toluidine blue surface staining on thick PMMA sections:

1. Etching with 0.7% formic acid	5 min
2. Wash with deionized (DI) water	
3. 0.1% toluidine blue working solution	10–15 min
4. Wash with tap water	
5. Wash with DI water	
6. Dab the section dry and leave to air-dry	30 min

The following protocol can be used to perform a toluidine blue stain on a non-decalcified PMMA thin section, which has already been deplasticized and rehydrated (Subheading 3.1.2, **step 12**):

1. Etching with 0.7% formic acid	5 min
2. Wash with deionized (DI) water	
3. 0.1% toluidine blue working solution	10 min
4. Wash with tap water	
5. Wash with DI water	
6. Differentiate in 70% ethanol	
7. Differentiate in 96% ethanol	
8. Wash with absolute ethanol	
9. Absolute ethanol	5 min
10. Wash with xylene	
11. Xylene	5 min.
12. Mount coverslip using mounting medium	

Special care must be taken for washing steps so that the cartilage layer does not detach from the slide or delaminate from the bone.

3.1.3.2 Alcian Blue

Alcian blue staining allows the identification of glycosaminoglycan-rich tissues by staining most acid mucins blue (*see* **Note 6**).

The following protocol can be used to perform an alcian blue stain on a non-decalcified PMMA thin section, which has already been deplasticized and rehydrated (Subheading 3.1.2, **step 12**):

1. Alcian blue 8GX (pH 2.5) warmed to 60 °C	30–60 min
2. Wash with tap water	5 min
3. 0.5% periodic acid	10 min
4. Wash with tap water	5 min
5. Schiff's reagent	15 min
6. 0.5% sodium pyrosulfate	6 min
7. Wash with tap water	10 min
8. Weigert's hematoxylin	15 min
9. Wash with tap water	5 min
10. Wash with DI water	
11. Wash with 80% ethanol	
12. Wash with 96% ethanol	
13. Wash with absolute ethanol	
14. Absolute ethanol	5 min

(continued)

15. Wash with xylene	
16. Xylene	5 min
17. Mount coverslip using mounting medium	

Special care must be taken for washing steps so that the cartilage layer does not detach from the slide or delaminate from the bone.

3.1.3.3 Safranin O

Safranin O is another stain that particularly highlights glycosaminoglycan-rich tissues such as cartilage in red, with the intensity proportional to the amount of sulfated glycosaminoglycans present in the matrix (*see* **Note** 7). In osteochondral sections, it can also highlight the calcified cartilage layer in a darker red, the bone in green (by using fast green as a counterstain), and cell nuclei in blue violet (by using Mayer's hematoxylin).

The following protocol can be used to perform a safranin O stain on a non-decalcified PMMA thin section, which has already been deplasticized and rehydrated (Subheading 3.1.2, **step 12**):

1. Mayer's hematoxylin	10 min
2. Bluing step under running tap water	10 min
3. 0.1% fast green	10 min
4. Wash with 1% acetic acid	10 s
5. 0.1% safranin O	15 min
6. 0.5% sodium pyrosulfate	6 min
7. Wash with DI water	
8. Wash with 80% ethanol	
9. Wash with 96% ethanol	
10. Wash with 96% ethanol	
11. Wash with absolute ethanol	
12. Absolute ethanol	2 min
13. Wash with xylene	
14. Xylene	5 min
15. Mount coverslip using mounting medium	

Special care must be taken for washing steps so that the cartilage layer does not detach from the slide or delaminate from the bone.

3.2 Processing Decalcified Samples Embedded in Paraffin

Preparing samples for fixation, decalcification, and paraffin embedding is similar to the procedure described above for PMMA embedded samples, with a few notable differences.

3.2.1 Sample Preparation, Fixation, and Decalcification

1. As with non-decalcified samples, it is recommended to retain the surrounding cartilage tissue around a cartilage defect or a repaired cartilage defect, with a margin of about 10 mm if possible. This allows the adequate evaluation of lateral integration and surrounding cartilage tissue health, thereby assessing any potential protective effects treatments may have. The bone in osteochondral paraffin embedded samples, however, must be trimmed further than with PMMA embedded samples, leaving only about 5 mm of subchondral bone in thickness (*see* **Note 8**).
2. Fixation is most often achieved using a 4% buffered formalin solution, placing the samples in 10–20 times their volume of fixation solution, for up to 72 h maximum, in order not to damage any epitopes which may have to be detected with immunohistochemistry later in the process (*see* **Note 9**).
3. After fixation, osteochondral samples must be thoroughly washed (using tap water, saline, or phosphate-buffered saline) and then decalcified prior to dehydration and paraffin embedding. Three main methods of decalcification are used, with the main trade-off being speed of decalcification versus preservation of epitopes for immunohistochemical detection.
4. The mildest solution for decalcification is a 25% solution of ethylenediaminetetraacetic acid or EDTA (pH 7.0–7.4), which captures calcium ions, thereby softening the bone and calcified cartilage. Osteochondral samples are placed in cassettes or wrapped in gauze, if too big for standard cassettes, and then placed in a container with the 25% EDTA solution, protected from light using aluminum foil. It is recommended to place the decalcifying samples on a shaker or stir the solution using a magnetic stirrer during the decalcification process. EDTA decalcification can take several months if the calcified tissue volume in the samples is large. For each cm^3 of bone tissue, ca. 80 mL of EDTA solution is needed (*see* **Notes 10** and **11**).
5. The decalcifying solution needs to be exchanged every 7–14 days for the decalcification process to remain efficient (*see* **Note 12**).
6. After decalcification, the sample can be washed (e.g., under running tap water) and further trimmed to fit into cassettes.

3.2.2 Dehydration, Paraffin Embedding, and Sectioning

Fixed, decalcified osteochondral tissue samples are placed in histology cassettes and processed through an ascending ethanol series in order to dehydrate the samples. Specifically, the process starts in a solution of 70% ethanol, progressing up to 100% ethanol, ensuring water is progressively removed from the samples. An automated processor can be used for this step to facilitate the procedure (e.g., Leica histoprocessor). Such a device not only dehydrates samples through an ethanol series but also can be programmed to perform the subsequent defatting procedure in xylene as well as the liquid paraffin infiltration step, prior to embedding. Most samples will require an overnight program (total ca. 15 h, each step automatized ca. 2–3 h) to go through all steps (*see* **Notes 13** and **14**).

1. 1 × 70% ethanol step.
2. 1 × 90% ethanol step.
3. 4 × 100% ethanol steps.
4. 3 × xylene steps.
5. 3 × liquid paraffin steps.

Once the samples are infiltrated with liquid paraffin, they are then properly positioned and and embedded in paraffin blocks. Usually, it is recommended to position osteochondral samples in such a way to cut through both the cartilage and bone tissue when producing sections, in a plane orthogonal to the surface of the cartilage, so that the bone/cartilage unit can be adequately assessed. Afterward, a microtome is used to produce sections of 2–4 microns. If hard calcified areas are still found while cutting, the block can be immersed overnight in a solution of 25% EDTA, making sure the cut plane is fully submerged, in order to decalcify these regions and allow for a smoother cutting process.

After cutting, sections are first placed in a room temperature water bath. This step is important for osteochondral samples, as the difference in water content in the native tissue between cartilage and bone is very different and therefore causes many folds in the cartilage part of the section as well as breaks at the cartilage/bone interface. Sections in the room temperature water bath can be carefully manipulated to reduce such folds, before transferring the sections to a warm water bath (42 °C) and then placing them on pre-treated slides such as SuperFrost™ Plus, which further ensures the adhesion of the section to the slide. As cartilage tissue can be particularly slippery, carefully performing these steps is important to ensure a good quality and allow for an adequate osteochondral tissue assessment.

Prior to staining the slides, sections need to be dewaxed and rehydrated:

1. Place slides two times xylene (5 min each).
2. Followed by 2× absolute ethanol (3 min each).
3. 2 × 96% ethanol (3 min each).
4. 1 × 70% ethanol (3 min each).
5. Then transfer to deionized water.
6. Once in water, the section is ready to be stained. This procedure must be performed immediately prior to staining.

3.2.3 Staining Decalcified Paraffin Sections

3.2.3.1 Toluidine Blue

Toluidine blue is a simple yet versatile stain which can differentiate important structures in an osteochondral sample by showing, among other things:

- Calcified tissues in various shades of blue depending on their maturity.
- Cell nuclei in dark blue.
- Glycosaminoglycan-rich matrix in purple.

The following protocol can be used to perform a toluidine blue stain on a decalcified section, which has already been dewaxed and rehydrated:

1. Etching with 0.7% formic acid	5 min
2. Wash with deionized (DI) water	
3. 0.1% toluidine blue working solution	8–10 min
4. Wash with tap water	
5. Wash with DI water	
6. Differentiate in 80% ethanol (wash shortly)	
7. Differentiate in 96% ethanol (wash shortly)	
8. Wash with absolute ethanol	
9. Absolute ethanol	2 min
10. Wash with xylene	
11. Xylene	5 min
12. Mount coverslip using mounting medium	

Special care must be taken for washing steps so that the cartilage layer does not detach from the slide or delaminate from the bone.

3.2.3.2 Alcian Blue

Alcian blue staining allows the identification of glycosaminoglycan-rich tissues, by staining most acid mucins blue. At the standard pH of 2.5, most acid mucins will be stained blue, including some sulfated acid mucins. However, by using a more acidic alcian blue solution (pH 1.0), strongly sulfated glycosaminoglycans will be stained more prominently. This is therefore recommended for osteochondral tissues, in order to adequately evaluate glycosaminoglycan content in the native tissue as well as the repair or damaged tissue. A standard counterstain such as nuclear fast red-aluminum sulfate can be used to stain cell nuclei red and other tissues various pink/red.

The following protocol can be used to perform an alcian blue/nuclear fast red stain on a decalcified section, which has already been dewaxed and rehydrated:

1. Alcian blue 8GX	30 min
2. Wash with tap water	5 min
3. 0.5% periodic acid	10 min
4. Wash with tap water	5 min
5. Schiff's reagent	10 min
6. 0.5% sodium pyrosulfate	6 min
7. Wash with tap water	10 min
8. Nuclear fast red	5 min
9. Wash with tap water	2 min
10. Wash with DI water	
11. Wash with 80% ethanol	
12. Wash with 96% ethanol	
13. Wash with absolute ethanol	
14. Absolute ethanol	5 min
15. Wash with xylene	
16. Xylene	5 min
17. Mount coverslip using mounting medium	

Special care must be taken for washing steps so that the cartilage layer does not detach from the slide or delaminate from the bone.

3.2.3.3 Safranin O

Safranin O staining highlights glycosaminoglycan-rich tissues such as cartilage in red, with the intensity proportional to the amount of sulfated glycosaminoglycans present in the matrix. That is why this staining method has also been used to quantitatively assess glycosaminoglycan content by evaluating the intensity of red staining in safranin O-stained cartilage sections. In osteochondral sections, it can also highlight the calcified cartilage layer in a darker red, the bone in green (by using fast green as a counterstain), and cell nuclei in blue violet (by using Gill No. 2 hematoxylin).

The following protocol can be used to perform a safranin O stain on a decalcified section, which has already been dewaxed and rehydrated:

1. Hematoxylin, Gill No. 2	10 min
2. Bluing step under running tap water	5 min
3. 0.1% fast green	5 min
4. Wash with 1% acetic acid	10 s
5. 0.1% safranin O	5–10 min

(continued)

6. Wash with 96% ethanol	
7. Wash with 96% ethanol	
8. Wash with absolute ethanol	
9. Absolute ethanol	2 min
10. Wash with xylene	
11. Xylene	5 min
12. Mount coverslip using mounting medium	

Special care must be taken for washing steps so that the cartilage layer does not detach from the slide or delaminate from the bone.

3.2.3.4 Immunohistochemistry

Decalcified sections may be used to perform immunohistochemical stains, thereby identifying important epitopes in osteochondral tissue. Standard targets include hallmarks of native hyaline cartilage such as collagen type II and aggrecan and collagen type X in calcified cartilage. Other markers include collagen type I in bone, as well as in fibrocartilaginous repair tissue (e.g., after microfracture repair of cartilage defects) as well as indicators of osteoarthritis such as MMP-13, ADAMTS-5, and other matrix components such as COMP and perichondral matrix components like collagen types IV and IX. While the potentially interesting targets may be many, the fundamental methodology remains the same, as well as the important aspects to consider when processing osteochondral tissue in order to preserve the bone/cartilage unit.

While matrix components are present relatively in abundance compared to, for example, intracellular epitopes, cross-species reactivity is a significant hurdle to overcome. Indeed, most preclinical evaluations of osteochondral repair techniques are performed on sheep or goat tissue, which may reduce the efficacy of antibodies originally designed to target human proteins. The antigen retrieval step is therefore important to perform, in order to maximize the chances of success. Many antibodies recommend heat-mediated antigen retrieval at 95 °C for 20 min in trisodium citrate buffer (pH = 6) or in Tris-EDTA buffer (pH = 9.0), in order to break the methylene bonds formed during formalin fixation and expose epitopes for antibody detection. With osteochondral tissues, however, the risk of losing the cartilage part during processing is high because of the slippery nature of cartilage. Indeed, despite using SuperFrost™ Plus slides, the cartilage tissue will often delaminate or become shriveled and folded after heat-mediated antigen retrieval, which makes evaluation impossible. It is therefore not recommended to perform heat-mediated antigen retrieval with osteochondral tissue, for the reasons listed above. Instead, enzymatic antigen retrieval is recommended, using, for example, proteinase K, hyaluronidase, or chondroitinase ABC to make the extracellular matrix more accessible to primary antibodies. It is important to note that engineered osteochondral tissue may require shorter enzymatic antigen retrieval treatments compared to native tissue, as the matrix would be less mature and therefore less dense.

Besides the risk posed by antigen retrieval, care must be taken during the many steps of a standard immunohistochemical protocol (endogenous peroxidase block, protein block, wash steps, primary antibody and secondary antibody incubations) in order not to delaminate the cartilage tissue from the osteochondral section. Furthermore, fresh sections must be prepared 24 h in advance of an immunohistochemical staining or kept at 4 °C if prepared a few days in advance, in order to retain not only epitope viability bit also tissue structure and integrity on the slide. This can be especially important with osteochondral tissue in order to maintain the osteochondral interface during staining.

Matrix tissue permeability is an important aspect to consider when evaluating sections that may contain immature repair tissue, damaged tissue, as well as dense, healthy native tissue. The penetration of antibodies into damaged or immature tissue is likely to be higher than in dense mature native tissue, possibly creating a bias in interpretation. That is why antigen retrieval must be tested with known control sections, with every immunohistochemical run, in order to ensure a homogenous and reliable staining throughout.

3.3 Evaluation of Cartilage Histological Sections

The morphology of repair tissue in cartilage predicts its functionality. For the microscopical evaluation of histological sections, several scores are currently used to evaluate either the OA grade or the stage of regeneration of cartilage.

To compare stages of OA or the success of an OA treatment, a reliable and reproducible (semi-)quantitative grading system has to be chosen. Although there are attempts to use automated histomorphometric analyses to objectify the assessment, a more subjective or semiquantitative scoring system may still be an appropriate tool [6].

The ideal histopathology system should fulfill the need for [6, 7]:

- *Simplicity*: Applicable by investigators with different levels of experience.
- *Utility*: Useful for clinical *and* experimental OA.
- *Scalability*: Easy correlation of the macroscopic assessment with the histopathological findings.
- *Extendibility*: Extendable to advanced applications.
- *Comparability*.

The use of a validated scoring system makes own observations comparable to those of other researchers. Depending on the research question, a suitable or even several score systems have to be chosen, or scores need to be slightly modified. For instance, if only the degree of OA needs to be assessed, a different score system makes sense as in cases where a cartilage defect was created and defect repair with or without biomaterials has to be evaluated. In this chapter only the most commonly used score systems are presented. The score systems themselves are found in the appendix, and the text highlights their advantages and disadvantages.

3.3.1 Scores for the Evaluation of Osteoarthritis

OA scores should focus on degenerative features of the hyaline cartilage. Several different scores with advantages and disadvantages are available, such as the OARSI Mankin or Little score system.

3.3.1.1 OARSI

The OARSI score (Table 1 in appendix) [8–11] includes the evaluation of subchondral bone, synovium, and meniscus and obviously is, therefore, prone for evaluation of OA in the knee. The scoring system is able to assess early stages of OA, because, e.g., cartilage matrix swelling as an early change is included.

In contrast to other scoring systems and as an advantage, the OARSI score not only focuses on the *severity* but also on the *extent* of the lesion, comparable to the grade and stage cancer pathology assessment: Lesion depth into cartilage area represents more severe arthritis, and lesion extent over cartilage surface represents extent of disease [7].

The limitations of the OARSI score system, however, is that lesions of the subchondral bone are not included in the OARSI grades 1–4 (early changes) but only in scores 5–6. This suggests a development of pathologies from cartilage damage to bone, which may not be true for all cases of clinical and experimental OA. Furthermore, a staging component may be difficult to determine, because of the difficulty to estimate the surface extent of OA damage in histological sections, where only parts or one plane is available on sections [12].

3.3.1.2 Mankin

The Mankin OA score (Table 2 in appendix, schematic representation in Table 3 in appendix) [13] was originally developed for the assessment of human hyaline cartilage of the pelvis but can be applied for other joints and also for the evaluation of animal cartilage. It represents the most widely used cartilage histology score. The 14-point score includes cellular changes; histochemical presence of safranin O matrix staining, or more generally called metachromasia; and architectural changes like erosion and tidemark integrity. The system can also be used with sections stained with toluidine or alcian blue as they also change staining intensity in accordance to PG content of the matrix.

Table 1
OARSI cartilage score, as described by Little et al. [11]

OARSI score scale	
A structure (score the worst area in field of view)	
Description	Grade
Normal	0
Slight surface irregularities (surface barely disturbed)	1
Moderate surface irregularities (surface roughened)	2
Severe surface irregularities (disruption, fissuring/fibrillation to <10% depth)	3
Fissures to transitional zone (1/3 depth)	4
Fissures to radial zone (2/3 depth)	5
Fissures to calcified zone (full depth)	6
Erosion or severe fibrillation to mid zone (1/3 depth)	7
Erosion or severe fibrillation to deep zone (2/3 depth)	8
Erosion or severe fibrillation to calcified zone (full depth)	9
Erosion or severe fibrillation to subchondral bone	10
B Chondrocyte density ("average" score for whole field of view in non-calcified cartilage)	
Description	Grade
Normal	0
Increase or slight decrease	1
Moderate decrease	2
Severe decrease	3
No cells	4
C Cell cloning (score the whole field of view)	
Description	Grade
Normal	0
Several doublets	1
Many doublets	2
Doublets and triplets	3
Multiple cell nests or no cells in section	4
D Interterritorial toluidine blue (score the worst area in field of view working from AC surface down)	
Description	Grade
Normal	0
Decreased staining to mid zone (1/3 depth)	1
Decreased staining to deep zone (2/3 depth)	2
Decreased staining to calcified zone (full depth)	3
No staining	4

(continued)

Table 1
(continued)

OARSI score scale	
E Tidemark/calcified cartilagesubchondral bone (score the worst area in field of view)	
Description	Grade
Intact subchondral bone plate, single tidemark	0
Intact subchondral bone plate, duplicated tidemark	1
Blood vessels penetrate through subchondral bone plate to calcified cartilage	2
Tidemark penetrated by blood vessels	3

Table 2
Mankin cartilage score, as described by Mankin et al. [13]

Mankin score scale	
I Structure	*Grade*
Normal	0
Surface irregularities	1
Pannus and surface irregularities	2
Clefts to transitional zone	3
Clefts to radial zone	4
Clefts to calcified zone	5
Complete disorganization	6
II Cells	*Grade*
Normal	0
Diffuse hypercellularity	1
Cloning	2
Hypocellularity	3
III Safranin O staining	*Grade*
Normal	0
Slight reduction	1
Moderate reduction	2
Severe reduction	3
No dye noted	4
IV Tidemark integrity	*Grade*
Intact	0
Crossed by blood vessels	1

Table 3
Schematic illustration of the Mankin score

Score	Description	Sketch	Histological sample
I Structure			
0	Normal	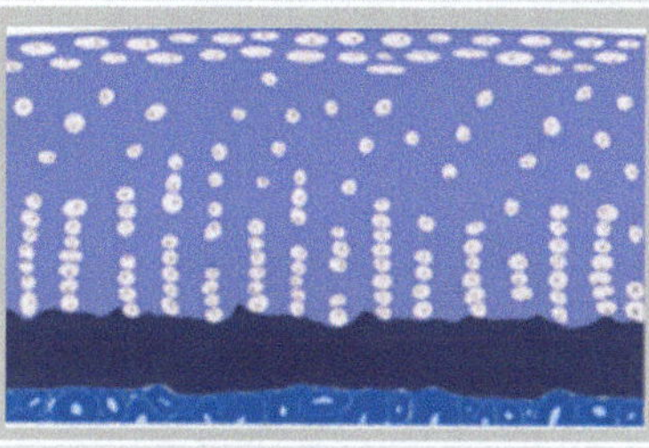	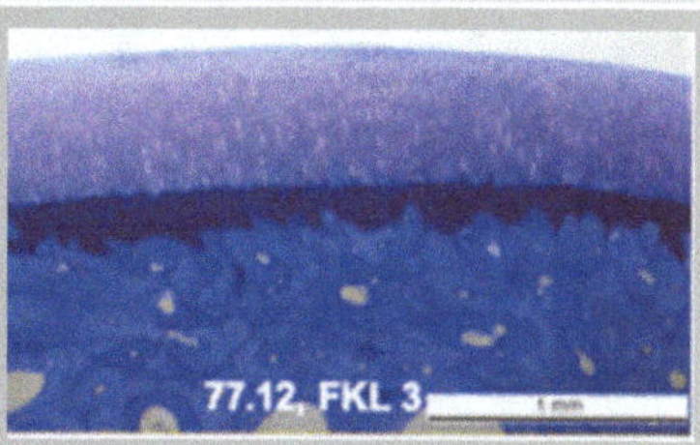
1	Surface irregularities	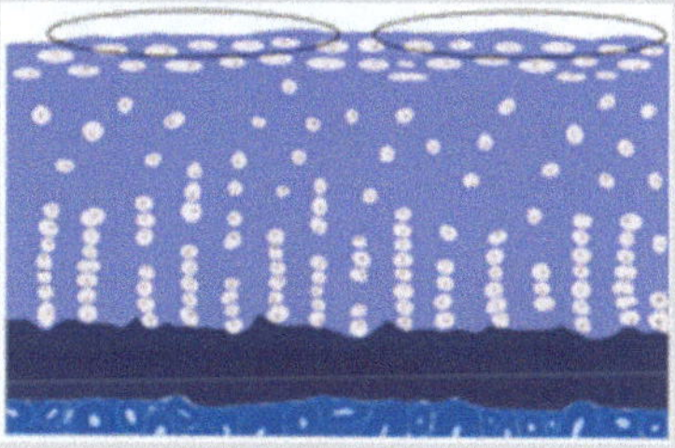	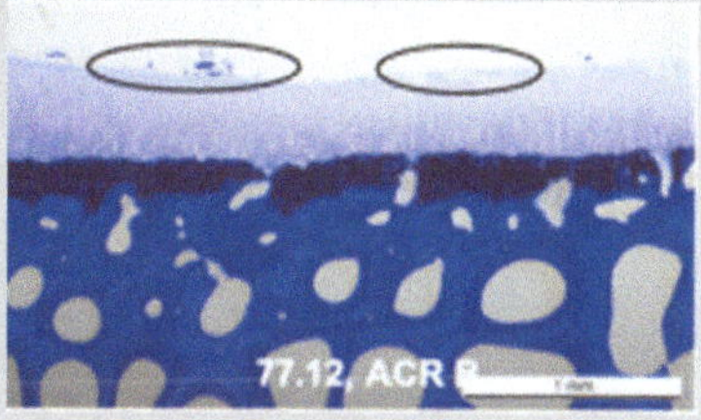
2	Pannus and surface irregularities	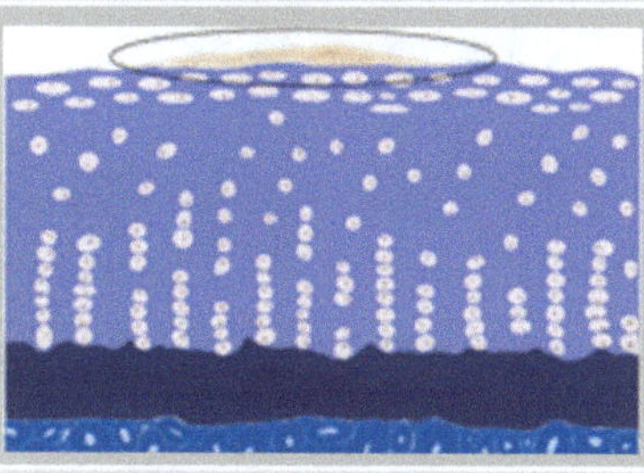	
3	Clefts to transitional zone	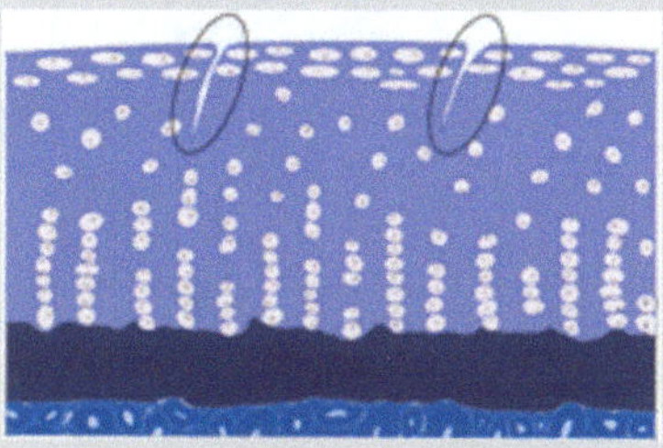	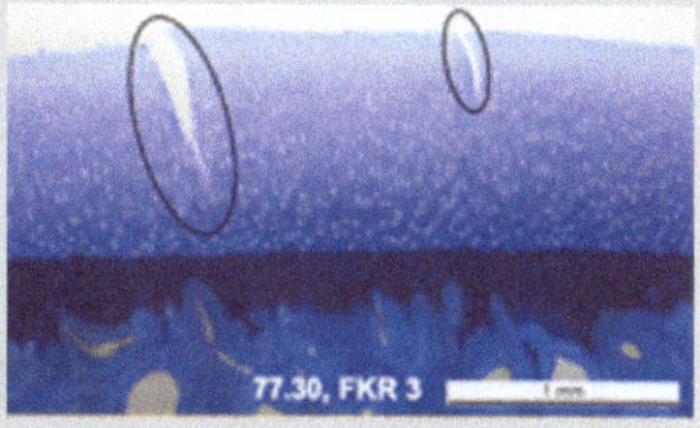
4	Clefts to radial zone	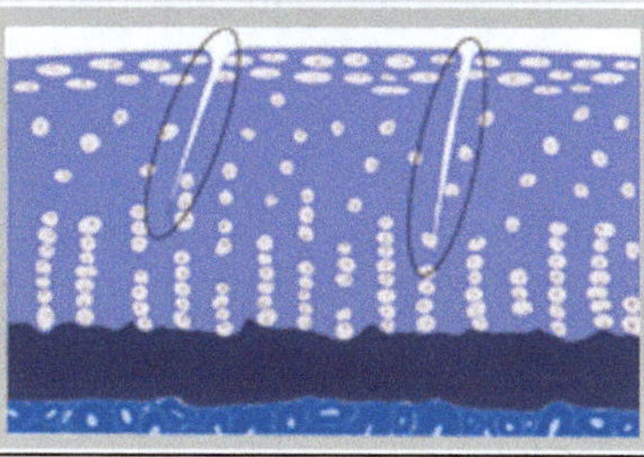	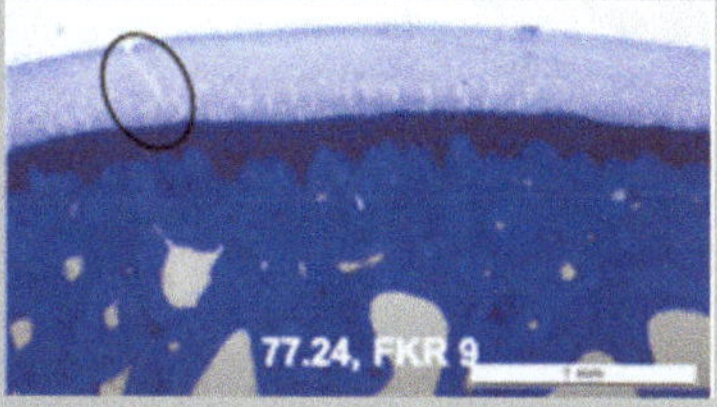

(continued)

Table 3
(continued)

Score	Description	Sketch	Histological sample
5	Clefts to calcified zone	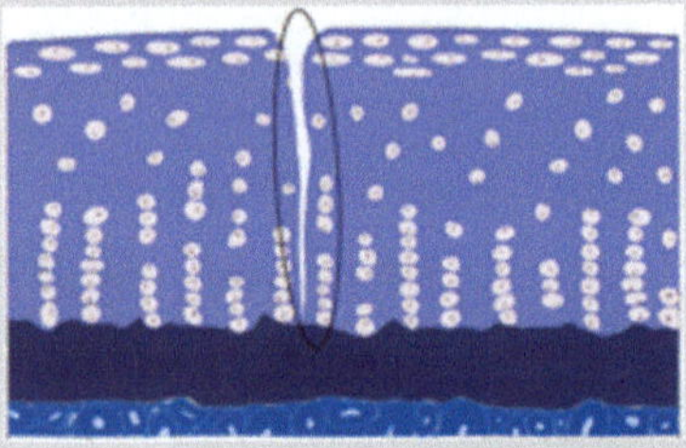	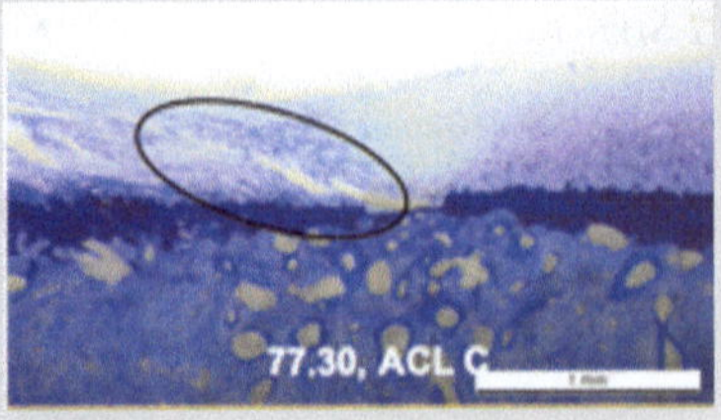
6	Complete disorganization	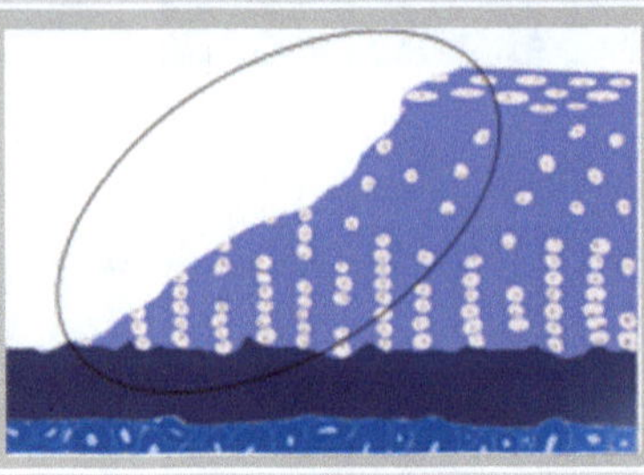	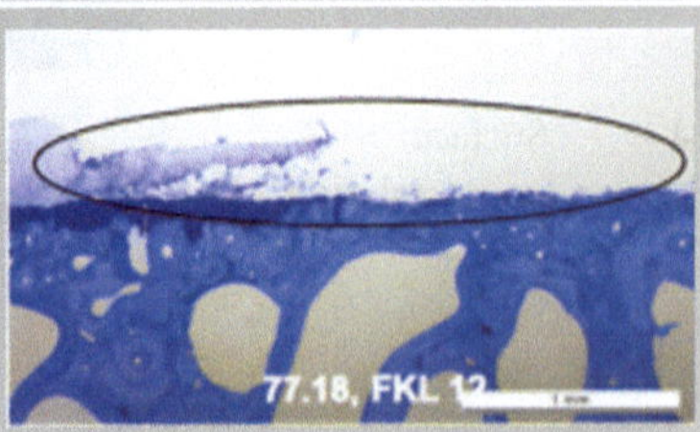
II Cells			
0	Normal		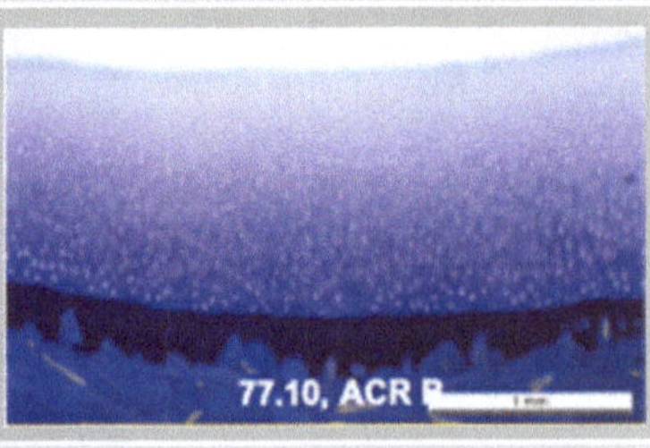
1	Diffuse hypercellularity	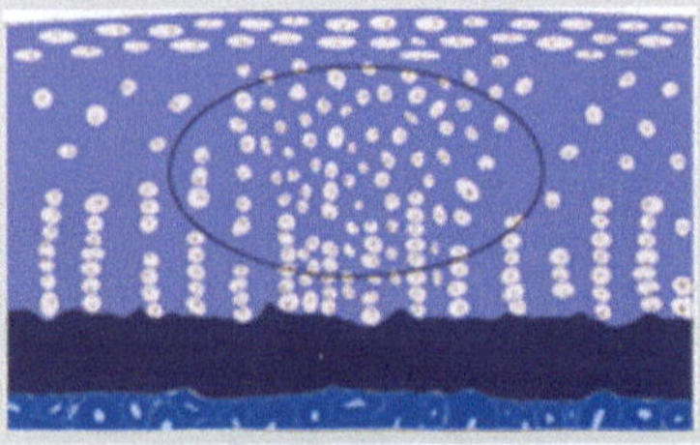	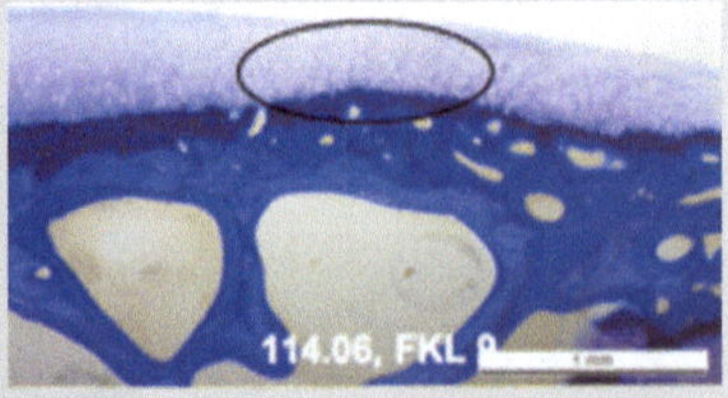
2	Cloning, formation of cell clusters	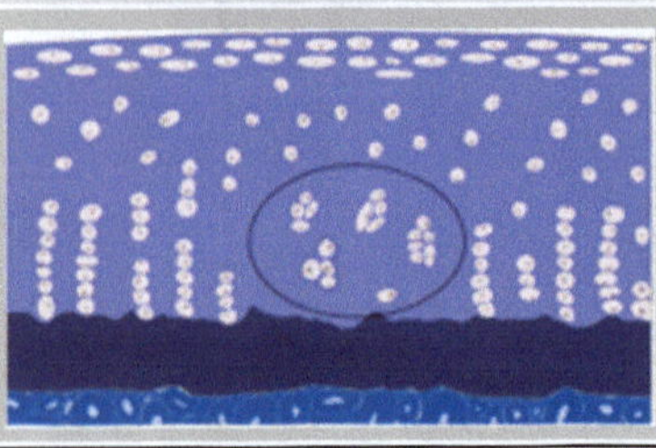	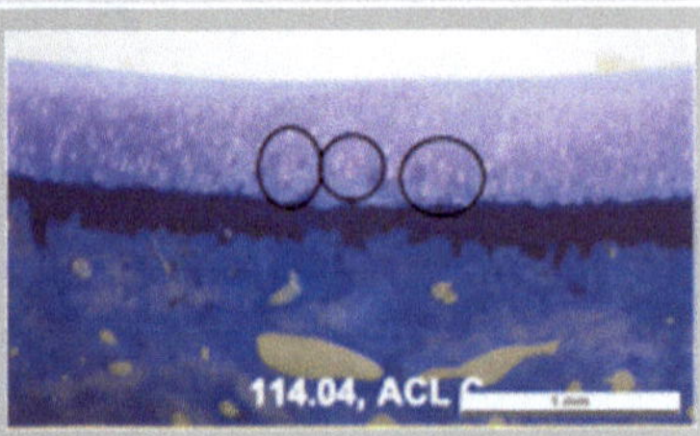

(continued)

Table 3 (continued)

Score	Description	Sketch	Histological sample
3	Hypocellularity	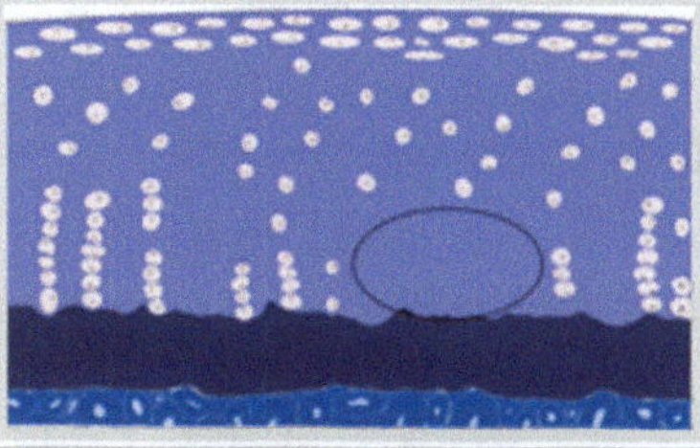	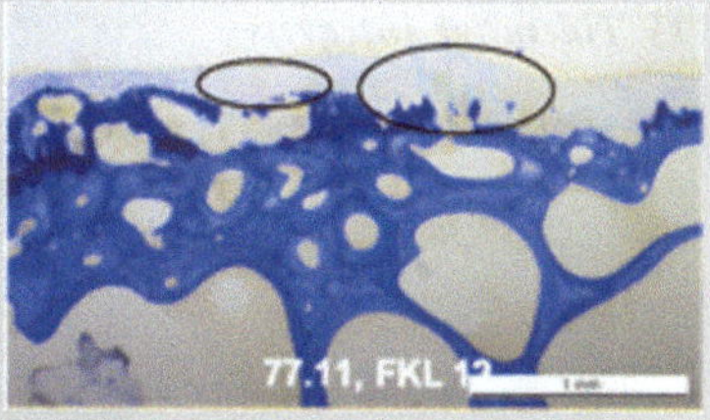
III Matrix staining			
0	Normal	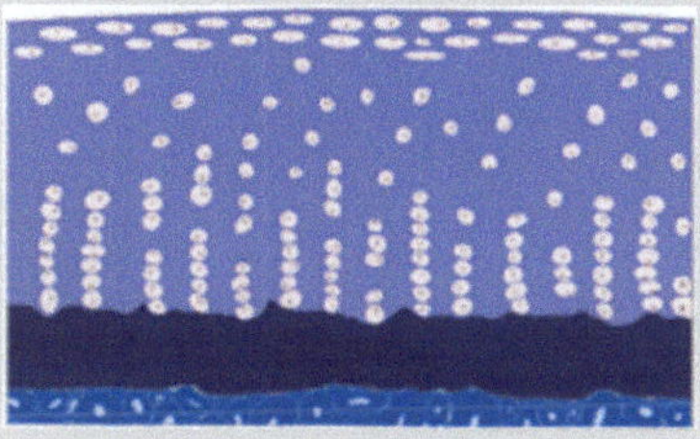	
1	Slight reduction	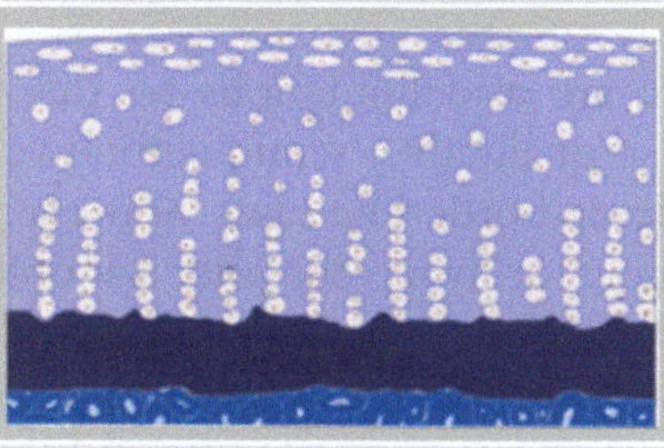	
2	Moderate reduction	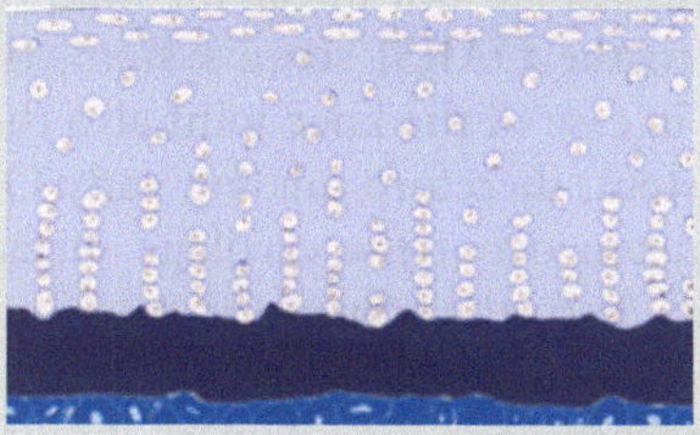	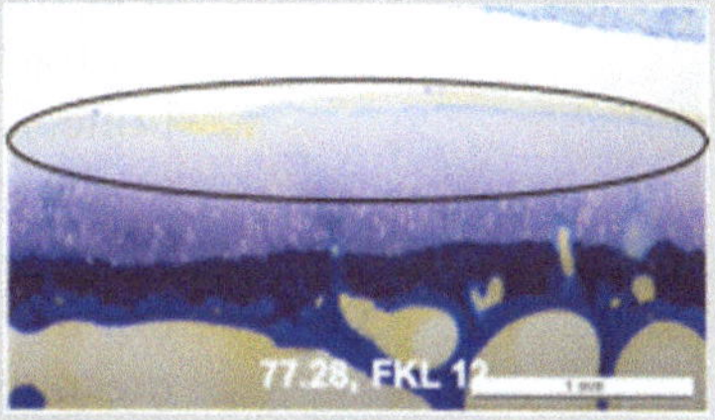
3	Severe reduction	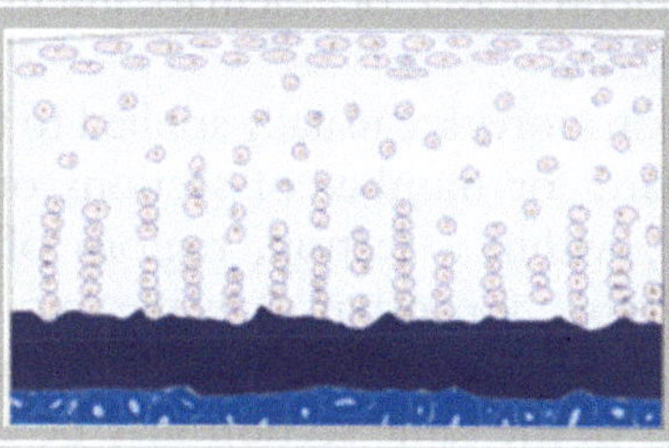	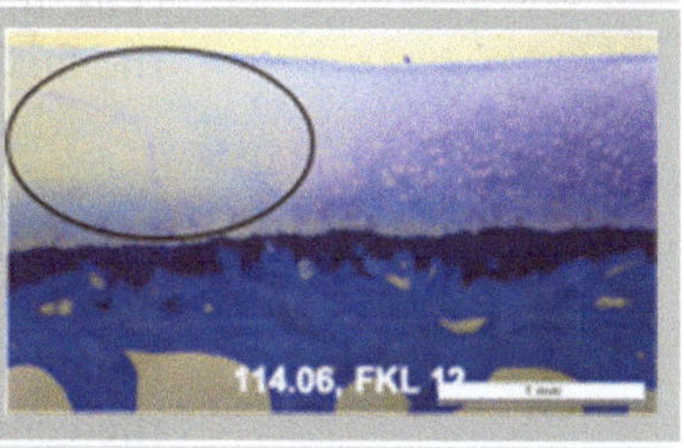
4	No dye noted	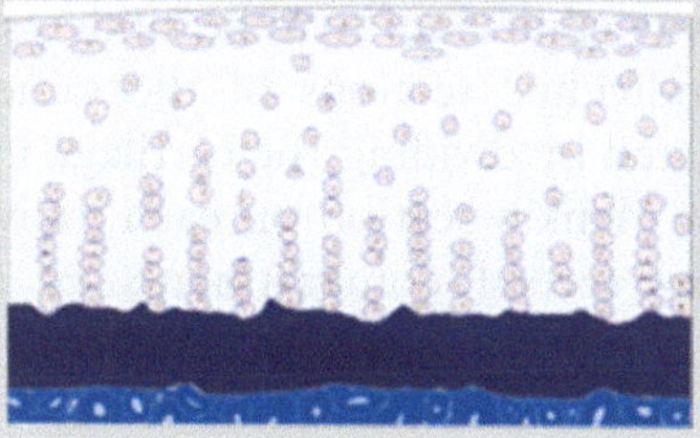	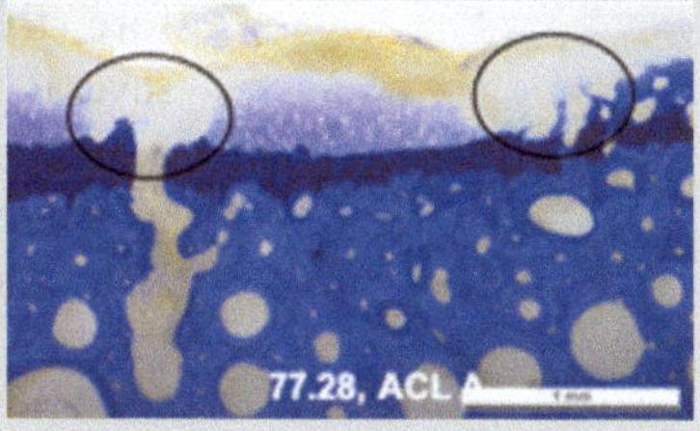

(continued)

Table 3
(continued)

Score	Description	Sketch	Histological sample
IV Tidemark integrity			
0	Intact	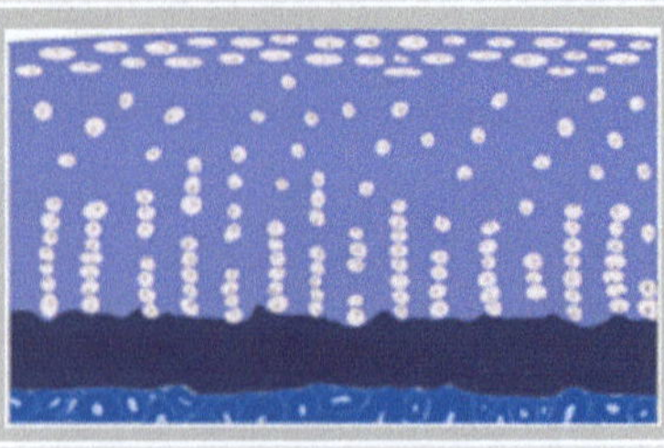	
1	Crossed by blood vessels	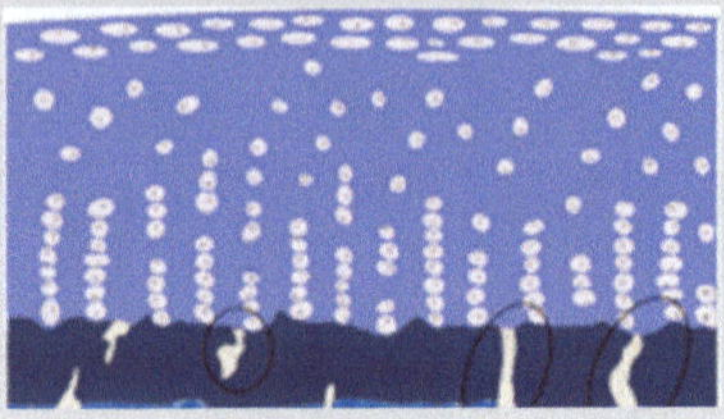	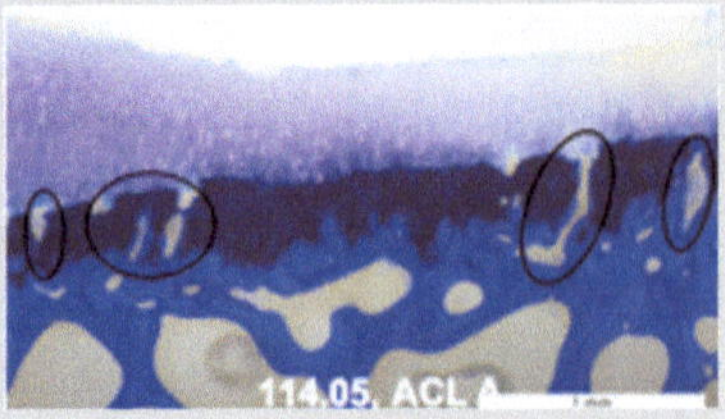

There are also limitations to this system [8, 14]. The low number of qualitative/semiquantitative units can induce scoring variabilities in borderline cases. Some scores such as pannus formation, cloning, or hypercellularity not only describe primary OA but also secondary cartilage formation resp. repair functions. This may lead to an inadequate assessment of mild and moderate OA, such that the scores are not linear for mild or earlier phases of OA. In contrast to the OARSI score system, the synovium, subchondral bone, or in case of the knee joint also the menisci are not included in the system, therefore missing early OA cases, where the overall joint is inflamed but the cartilage surface is not yet destroyed or fibrillated. Also, the scores for each category are not "weighed" meaning that in the final score number the section of structure, cells, metachromasia, or tidemark integrity value is equally high in the final score.

Mankin scores are usually applied to the overall appearance, and no standard for number of sections or their location is scored. However, within a section, regions of different severity of OA may be visible, for instance, in the impingement syndrome of the hip joint, only focal parts in the periphery may be damaged whereas the center of the same section may be completely normal. Therefore, this overall score may be underestimated if not applied on several individual sections of the same slide section. It can be counteracted by dividing the cartilage in several sections and then using the Mankin section on each individual section. This would also allow several compartments of a joint to be compared individually and the average of all calculated. Last but not least, there may be inter- and intra-observer variability for cellularity, metachromasia, and tidemark integrity. Especially for metachromasia it is important that all histology sections should be stained in the same batch.

3.3.1.3 Little

Similarly to the Mankin score, the Little score (Table 4 in appendix) [10, 11] assesses only the hyaline cartilage component and does not include synovial membrane or menisci. Its focus is similar to the Mankin score on cartilage degradation looking at structure, cellularity, metachromasia, and subchondral bone remodeling including tidemark integrity. In contrast to the Mankin score, the Little score system distinguishes the cartilage state in a more detailed way, such that each component of structure, cellularity, metachromasia, and subchondral bone remodeling is better staged, showing also higher numbers to each component. There is even a separate compartment for the assessment of cloning. High cloning activity is clearly assessed as degenerative sign, which may be questioned since it may also be a sign for regeneration or repair attempts by the tissue. However, the score system allows to identify better details of the cartilage stage. As in the Mankin score, the system should be applied for individual compartments of the histology section separately to allow for a more accurate assessment.

3.3.2 Score Systems for Cartilage Repair

Scores for cartilage repair or regeneration differ from those for OA such that they concentrate on type of tissue repair, stage of cell differentiation, and quality of cartilage repair [6]. Cartilage surface and structure as well as remodeling of calcified cartilage zone and tidemark are also considered, as well as integration into the adjacent cartilage. The most commonly used are the O'Driscoll score and the ICRS I and ICRS II. The latter score was developed by an international panel of cartilage experts within the frame of the International Cartilage Regeneration and Joint Preservation Society (ICRS).

3.3.2.1 O'Driscoll

The O'Driscoll score (Table 5 in appendix, [15]) concentrates on the percent filling of the original defect and the reconstitution of the osteochondral junction. Cellular morphology as for type of tissue, cartilage structure as for surface regularity and integrity, percentage of cartilage filling, and freedom of degenerative signs of the adjacent cartilage are respected. As in previous OA scores, signs of cluster formation are considered to be degenerative, which may be debated. The score system has a relatively high intra- and interobserver reproducibility. Differences in matrix staining intensity and, thus, biochemical composition (PG, collagen) of the new repair tissue are not considered with this score system as a major

Table 4
Little cartilage score, as described by Little et al. [10]

Little score scale	
Structure	
Description	Grade
Normal	0
Slight surface irregularities	1
Moderate surface irregularities	2
Severe surface irregularities	3
Clefts into transitional zone (1/3 depth)	4
Clefts into radial zone (2/3 depth)	5
Clefts into calcified zone (full depth)	6
Fibrillation and/or loss to transitional zone (1/3 depth)	7
Fibrillation and/or loss to radial zone (2/3 depth)	8
Fibrillation and/or loss to calcified zone (full depth)	9
Fibrillation and/or loss to subchondral bone	10
Cellularity	
Description	Grade
Normal	0
Increase or slight decrease	1
Moderate decrease	2
Severe decrease	3
No cells	4
Cell cloning	
Description	Grade
Normal	0
Several doublets	1
Many doublets	2
Doublets and triplets	3
Multiple cell nests	4
Territorial toluidine blue	
Description	Grade
Normal	0
Increase or slight decrease in staining	1
Moderate decrease	2

(continued)

Table 4
(continued)

Little score scale	
Severe decrease	3
No staining	4
Interterritorial toluidine blue	
Description	Grade
Normal	0
Loss of staining in tangential zone (1/3 depth)	1
Loss of staining in transitional zone (2/3 depth)	2
Loss of staining in radial zone (full depth)	3
No staining	4
Tidemark/calcified cartilage/subchondral bone	
Description	Grade
Intact subchondral bone plate and single tidemark	0
Intact subchondral bone plate and multilayered tidemark	1
Blood vessels through subchondral bone to calcified cartilage	2
Tidemark crossed by blood vessels	3

Table 5
O'Driscoll score for evaluating cartilage defect filling, as described by O'Driscoll et al. [15]

O'Driscoll score scale	
Nature of the predominant tissue	
Cellular morphology	Grade
Hyaline articular cartilage	2
Incompletely differentiated mesenchyme	1
Fibrous tissue or bone	0
Safranin O staining of the matrix	Grade
Normal	3
Moderate	2
Slight	1
None	0
Structural changes	

(continued)

Table 5
(continued)

O'Driscoll score scale	
Surface regularity	Grade
Smooth and intact	3
Superficial horizontal lamination	2
Fissures 25–100% of the thickness	1
Severe disruption, including fibrillation	0
Structural integrity	Grade
Normal	2
Slight disruption, including cysts	1
Severe disintegration	0
Thickness	Grade
100% of normal adjacent cartilage	2
50–100% of normal cartilage	1
0–50% of normal cartilage	0
Bonding to adjacent cartilage	Grade
Bonded at both ends	2
Bonded at one end or partially at both ends	1
Not bonded	0
Freedom from cellular changes of degeneration	
Hypocellularity	Grade
Normal cellularity	3
Slight hypocellularity	2
Moderate hypocellularity	1
Severe hypocellularity	0
Chondrocyte clustering	Grade
No clusters	2
<25% of the cells	1
25–100% of the cells	0
Freedom from degenerative changes in adjacent cartilage	Grade
Normal cellularity, no clusters, normal staining	3
Normal cellularity, mild clusters, moderate staining	2
Mild of moderate hypocellularity, slight staining	1
Severe hypocellularity, poor or no staining	0

disadvantage. Also, the differentiation possibility of structural integrity with normal surface and slight and severe disruption is low.

To have unity there, our laboratory reversed the scale (*see* Table 6 in appendix). Originally, high numbers are "good" scores, and low numbers are "bad" scores, which is the opposite of the Mankin score system, where high scores are "bad" scores.

3.3.2.2 ICRS I and II

The ICRS score systems I (Table 7 in appendix) and II (Table 8 in appendix) [16, 17] concentrate also on the repair of cartilage lesions. Scores are based on surface roughness, nature of repair tissue, cell distribution, viability of cell population, subchondral bone remodeling, and cartilage mineralization. As for the matrix composition, no staining intensity resp. biochemical evaluation of PG content is integrated in this score system. Instead, typification of matrix type, such as hyaline, fibrocartilage, fibrous tissue, or mixtures thereof, is scored, which renders high inter-reader variability in these criteria. The exclusion of the evaluation of the integration to the adjacent cartilage matrix is a second major disadvantage of the ICRS score system I.

The ICRS score system II addresses this critique such that more emphasis on the biological significance of the repair tissue is placed. Instead of semiquantitative scores, numbers reflect the percentage of the repair tissue. Vascularization, basal integration of repaired tissue, formation of a new tidemark, and inflammation are also considered. Interobserver variability seems to be lower, compared to the ICRS I score system. However, out of our own experience, giving percentages of the various parameters is not easy and is very much dependent on the expertise of the observers. Therefore, the O'Driscoll system is more reproducible.

3.3.2.3 MSRU Subchondral Bone Cyst Score

As in cartilage repair and regeneration, the development of subchondral bone cysts is a common feature deciding the success of the repair; therefore the MSRU has developed its own score system for this particular aspect (Table 9 in appendix). The type of cyst, its connection to the synovial space, and type of tissue for cyst filling (fibrous tissue, fluid, etc.) are graded. These parameters may decide whether the cartilage repair tissue may hold its integrity or whether it collapses into the cyst leaving a greater defect filled with only fibrous tissue or even left empty forming a crater in the original defect area. Subchondral cysts appear within the first 6 months of repair in animals and also humans, but may later disappear and subchondral bone formation may take place if collapse of grafts can be avoided.

Table 6
Modified O'Driscoll score for evaluating cartilage defect fillings, with low scores indicating more physiological findings and higher scores indicating abnormalities

Modified O'Driscoll score scale (MSRU)	
Nature of the predominant tissue	
Cellular morphology in defect filling	Grade
Normal hyaline articular cartilage/columnar structures	0
Hyaline-like cartilage cells	1
Incompletely differentiated mesenchyme	2
Flattened/elongated cells	3
Fibrous tissue/no filling/no cells	4
Safranin O staining of the filling matrix	Grade
Normal	0
Moderate	1
Slight	2
None	3
Structural changes	
Surface coverage (%)	Grade
100% coverage	0
>75%	1
75–25%	2
Severe disruption (<25%)	3
Structural integrity of tidemark/subchondral bone plate	Grade
Intact tidemark under defect filling	0
Sightly disrupted	1
Severely disrupted/cyst formation	2
Thickness	Grade
75–100% of normal adjacent cartilage	0
50–75% of normal adjacent cartilage	1
25–50% of normal adjacent cartilage	2
0–25% of normal adjacent cartilage	3
Bonding of new tissue to adjacent cartilage	Grade
Fully bonded	0
Partially bonded	1
Not bonded	2
Cellularity of defect filling	

(continued)

Table 6
(continued)

Modified O'Driscoll score scale (MSRU)	
Cellularity of defect filling	Grade
Normal/increased	0
Slight hypocellularity	1
Moderate hypocellularity	2
Severe hypocellularity	3
Chondrocyte clustering	Grade
No clusters	2
<25% of the cells	1
25–100% of the cells	0

Table 7
ICRS I score scale, as described by Mainil-Varlet et al. [16]

ICRS I score scale	
Features	**Grade**
I. Surface	
Smooth/continuous	3
Discontinuities/irregularities	0
II. Matrix	
Hyaline	3
Mixture: hyaline/fibrocartilage	2
Fibrocartilage	1
Fibrous tissue	0
III. Cell distribution	
Columnar	3
Mixed/columnar clusters	2
Clusters	1
Individual cells/disorganized	0
IV. Cell population viability	
Predominantly viable	3
Partially viable	1
<10% viable	0

(continued)

Table 7
(continued)

ICRS I score scale	
Features	**Grade**
V. Subchondral bone	
Normal	3
Increased remodeling	2
Bone necrosis/granulation tissue	1
Detached/fracture/callus at base	0
VI. Cartilage mineralization (calcified cartilage)	
Normal	3
Abnormal/inappropriate location	0

Table 8
ICRS II score scale, as described by [17]

ICRS II score scale	
Histological parameters	**Score**
1. Tissue morphology (viewed under polarized light)	0%: Full-thickness collagen fibers 100%: Normal cartilage birefringence
2. Matrix staining (metachromasia)	0%: No staining 100%: Full metachromasia
3. Cell morphology	0%: No round/oval cells 100%: Mostly round/oval cells
4. Chondrocyte clustering (four or more grouped cells)	0%: Present 100%: Absent
5. Surface architecture	0%: Delamination or major irregularity 100%: Smooth surface
6. Basal integration	0%: No integration 100%: Complete integration
7. Formation of a tidemark	0%: No calcification front 100%: Tidemark
8. Subchondral bone abnormalities/marrow fibrosis	0%: Abnormal 100%: Normal marrow
9. Inflammation	0%: Present 100%: Absent
10. Abnormal calcification/ossification	0%: Present 100%: Absent

(continued)

Table 8
(continued)

ICRS II score scale	
Histological parameters	**Score**
11. Vascularization (within the repaired tissue)	0%: Present 100%: Absent
12. Surface/superficial assessment	0%: Total loss or complete disruption 100%: Resembles intact articular cartilage
13. Mid/deep zone assessment	0%: Fibrous tissue 100%: Normal hyaline cartilage
14. Overall assessment	0%: Bad (fibrous tissue) 100%: Good (hyaline cartilage)

Table 9
Subchondral bone cyst score, designed by the Musculoskeletal Research Unit (MSRU) at the University of Zürich

MSRU subchondral bone cyst score	
Size	*Grade*
Shallow	1
Mid-size (depth and width equal to defect size)	2
Deep	3
Isolation	*Grade*
Complete isolation from synovial space by cartilage/bone	0
Isolation by fibrous tissue	1
Cyst in contact with synovial space	2
Filling	*Grade*
Completely/predominantly filled with bone	0
Mix of bone/cartilage-like tissue	1
Mix of fibrotic/bone/cartilage-like tissue	2
Predominantly fibrotic tissue	3
Predominantly fluid	4
Bone tissue activity (remodeling)	*Grade*
More than 50% of cyst	0
25–50%	1
Less than 25%	2
None	3

(continued)

Table 9
(continued)

MSRU subchondral bone cyst score	
Non-fibrotic soft tissue activity	*Grade*
More than 50% of cyst	0
25–50%	1
Less than 25%	2
None	3

4 Notes

1. Samples can therefore be radiographed (e.g., with a Faxitron unit to show high details) at this stage to better estimate the depth of subchondral bone disruptions.
2. The standard fixation solution to be used in most cases is 4% buffered formalin. If the sample contains fluorochromes such as cell tracking fluorochromes or fluorochromes used to track bone deposition kinetics, a 40% ethanol solution is used for fixation instead of formalin, as it preserves the fluorescent signal. In this case, the sample processing must take into account fluorescent bleaching and reduce exposure to light as much as possible (e.g., by wrapping the sample containers with foil).
3. Ethanol-fixed samples can be directly moved to a 50% ethanol solution, skipping the washing step in order to not disrupt the fluorochromes in the sample.
4. Caution must be taken when handling the PMMA solution at room temperature as the polymerization process is exothermic. For safety reasons it is therefore recommended to work under a chemical hood (eosinophilic reaction of lung tissue may occur when chronically exposed to these fumes).
5. It is important to note that the saw blade itself will cut an additional 300–600 microns of tissue (depending on the saw size) which will be lost during the cutting process. It is therefore useful to account for 700–1300 microns for each thick section produced, especially if regions of interest are small and multiple sections are needed.
6. Weigert's hematoxylin may be used as a counterstain to highlight nuclei blue and osteoids pale pink.
7. This staining method has also been used to quantitatively assess glycosaminoglycan content by evaluating the intensity of red staining in safranin O-stained cartilage sections.

8. Indeed, as the subchondral bone plate is made of dense calcified tissue, the decalcification process would take too long if deeper bone layers were retained. The trade-off, in this case, is that deeper subchondral bone cysts cannot be assessed histologically in the same block. It is therefore recommended, in case of deep cyst formation (deeper than ca. 5 mm), to split the osteochondral tissue in two parts: one containing the osteochondral unit extending to about 5 mm below the calcified cartilage layer and another containing only the deeper bone.
9. Normally, for small samples, fixation may be adequately achieved within 1 h, but in the case of osteochondral tissue, as the tissue architecture is very dense, especially at the interface between cartilage and bone, it is recommended to keep samples in the fixation solution for 24 h at least. It is recommended to place samples in their fixation solutions on a shaker to allow better infiltration.
10. An alternative to EDTA would be to use the RDF™ mild decalcifier. This formic acid-based mild decalcifying solution allows for faster decalcification of osteochondral samples. However, although labeled "mild" compared to its harsher counterpart, RDC™, which is a hydrochloric acid-based solution, RDF™ could still damage some sensitive epitopes for immunohistochemistry. Testing both the EDTA and RDF methods side by side when establishing an immunohistochemical detection protocol is important. Optimizing the process requires balancing on one hand an adequate detection of the epitope of interest and on the other hand a shortened time of decalcification. Note that samples decalcified with RDF™ must be placed in glass containers and not in plastic cassettes or gauze.
11. A needle can be used to probe the samples to verify if the decalcification is sufficient, prior to further processing. As a rule, if the needle can go through the tissue without much resistance, so will the blade of the microtome. It is recommended to probe an area away from the region of interest to not disrupt the structures to be evaluated.
12. For large osteochondral samples, EDTA decalcification can take months, whereas RDF™ decalcification can take weeks.
13. For routine decalcified osteochondral tissue processing, these steps are performed with the pressure (infiltration pressure at +35 kPa) and vacuum (impregnation and filling vacuum at −70 kPa) function turn on in order to facilitate the homogenous infiltration of solutions (ethanol, xylene, paraffin) into the dense osteochondral tissue matrix.

14. Unlike with native tissues, in the case of tissue engineered cartilage, or osteochondral tissue with an immature defect matrix filling, the pressure/vacuum function can cause glycosaminoglycans loosely attached to an immature cartilage matrix to leach out during processing. An analysis of such samples would then underestimate the amount of glycosaminoglycans in the extracellular matrix of an engineered osteochondral tissue or immature cartilage defect filling. It is therefore recommended to modify the standard program on a histoprocessor by turning the pressure/vacuum function off when dehydrating and infiltrating such immature tissue samples.

Acknowledgments

The authors thank Aymone Lenisa for the helpful discussion and her technical insights.

Appendix: Score Scales for Cartilage Evaluation

References

1. Schmitz N, Laverty S, Kraus VB et al (2010) Basic methods in histopathology of joint tissues. Osteoarthr Cartil 18(Suppl 3):S113–S116. https://doi.org/10.1016/j.joca.2010.05.026
2. von der Mark K, Kirsch T, Nerlich A et al (1992) Type X collagen synthesis in human osteoarthritic cartilage. Indication of chondrocyte hypertrophy. Arthritis Rheum 35(7):806–811. https://doi.org/10.1002/art.1780350715
3. Fuerst M, Bertrand J, Lammers L et al (2009) Calcification of articular cartilage in human osteoarthritis. Arthritis Rheum 60(9):2694–2703. https://doi.org/10.1002/art.24774
4. Zhang Y, Wang F, Tan H et al (2012) Analysis of the mineral composition of the human calcified cartilage zone. Int J Med Sci 9(5):353–360. https://doi.org/10.7150/ijms.4276
5. Goldstein DJ, Horobin RW (1974) Surface staining of cartilage by Alcian blue, with reference to the role of microscopic dye aggregates in histological staining. Histochem J 6(2):175–184. https://doi.org/10.1007/BF01011805
6. Rutgers M, van Pelt MJ, Dhert WJ et al (2010) Evaluation of histological scoring systems for tissue-engineered, repaired and osteoarthritic cartilage. Osteoarthr Cartil 18(1):12–23. https://doi.org/10.1016/j.joca.2009.08.009
7. Moskowitz RW (2006) Osteoarthritis cartilage histopathology: grading and staging. Osteoarthr Cartil/OARS Osteoarthr Res Soc 14(1):1–2. https://doi.org/10.1016/j.joca.2005.08.015
8. Cook JL, Kuroki K, Visco D et al (2010) The OARSI histopathology initiative - recommendations for histological assessments of osteoarthritis in the dog. Osteoarthr Cartil/OARS Osteoarthr Res Soc 18(Suppl 3):S66–S79. https://doi.org/10.1016/j.joca.2010.04.017
9. Custers RJ, Creemers LB, Verbout AJ et al (2007) Reliability, reproducibility and variability of the traditional Histologic/Histochemical Grading System vs the new OARSI Osteoarthritis Cartilage Histopathology Assessment System. Osteoarthr Cartil/OARS Osteoarthr Res Soc 15(11):1241–1248. https://doi.org/10.1016/j.joca.2007.04.017
10. Little C, Smith S, Ghosh P et al (1997) Histomorphological and immunohistochemical evaluation of joint changes in a model of

osteoarthritis induced by lateral meniscectomy in sheep. J Rheumatol 24(11):2199–2209

11. Little CB, Smith MM, Cake MA et al (2010) The OARSI histopathology initiative - recommendations for histological assessments of osteoarthritis in sheep and goats. Osteoarthr Cartil/OARS Osteoarthr Res Soc 18(Suppl 3):S80–S92. https://doi.org/10.1016/j.joca.2010.04.016
12. Pritzker KP, Gay S, Jimenez SA et al (2006) Osteoarthritis cartilage histopathology: grading and staging. Osteoarthr Cartil/OARS Osteoarthr Res Soc 14(1):13–29. https://doi.org/10.1016/j.joca.2005.07.014
13. Mankin HJ, Dorfman H, Lippiello L et al (1971) Biochemical and metabolic abnormalities in articular cartilage from osteo-arthritic human hips. II. Correlation of morphology with biochemical and metabolic data. J Bone Joint Surg Am 53(3):523–537
14. Aigner T (2012) Osteoarthritis histopathology grading criteria - a never ending story? Osteoarthr Cartil/OARS Osteoarthr Res Soc 20(6):469–470. https://doi.org/10.1016/j.joca.2012.02.008
15. O'Driscoll SW, Keeley FW, Salter RB (1988) Durability of regenerated articular cartilage produced by free autogenous periosteal grafts in major full-thickness defects in joint surfaces under the influence of continuous passive motion. A follow-up report at one year. J Bone Joint Surg Am 70(4):595–606
16. Mainil-Varlet P, Aigner T, Brittberg M et al (2003) Histological assessment of cartilage repair: a report by the Histology Endpoint Committee of the International Cartilage Repair Society (ICRS). J Bone Joint Surg Am 85-A(Suppl 2):45–57
17. Mainil-Varlet P, Van Damme B, Nesic D et al (2010) A new histology scoring system for the assessment of the quality of human cartilage repair: ICRS II. Am J Sports Med 38(5):880–890. https://doi.org/10.1177/0363546509359068

Chapter 18

Articular Cartilage Imaging in the Context of the Superficial Chondrocyte Spatial Organization (SCSO) as a Surrogate Marker for Functional Pathology

Bodo Kurz, Thomas Lange, Marita Voelker, and Bernd Rolauffs

Abstract

Articular cartilage imaging has undergone tremendous changes and nowadays enables functionally relevant quantitative information useful for detecting the early, preclinical state of articular cartilage degeneration as seen in primary or post-traumatic osteoarthritis (OA/PTOA) to be obtained. In this context, we describe the necessary steps for articular cartilage imaging with the goal to utilize the superficial chondrocyte spatial organization (SCSO) as a score that is responsive to its environment and dynamically changes during the lifetime of an individual and that can be used as a surrogate marker for loss of articular cartilage surface stiffness on the nanoscale.

Key words Articular cartilage, Imaging, AI, A.I., Optical biopsy

1 Introduction

One goal of modern articular cartilage imaging is to visualize structure-function relationships in the context of tissue or joint health and early or generalizing degeneration. In this context, novel imaging methods or data analysis tools that generate functional quantitative information would be helpful for developing preventive and/or early therapeutic interventions that are not yet in reach. Our contribution to this field is the development of a future-oriented, machine learning-supported, nondestructive quantitative optical biopsy for early disease detection. This optical biopsy visualizes and interprets the superficial chondrocyte spatial organization (SCSO) as a digital tissue architecture fingerprint of tissue ultrastructure and uses the association(s) of the SCSO with functional pathology [1]. For example, the SCSO changes dynamically during the lifetime of a hypothetical individual [2], which illustrated that the SCSO is responsive to its environment. Moreover, we have demonstrated SCSO-dependent nanoscale stiffness

Martin J. Stoddart et al. (eds.), *Cartilage Tissue Engineering*,
Methods in Molecular Biology, vol. 2598, https://doi.org/10.1007/978-1-0716-2839-3_18,

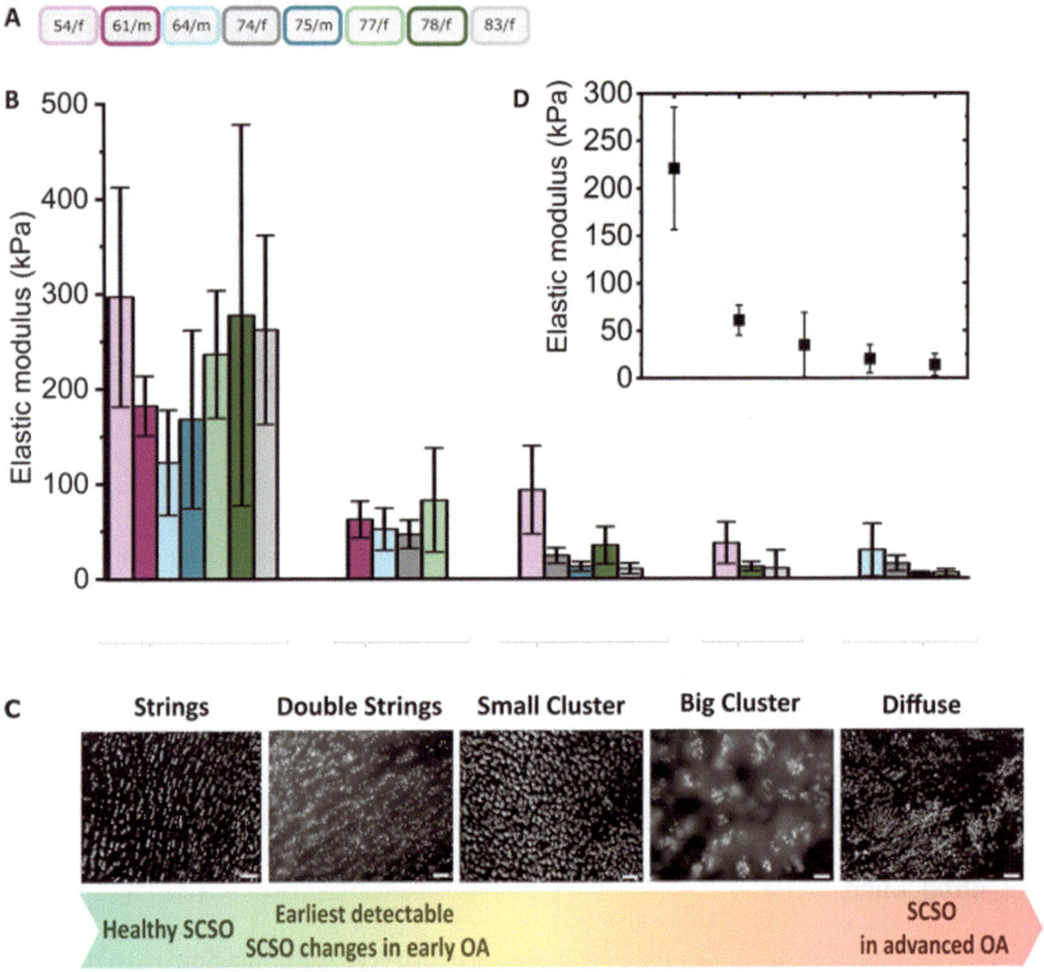

Fig. 1 Correlation of fluorescence and AFM-based data. (**a**) Color-coded patient age and gender. (**b**) The elastic moduli of eight OA patients. Each bar represents the mean elastic modulus of the articular surface of an AC disc (three regions of interest (ROI) per disc, 256 force vs. indentation curves per ROI). The respective error bars indicate the standard deviation. (**c**) The corresponding predominant SCSO of each AC disc. Given are fluorescence images of DAPI-stained nuclei of superficial zone chondrocytes displaying distinct stages of healthy (strings) and OA-related SCSOs. Scale bars: 100 μm. (**d**) The patient-averaged elastic modulus. AFM indentation data showed a strong correlation with the SCSO ($r_{rm} = -0.91$, 95%CI: −0.97, −0.73, $p < 0.000002$ for linearly transformed disc averaged data elastic modulus vs. SCSO data). The elastic moduli differed between the string SCSO and all other SCSO stages ($P \leq 0.001$). (This figure is a reprint of Fig. 2 published in [1]. For this reprint permission has been obtained)

changes of the articular cartilage surface (Fig. 1), which established the SCSO as a surrogate marker of loss of articular cartilage surface stiffness [1]. Thus, we have used the SCSO as a nondestructively visualizable surrogate marker for not directly visualizable functional pathology. The present chapter describes SCSO imaging and classification procedures and the testing for potential correlations and/or differences between SCSO stages and ones choice of tissue pathology analysis.

2 Materials

1. Standard medium: DMEM low glucose, Sigma Aldrich, D5546; 4% PenStrep, Thermo Fisher Scientific, 15140-122; 1% amphotericin B, Sigma Aldrich, A2612).
2. Nucleus stain: DAPI (Sigma Aldrich, D8417-5 mg).
3. Live cell imaging: Hoechst 33342 staining dye solution (ThermoFisher, H1399).
4. Fluorescence microscopy of the SCSO: for example, an Axio Observer A1, Zeiss, Germany.
5. Fluorescence microscopy of the SCSO together with nanomechanical analyses: MFP-3D AFM (Asylum Research, an Oxford Instruments Company, USA) with the AFM head mounted onto the fluorescence microscope platform (*see* **Notes 1–3**).
6. Statistical analysis software: for example, SigmaStat 12.5 (Systat Software, USA), which has the advantage that a statistical advisor assesses the validity of the chosen test, avoiding potential statistical errors of inexperienced users.

3 Methods

1. Obtain articular cartilage with the approval of the local research ethics committee and with the informed consent from OA patients, e.g., during total joint arthroplasty, from healthy organ/tissue donors, or from animal joints from the local slaughterhouse. Avoid drying out by placing the articular cartilage in the operating/dissecting room as soon as possible into standard medium (see above), and immediately transport the sample to the laboratory.
2. Keep the sample(s) at 4 °C for same or next day processing.
3. Process the sample(s) into the desired geometry, e.g., into discs with a specific diameter and thickness, and retain the articular surface.
4. Stain the sample(s) with a nucleus stain, e.g., with DAPI (1 h RT, Sigma Aldrich, D8417) or a live cell nuclear stain Hoechst 33342 (1 h 37 °C, Thermo Fisher Scientific, H1399) according to the company's protocol.
5. Image each disc's articular surface in a top-down view at 10×. Classify the SCSO state according to the sample's *predominant* SCSO type within a given sample [1–4]:

   (i) A healthy SCSO, for example, chondrocyte strings, in addition to a few clusters, pairs, or single chondrocytes, as observed in the intact cartilage of joints with a rotational component such as the human knee joint [5]. For more details on the type and frequency of healthy SCSO patterns in various large human joints, please see Fig. 3 in reference [5].

(ii) Double strings as observed in the macroscopically intact cartilage of joints with focal OA [6]. Those double strings are the earliest identifiable OA-induced changes in the SCSO, according to our model that summarizes the SCSO changes that occur during the lifetime of a hypothetical individual [2].

(iii) Small and big clusters [2].

(iv) The absence of any discernible organization based on randomly aggregating cells within OA lesions, termed "diffuse" arrangement [4].

6. This classification of the SCSO might be sufficient for correlation analyses with other scores such as the OARSI scale [7, 8] or the 5-point scale established by Collins-Muehleman [30], biochemical read-outs such as the glycosaminoglycan (GAG) content in a tissue trauma context [9, 10] using the dimethylmethylene blue dye binding assay [11], or any quantitative read-outs on matched samples that are established in the laboratory.
7. For testing for correlations between the SCSO and the to-be-tested pathology, numerically code the predominant SCSO of each individually assessed tissue sample, e.g., predominantly strings as 1, predominantly double strings as 2, predominantly small clusters as 3, predominantly big clusters as 4, and the absence of any organizational patterns as 5.
8. For correlation analyses, we suggest considering the Spearman rank-order correlation test because of the usage of classified SCSO data or a repeated measures correlation test, which would potentially require data transformation to achieve dataset linearity.
9. To test for significant differences between different SCSO stages across patients/animals/joints, we suggest one-way repeated measures ANOVA with patients/animals/joints as subjects and the numerically coded SCSO as level/treatment and the Holm-Sidak post hoc test.
10. For generating quantitative SCSO data as demonstrated in [1], one can define the SCSO patterns as spatial point patterns, as described previously [1–4, 12].

4 Notes

1. For measuring nanoscale stiffness changes of the articular surface in relation to SCSO-classified discs, the AFM head needs to be mounted onto a fluorescence microscope platform to exactly correlate cantilever tip position, force data, and local SCSO [13].
2. We recommend recording multiple three force maps per region of interest with each recorded force map consisting of a larger number of individual measurement points to utilize high-resolution force mapping to account for articular cartilage local inhomogeneity.
3. The chosen AFM indentation depth should be small, compared to the sample thickness.

References

1. Tschaikowsky M, Selig M, Brander S et al (2021) Proof-of-concept for the detection of early osteoarthritis pathology by clinically applicable endomicroscopy and quantitative AI-supported optical biopsy. Osteoarthr Cartil 29(2):269–279. https://doi.org/10.1016/j.joca.2020.10.003
2. Felka T, Rothdiener M, Bast S et al (2016) Loss of spatial organization and destruction of the pericellular matrix in early osteoarthritis in vivo and in a novel in vitro methodology. Osteoarthr Cartil 24(7):1200–1209. https://doi.org/10.1016/j.joca.2016.02.001
3. Aicher WK, Rolauffs B (2014) The spatial organisation of joint surface chondrocytes: review of its potential roles in tissue functioning, disease and early, preclinical diagnosis of osteoarthritis. Ann Rheum Dis 73(4):645–653. https://doi.org/10.1136/annrheumdis-2013-204308
4. Rolauffs B, Rothdiener M, Bahrs C et al (2011) Onset of preclinical osteoarthritis: the angular spatial organization permits early diagnosis. Arthritis Rheum 63(6):1637–1647. https://doi.org/10.1002/art.30217
5. Rolauffs B, Williams JM, Grodzinsky AJ et al (2008) Distinct horizontal patterns in the spatial organization of superficial zone chondrocytes of human joints. J Struct Biol 162(2):335–344. https://doi.org/10.1016/j.jsb.2008.01.010
6. Rolauffs B, Williams JM, Aurich M et al (2010) Proliferative remodeling of the spatial organization of human superficial chondrocytes distant from focal early osteoarthritis. Arthritis Rheum 62(2):489–498. https://doi.org/10.1002/art.27217
7. Custers RJ, Creemers LB, Verbout AJ et al (2007) Reliability, reproducibility and variability of the traditional histologic/histochemical grading system vs the new OARSI osteoarthritis cartilage histopathology assessment system. Osteoarthr Cartil 15(11):1241–1248. https://doi.org/10.1016/j.joca.2007.04.017
8. Waldstein W, Perino G, Gilbert SL et al (2016) OARSI osteoarthritis cartilage histopathology assessment system: a biomechanical evaluation in the human knee. J Orthop Res 34(1):135–140. https://doi.org/10.1002/jor.23010
9. Rolauffs B, Kurz B, Felka T et al (2013) Stress-vs-time signals allow the prediction of structurally catastrophic events during fracturing of immature cartilage and predetermine the biomechanical, biochemical, and structural impairment. J Struct Biol 183(3):501–511. https://doi.org/10.1016/j.jsb.2013.06.011
10. Rolauffs B, Muehleman C, Li J et al (2010) Vulnerability of the superficial zone of immature articular cartilage to compressive injury. Arthritis Rheum 62(10):3016–3027. https://doi.org/10.1002/art.27610

11. Farndale RW, Buttle DJ, Barrett AJ (1986) Improved quantitation and discrimination of sulphated glycosaminoglycans by use of dimethylmethylene blue. Biochim Biophys Acta 883(2):173–177
12. Meinhardt M, Luck S, Martin P et al (2012) Modeling chondrocyte patterns by elliptical cluster processes. J Struct Biol 177(2): 447–458. https://doi.org/10.1016/j.jsb.2011.11.023
13. Tschaikowsky M, Brander S, Barth V et al (2022) The articular cartilage surface is impaired by a loss of thick collagen fibers and formation of type I collagen in early osteoarthritis. Acta Biomater 146:274–283. https://doi.org/10.1016/j.actbio.2022.04.036

Chapter 19

Unconfined Compression Experimental Protocol for Cartilage Explants and Hydrogel Constructs: From Sample Preparation to Mechanical Characterization

Seyed Ali Elahi, Rocío Castro-Viñuelas, Anke Govaerts, Rik Lories, Nele Famaey, and Ilse Jonkers

Abstract

Mechanical characterization of articular cartilage and cell-seeded hydrogel constructs is a challenging task due to the complex biphasic behavior of these materials. Here we describe a step-by-step unconfined compression testing protocol for inverse mechanical characterization of these materials from sample preparation to parameter identification. Examples from our ongoing experiments on alginate hydrogel constructs and preserved and damaged cartilage explants obtained from human hip samples are presented.

Key words Articular cartilage, Hydrogel construct, Mechanical characterization, Unconfined compression, Mechanical testing, Experimental protocol

1 Introduction

Articular cartilage plays a key mechanical role during movement of diarthrodial joints by allowing almost frictionless motion between the articular surfaces, minimizing the contact stresses by distributing articulation loads over a larger contact area and dissipating the energy associated with the load [1]. Articular cartilage degeneration is the hallmark of osteoarthritis (OA), with associated changes in the subchondral bone, synovium, bone marrow, and ligaments, positioning it as a whole joint disease [2]. Tissue engineering tools provide a potential solution for repairing cartilage defects by implantation of chondrocyte-seeded scaffolds, such as hydrogel constructs, within the defect [3, 4]. Furthermore, cell-seeded models are often used to elucidate the role of molecular pathways—in particular those relevant to biological cell function in response to

Nele Famaey and Ilse Jonkers have contributed equally to this work and share senior authorship.

Martin J. Stoddart et al. (eds.), *Cartilage Tissue Engineering*,
Methods in Molecular Biology, vol. 2598, https://doi.org/10.1007/978-1-0716-2839-3_19,

mechanical loading. Given the mechanical function of articular cartilage, any tissue engineering treatment aims at restoring not only the biochemical composition but also the mechanical properties of the tissue. Therefore, mechanical characterization of cartilage and cell-seeded hydrogel constructs is an absolute necessity to assess the quality of the engineered tissue [5], evaluate the changes in tissue mechanics due to disease onset and progression [6], and calibrate in silico models for optimization of the treatment methods [7].

The complex anisotropic and heterogeneous structure of articular cartilage makes it highly challenging to characterize its mechanical behavior [8, 9]. Inverse mechanical characterization methods are typically used to identify the properties of such complex soft tissues by answering three interrelated questions [10]: (i) How does the tissue behave when subjected to mechanical loading? (ii) Which mathematical model can simulate the behavior of tissue under mechanical loading? (iii) What are the parameters describing the mechanical properties (i.e., constants of the mathematical model)? To answer the question, (i) a mechanical loading experiment is required to apply a known/specific displacement to a sample and measure the reaction force (or vice versa). Question (ii) can be answered by simulating the observed mechanical behavior in the experiment using an analytical or numerical (such as finite element) solution. Finally, the mechanical properties of the material can be obtained by answering the question (iii) through minimizing the difference between the mathematical simulations of step (ii) and the tissue behavior observed in step (i).

A variety of mechanical testing methods have been used by different research groups for in vitro mechanical characterization of cartilage and hydrogel constructs [11–17]. Lack of a standardized mechanical test protocol highlights the necessity of introducing a unified simple experimental protocol that is applicable in most cartilage tissue engineering laboratories and that simplifies the mathematical modeling for inverse characterization. Therefore, this chapter aims to provide a practical testing protocol for in vitro mechanical characterization of cartilage explants and hydrogel constructs using unconfined compression. An unconfined compression testing setup is chosen due to its simplicity in comparison with the other prevalent testing setups (e.g., indentation and confined compression) and as it allows lateral deformation of the sample to be measured using optical techniques, in addition to the measured reaction force in the experiments. Obtaining this extra experimental output will increase the accuracy of mechanical property estimation.

2 Materials

1. 0.9% NaCl solution in Milli-Q water.
2. 102 mM $CaCl_2$ in Milli-Q water.
3. 0.22 μm pore size filters.
4. Alginic acid sodium salt from brown algae (A0682, Sigma).
5. Phosphate-buffered saline (PBS).
6. PBS supplemented with 1% antibiotic/antimycotics (15240062, Gibco).
7. Sterile scalpel.
8. Surgical clamps.
9. Sterile tweezers.
10. Petri dish.
11. 8 mm diameter biopsy punch.
12. DMEM/F12.
13. Fetal bovine serum.
14. Antibiotics/antimycotics (15240062, Gibco)
15. L-Glutamine (25030081, Gibco)

3 Methods

3.1 Hydrogel Materials and Preparation

1. Prior to hydrogel making, prepare the following solutions: (i) 1 L of 0.9% NaCl solution in Milli-Q water and (ii) 500 mL 102 mM $CaCl_2$ in Milli-Q water. The final volume of each solution can be adjusted depending on the number of hydrogels to be made. Solutions can be prepared in advance and stored at room temperature. Check that precipitates are not formed before using the solutions. In case sterility is needed, solutions can be autoclaved or filtered with a 0.22 μm pore size filter.
2. Prepare 2.2% alginate solution (*see* **Note 1**). Since the use of freshly made alginate solution is recommended, the final volume of this solution must be adjusted depending on the number of hydrogels needed. In this protocol, 0.5 mL of alginate solution are used per hydrogel. Weigh the desired amount of alginate powder, and dissolve it in the adequate amount of 0.9% NaCl (*see* **Note 2**). In case sterility is needed, the alginate solution can be autoclaved for 20 min at 120 °C.
3. Prepare the molds for hydrogel casting. We use cylindrical molds with 8 mm diameter and height for our hydrogels (Fig. 1a). Put filter paper in the bottom of the mold. Then, cut small pieces of the filter paper (10 mm high × 30 mm long approximately), fold them in a cylinder, and put in the holes of

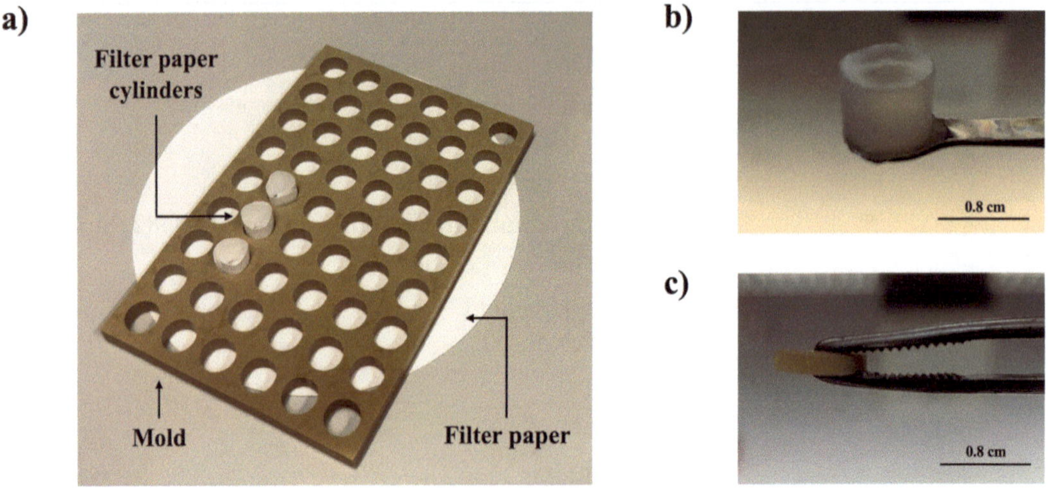

Fig. 1 (**a**) Picture showing the custom-made mold as well as the filter papers used for alginate hydrogel preparation. (**b**) Picture showing a representative alginate hydrogel. (**c**) Picture showing a representative cartilage explant in lateral view

the mold (*see* **Note 3**). These filter papers are helpful to remove the samples from the mold. In case sterility is needed, take the mold with the papers and place inside the cell culture hood. Turn on the UV light for at least 15 min.

4. To prepare the alginate hydrogels (Fig.1b), use a 10–25 mL pipette to cover the filter paper with 102 mM $CaCl_2$ until all filter papers are completely soaked (*see* **Note 4**). With a 1 mL micropipette, pipet 0.5 mL of the alginate solution into each well of the mold, without making any air bubbles (*see* **Note 5**). Polymerize the hydrogels for 30 min at room temperature. Carefully collect the hydrogels with sterile tweezers and place them in a petri dish filled with 102 mM $CaCl_2$ and let them cross-link for another 30 min at room temperature. Then, collect the hydrogels with a sterile spoon and place them in a 12-well plate containing phosphate-buffered saline (PBS) (*see* **Note 6** in case cell-seeded alginate hydrogels are required). We prepared five alginate hydrogel constructs for demonstration tests (*see* Subheading 3.7).

3.2 Cartilage Explant Preparation

In this protocol, cartilage explants are obtained from preserved and damaged areas of human femoral heads (hips) of OA patients undergoing hip arthroplasty surgery. Fresh human material should be handled as Biosafety class 2.

Place the femoral head in the petri dish and firmly grab it with the clamps. Peel off the cartilage by cutting with a scalpel parallel to and as close to the subchondral bone as possible. It is important to not take the mineralized cartilage, and scalpel off only the superficial layer. Take the cartilage pieces with the tweezers and place them in a petri dish containing PBS to avoid drying of the tissue. Once enough tissue has been harvested, cut small cylinders with a biopsy punch of the desired diameter (Fig. 1c) (*see* **Notes 7** and **8**). We used a biopsy punch with 8 mm diameter to prepare preserved and damaged areas of human femoral heads (hips). In total, five preserved and five damaged samples were prepared for demonstration tests (*see* Subheading 3.7). The thicknesses of the preserved and damaged samples were 1.20 ± 0.01 mm and 1.00 ± 0.01 mm, respectively.

3.3 Sample Preservation

If the samples are not immediately used for the mechanical testing, they can be stored using the following protocols:

- Hydrogels: these samples can be stored in a 50 mL tube or petri dish containing PBS at room temperature. In case of cell-seeded hydrogels, they can be kept inside an incubator at 37 °C in a humidified atmosphere with 5% CO_2 in the appropriate culture medium (*see* **Note 9**). To keep the hydrogel wet during the test, PBS or culture medium can be used depending on whether or not the testing is performed in a cell-seeded hydrogel.
- Cartilage explants: these samples can be kept inside an incubator at 37 °C in a humidified atmosphere with 5% CO_2 in the appropriate culture medium (*see* **Note 10**). Alternatively, cartilage explants can be stored at 4 °C in culture medium [18]. To keep the samples wet during the test, PBS or culture medium can be used (*see* Subheading 3.4).

3.4 Test Setup Preparation

1. For the unconfined compression test, a compression loading machine is required (Fig. 2). The machine should allow application of a specified compressive displacement to the sample by an actuator moving the upper or lower platen (*see* **Note 11**) and measuring the reaction force using a load sensor installed under the static platen if it is the lower platen or over the static platen, if it is the upper platen (*see* **Note 12**). The load sensor needs to be connected to a data acquisition system to sample the measured loading and store it to a computer unit.
2. Use a load sensor with a capacity corresponding to the sample stiffness. For this, a trial test can be helpful, in which a sample is compressed with the chosen loading protocol for the measurements (*see* Subheading 3.5), and the reaction force is measured. Always start with using a load sensor with a higher capacity and check the maximum of the measured reaction force. Then, select a proper load sensor based on the initial measurement of the maximum reaction force and capacity of the load sensor. Usually, the lower the capacity of the load sensor, the higher the accuracy of the force measurements. For instance, we used a load sensor with 2.5 N capacity for unconfined compression of our alginate hydrogel constructs (maximum measured load using the loading protocol in Subheading 3.5 was around

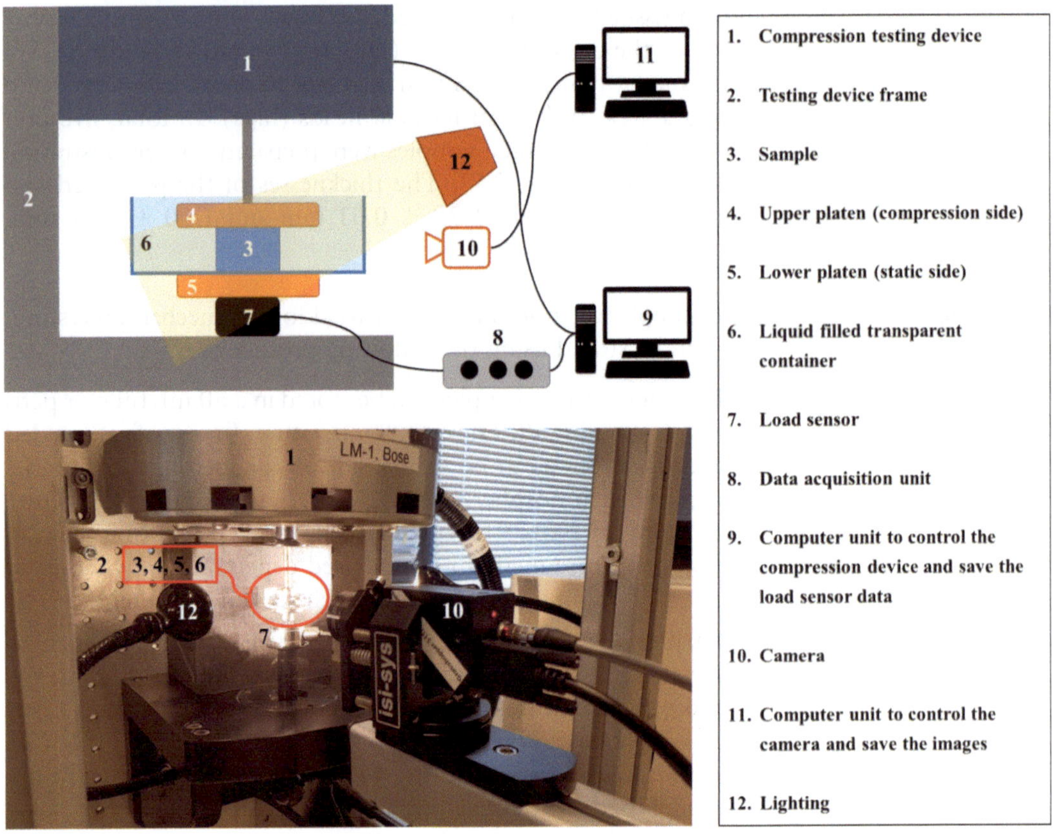

Fig. 2 Components of the unconfined compression testing setup

0.05 N) and a load sensor with 225 N capacity for our human cartilage explants (maximum measured load using the loading protocol in Subheading 3.5 was around 17 N).

3. Install two rigid nonporous platens on the static and the moving sides of the compression device (*see* Fig. 2 and **Notes 13** and **14**).
4. Use a transparent container with the proper dimensions and place the sample in the middle of the container (*see* Fig. 2 and **Notes 14** and **15**). Place the container on the lower platen and concentric with that.
5. Ensure contact between the upper platen and the sample. Slowly bring the loading platens close to each other until a compressive load is measured by the load sensor, and let the sample relax under the applied preload until no further change in the load is observed (*see* **Notes 16** and **17**).
6. Fill the container with a medium or liquid to the top of the upper platen (*see* **Note 18**). The samples should be kept hydrated during the test to avoid any change in mechanical properties that can consequently affect the measured reaction force.

3.5 Compression Loading Protocol

1. Measure the diameter and thickness of the sample accurately before the compression test. The sample dimensions are important to make the mathematical model of the experiments for inverse characterization (Subheading 3.7). Besides, accurate measurement of cartilage explant thickness and hydrogel construct height are crucial for applying a strain-controlled compression to the samples (detailed in the following steps). To this end, sample images with a length scale (*see* **Note 19**) or a laser measurement device can be used.
2. The choice of loading protocol in the experiments depends on the mechanical properties to be derived from the measurements. For example, if the stiffness change of the sample (e.g., during a treatment process) is of interest, a ramp loading (increasing compressive strain from zero to a defined strain in a given time) in an unconfined compression setup is adequate (*see* **Note 20**). However, to characterize material parameters related to the porous properties of the materials or collagen fibers' characteristics, different loading protocols are required (e.g., time-dependent loading protocols such as relaxation, creep, or dynamic loadings are required to characterize porous properties). In this chapter, we aim to propose two unconfined compression loading protocols for identifying the mechanical properties of hydrogels and cartilage using two commonly used material models: biphasic elastic material [19] (for use in both hydrogel and cartilage samples; *see* **Note 21** for more details about the material model and mechanical properties) and fibril-reinforced poroelastic (FRPE) material [6] (for use in cartilage samples, *see* **Note 21** for more details about the material model and mechanical properties).
3. Loading protocol to identify the material properties of the *biphasic elastic material*: the biphasic properties of the material cause it to have a time-dependent behavior that needs to be reflected in a time-dependent loading protocol. Based on our parameter sensitivity analysis results (reported in [20] for a porohyperelastic material but also likewise applicable for a biphasic material assuming the constant material properties, *see* **Note 21**), we recommend a compression loading until 20% of the sample thickness with the rate of 10%/s followed by 15 min of relaxation (*see* Fig. 3a and **Note 22**).
4. Loading protocol to identify the material properties of the *FRPE material*: since this material is more complex, a different loading protocol is required to identify the time-dependent properties related to both nonlinear biphasic and fibrillar matrix properties. Based on our parameter sensitivity analysis results (reported in [20]), we recommend a two-step compres-

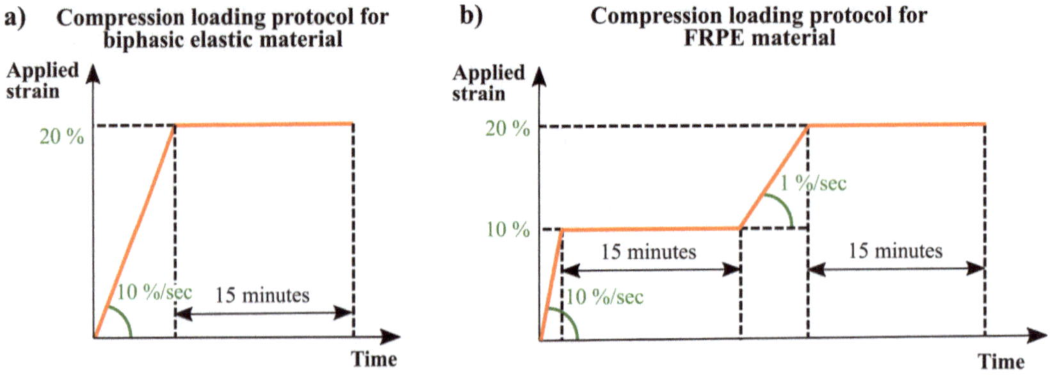

Fig. 3 The proposed unconfined compression loading protocols for (**a**) biphasic elastic material and (**b**) FRPE material. Strains are in percentage of the sample thickness

sion loading, first until 10% of the sample thickness at 10%/s followed by 15 min of relaxation, which is followed by 10% compression at 1%/s followed by 15 min of relaxation (Fig. 3b).

5. Select the appropriate loading protocol to be applied to the sample using the software associated with the loading device.

3.6 Experimental Data Acquisition and Testing

1. Reaction force: typically, the software associated with the compression loading device allows the sampling frequency for different parts of the loading protocol to be selected. Saving a high number of data points may increase the noise in the results, while saving a low number of data points may increase the risk of losing important data for mechanical characterization. Based on our experience we recommend the following optimized data acquisition frequencies for the ramp and relaxation parts of the loading protocols proposed in Subheading 3.5 (*see* **Note 23**):
   - Ramp loading: 100 data points.
   - Relaxation: 500 data points.
2. Optical edge detection (optional): the unconfined compression testing setup allows the deformation of the sample edges to be optically measured. Based on our parameter sensitivity analysis [20], these data can be used in addition to the measured reaction force for more accurate mechanical characterization of the samples (especially accurate identification of Poisson's ratio—*see* **Note 24**). However, collecting these data requires specific facilities and software that may not be available in every mechanical characterization laboratory. Therefore, we recommend it as an optional step to increase the accuracy of mechanical characterization. The steps to be followed for optical edge deformation measurement are:

- Provide a black background and proper lighting to increase the contrast between the sample and its background in the images.
- Use a camera capable of high-frequency imaging (minimum imaging frequency of 20 Hz, see the following steps for more details).
- Place the camera perpendicular and as close as possible to the sample (Fig. 2) to increase the number of pixels in the images at the zone of interest (edges of the sample). Adjust the focus of the camera and the lighting to reach an optimized view of the sample in the images.
- Use a software (e.g., Vic-Snap software provided by Correlated Solutions) to control the camera and acquire timely images of the sample during loading. Based on our experience we recommend a minimum of 20 Hz imaging frequency for the loading protocols proposed in Subheading 3.5 (*see* **Note 25**).

3. Run the test: if optical edge detection is used, run the imaging first and then run the loading to avoid losing the edge deformation data during loading. Extra initial images can be removed during experimental data analysis.
4. Repeat the tests for at least five samples of the same material to check the reproducibility of the experimental results.

3.7 Experimental Data and Inverse Mechanical Characterization

To estimate the mechanical properties of the samples using the inverse mechanical characterization method, three steps are required:

1. Obtain the experimental data: these are the measured reaction force versus time (obtained from load sensor measurements and the loading device software) and sample edge deformation versus time (obtained from images of the sample during loading—*see* **Note 26**). Figure 4a–c show examples of the reaction force versus time and maximum lateral deformation versus time curves for a hydrogel construct, a preserved, and a damaged human cartilage explant, respectively. The specifications of the hydrogel and cartilage samples are given in Subheadings 3.1 and 3.2, respectively. An example of the images used for the optical edge detection of the preserved cartilage explant is shown in Fig. 4d.
2. Mathematical modeling of the experimental loading:
   - For unconfined compression of a cylindrical biphasic elastic material, an analytical solution exists [19] that calculates the reaction force and lateral edge deformation of the sample subjected to a ramp loading followed by a relaxation.

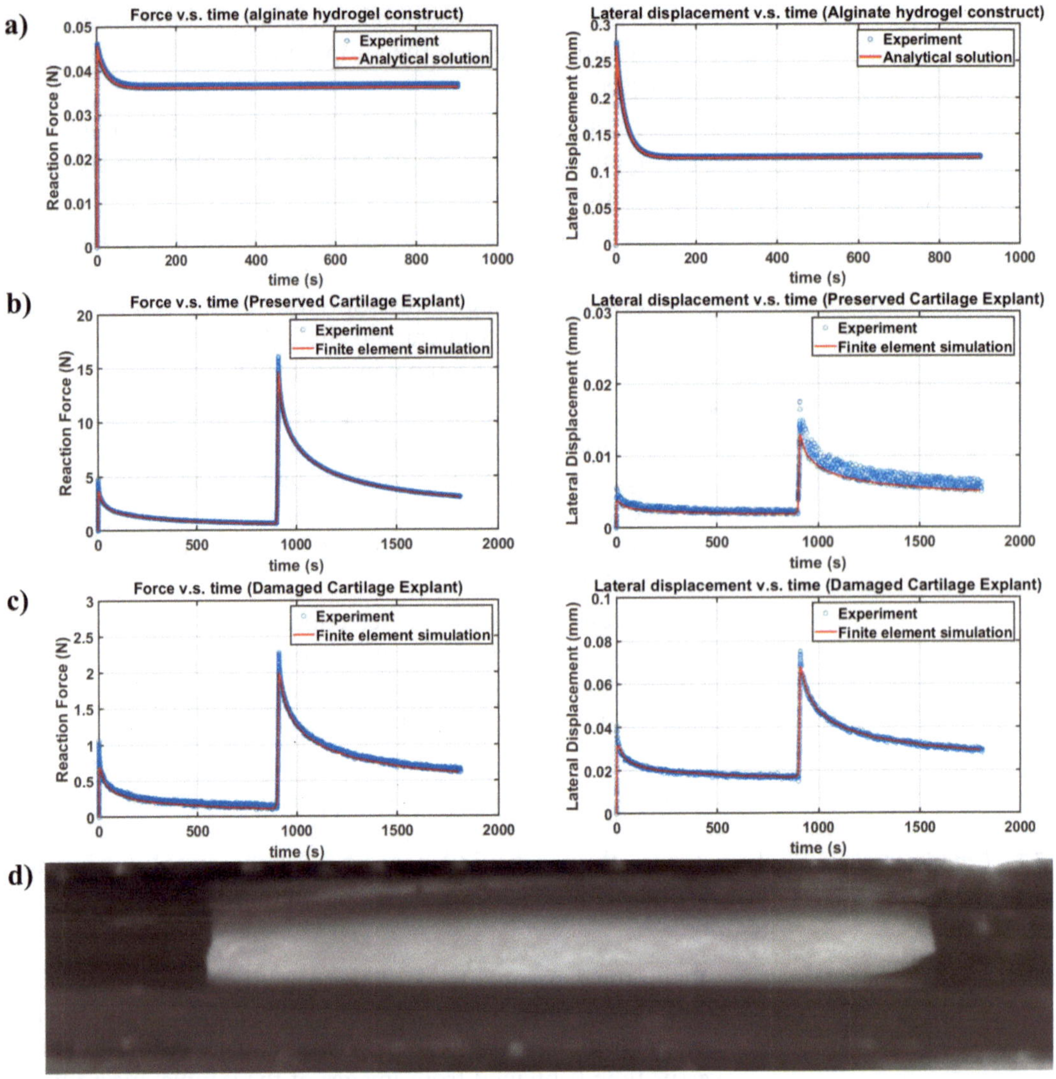

Fig. 4 Examples of the experimental curves of reaction force versus time and maximum lateral deformation versus time for (**a**) a hydrogel construct fitted with the curves obtained from the analytical solution of biphasic elastic material [19] and (**b**) a preserved and (**c**) a damaged cartilage explant fitted with the curves obtained from finite element models of FRPE material [20]. (**d**) An example of images used for optical edge detection of a preserved cartilage explant

- Due to the more complexity of the FRPE material, model finite element method can be used to simulate the experiments. The details for such a finite element model can be found in [6, 20]. We used ABAQUS 2017 software (provided by Dassault Systemes) for our finite element simulations.

3. A parameter fitting algorithm is required to fit the reaction force versus time (and lateral deformation versus time, in case measured) obtained from the mathematical models to the experimental data by modifying the materials' constants. To

Table 1
The estimated material properties (mean ± standard deviation) for the hydrogel construct, preserved, and damaged human cartilage explants

Biphasic elastic material parameters (obtained for 2.2% alginate hydrogel constructs)		
Young's modulus of solid matrix	$(3.59 \pm 0.11) \times 10^{-3}$ MPa	
Poisson's ratio of solid matrix	0.15 ± 0.05	
Permeability	$(4.84 \pm 0.42) \times 10^{-11}$ m^4/Ns	
FRPE material parameters (obtained for preserved and damaged human cartilage explants)		
	Preserved explant	Damaged explant
Young's modulus of non-fibrillar matrix	0.55 ± 0.11 MPa	0.22 ± 0.10 MPa
Poisson's ratio of non-fibrillar matrix	0.36 ± 0.11	0.21 ± 0.14
Initial modulus of fibril network	0.65 ± 0.25 MPa	0.26 ± 0.12 MPa
Strain-dependent modulus of fibril network	21.50 ± 7.61 MPa	8.69 ± 5.36 MPa
Initial permeability	$(0.89 \pm 0.36) \times 10^{-15}$ m^4/Ns	$(12.25 \pm 4.11) \times 10^{-15}$ m^4/Ns
Permeability void-ratio dependency constant	2.36 ± 1.27	5.69 ± 2.34

this end, we used the *lsqnonlin* function in MATLAB R2018a (provided by the MathWorks Inc.) and fitted the curves obtained from the analytical solution for the biphasic elastic material model [19] and finite element simulation for the FRPE material model [20] to the experimental curves of the hydrogel construct and cartilage explants, respectively (Fig. 4a–c). The MATLAB script for parameter fitting using the biphasic elastic material model is provided in the Supplementary Materials section. The same script can be used together with finite element simulations for more complex material models like FRPE material.

4. For demonstration, five samples were prepared for each of the hydrogel constructs, preserved, and damaged human cartilage explants (*see* Subheadings 3.1 and 3.2). These samples were tested using the proposed testing protocols in Subheading 3.5 for biphasic elastic material (hydrogel samples) and FRPE material (cartilage explants). For each category of the samples, the experimental results (reaction force and lateral deformation) of the five tests were used in the inverse identification. The average and standard deviation of the estimated material properties for the hydrogel construct, preserved, and damaged human cartilage explants are given in Table 1.

4 Notes

1. Read the specifications of the alginate beforehand to prepare the solution. The alginate used in this protocol is soluble in water, but some powders without sodium salt added must be prepared in other solvents.
2. Alginic acid sodium salt powder is hard to dissolve. The use of magnetic stirring bars and magnetic stir are recommended. Do not heat the solution during stirring or the concentration of the alginate may be affected due to evaporation.
3. Samples with cylindrical shape (*see* Fig. 1b) are easier remove from the mold than samples with multiple edges (e.g., rectangular samples).
4. Press thoroughly the mold against the filter paper to ensure vacuum so there is no solution leak. Be sure that there is no residual 102 mM $CaCl_2$ in the holes of the mold or the bottom of the hydrogels will not be flat.
5. Ensure flat surface with, for example, glass lids on top of the samples.
6. All solutions must be sterile and the protocol must be performed inside a laminar flow cabinet. Harvest the needed number of cells (we used 1 million cells/mL of alginate hydrogel), and centrifuge them to obtain a cell pellet. During the centrifugation step, cover the filter paper with 102 mM $CaCl_2$ until all filter papers are completely soaked as indicated in Subheading 3.1, **step 4**. After centrifugation, aspirate supernatant and mix the cell pellet with appropriate volume of freshly made and autoclaved 2.2% alginate solution without making any air bubbles. The number of cells per hydrogel must be adjusted for each specific experimental condition (we used 0.5 million cells for 0.5 mL of alginate hydrogel). Pipet 0.5 mL of the cell-alginate mixture into each well of the mold. Let the hydrogels set for 30 min at room temperature for cross-linking. Collect the hydrogels in a petri dish filled with 102 mM $CaCl_2$, and let them rest for another 30 min at room temperature. Then, collect the hydrogels in a 12-well plate containing warm cell culture medium.
7. The diameter of the biopsy punch can be selected as needed. Smaller diameters ensure a more regular shape of the explants.
8. Ensure a flatter surface by taking explants from the parts with a more regular shape. Constant thickness is also important.
9. Long-term storage periods are not recommended since the mechanical properties of the hydrogels may be altered.
10. We recommend DMEM F12 + 10% fetal bovine serum +1% antibiotics/antimycotics +1% L-glutamine.

11. Due to the small thickness of the cartilage samples (typically between 1 and 4 mm for human cartilage in different lower limb joints [21]), the compression testing machine should have the capability to apply compressive displacements as small as 0.2 mm (20% of minimum cartilage sample thickness) to use the proposed loading protocol here (Subheading 3.5).
12. The load sensor is recommended to be installed on the static side of the compression device to avoid the effects of load sensor movement and inertia on the measured loading. Since the load sensor will be placed under the lower platen or over the upper platen, the weight of the platen will be applied to the load sensor in addition to the applied loading during the test. Therefore, one should note that the sum of the applied loads to the load sensor will not pass the capacity of the sensor. To have an estimation of the applied loads to the load sensor, the trial test introduced in Subheading 3.4, **step 2** can be used. Usually, the compression devices are controlled using a software in which there is a possibility to set a limit for the load sensed by the load sensor and stop the machine if the limit is passed.
13. The diameter of the compression platens should be large enough to avoid contact loss with the sample during the compression when the sample expands laterally. At the same time, the platens should be as light as possible to decrease the unnecessary applied loading on the load sensor (*see* **Note 12**).
14. The surface of cartilage explants and hydrogel constructs may provide a low friction contact with the upper compression platen and bottom of the container (*see* Subheading 3.4, **step 4**). This may lead to the movement of the samples during the compression test. If such a movement is observed during the experiments, the friction between the sample and the upper platen or bottom of the container needs to be increased. This can be done by, e.g., slightly scratching surfaces of the platen and the container that are in contact with the sample. Avoid increasing the friction if it is not required to comply with the frictionless contact hypothesis in numerical simulations for inverse identification (Subheading 3.7).
15. The width of the container needs to be large enough to allow the upper platen to insert the container and compress the sample. The height of the container is recommended to be more than the sum of the sample and the upper platen thicknesses. This will allow filling the container with liquid in a way that the upper platen and the sample will be completely submerged within the liquid (*see* Subheading 3.4, **step 6**).
16. While bringing the loading platens close to each other, the first observed changes in the measured loading may be due to traction forces applied by a thin layer of liquid over the sample due to its liquid contents. At this point, the upper platen and

the sample are not yet in contact. Therefore, the platens should be further approximated until an initial compressive loading is measured by the load sensor. Typically, the compressive loading is shown as a negative value in the loading device software.

17. To have a perfect contact between the sample and loading platens in unconfined compression test, the samples are required to have flat surfaces. A few tips are given in Subheadings 3.1 and 3.2 to prepare hydrogel constructs and cartilage explants with flat surfaces. However, it is challenging and sometimes impossible to have a cartilage explant with a completely flat surface. For the samples with low surface roughness, the preloading in the unconfined compression test needs to be continued until the whole sample surface is in contact with the upper platen. This can be controlled using a camera and zooming in on the contact area for more accuracy. The applied preloading to the sample should be small compared to the maximum load applied during the test (less than 5% of the maximum load). In this case, more time is required for the sample to relax before running the test. If the roughness of the surface is too high, a higher preload is required to reach the complete contact (more than 5% of the maximum load); therefore, it is not recommended to perform an unconfined compression test to characterize the mechanical properties of such samples. Indentation can be an alternative test for these samples [6].

18. If the upper platen becomes in contact with the liquid surface, the surface traction of the liquid will affect the force measurement results. To avoid this effect, it is recommended to submerge the upper platen within the liquid before the start of the test.

19. To use images of a sample for measuring its dimensions, a high-quality image of the sample with the maximum possible number of pixels is required. The image can be taken using a normal camera (e.g., a cellphone camera). The camera should be positioned perpendicular to the sample for correct measurement of the sample dimensions (Fig. 2). A length scale is required to be placed at the same distance to the camera as the distance of the zone of interest. For example, if we need to measure the diameter of a cylindrical sample using the images taken from a side view of the sample, the length scale needs to be placed next to the sample. A proper background and lighting needs to be used for clear detection of the sample edges in the image. Based on the image, the length of each pixel can be calculated using the scale, and the sample dimension is determined by counting the pixels of the image (e.g., number of the image pixels in the thickness of the sample).

20. If ramp loading is used to identify the changes in stiffness of the sample or to compare the stiffness of different samples, the same loading rate (e.g., 1% of the sample thickness in second) needs to be used in all the tests. This is due to the effect of the loading rate on the measured stiffness of hydrogel constructs and cartilage samples with time-dependent mechanical properties. To avoid any damage of the hydrogel and cartilage samples, it is recommended to use a maximum strain of 20% (with respect to the sample thickness) in the compression tests [17].
21. The biphasic elastic material model accounts for the elastic properties of the solid matrix and the linear permeability of the material structure [19]. The material properties in this model are Young's modulus and Poisson's ratio of the solid matrix and permeability of the material. The FRPE material model accounts for the elastic properties of the non-fibrillar solid matrix, nonlinear elastic properties of the fibrillar solid matrix, and nonlinear permeability of the structure [6]. The material properties in this model are Young's moduli of fibrillar and non-fibrillar matrices, Poisson's ratio of the non-fibrillar matrix, initial permeability of the material, and a permeability strain-dependent coefficient.
22. The hydrogel samples may buckle under the compression due to the softness of the material and high thickness to diameter ratio. Such a buckling can be prevented by reducing the maximum applied strain (e.g., to 10% of the sample thickness) and/or reducing the strain rate (e.g., to 1%/s).
23. It is important to ensure saving the data of the maximum reaction force at the end of compression loading or a very close data point to that. This is a critical point in determining the mechanical properties of the material. In addition, initial reaction force data points in the 15 min of relaxation should be measured with a high frequency, as the change in the reaction force is relatively high in this part of loading cycle.
24. The deformation of sample edges can be used as additional experimental data for more accurate identification of mechanical parameters, especially the Poisson's ratio of the hydrogel constructs and cartilage explants. In our previous studies, these data were shown to be even more important than the measured reaction force for accurate mechanical identification of hydrogel constructs and cartilage explants [20]. Therefore, the optical edge deformation measurement results in combination with the reaction force measurements allow for more accurate mechanical characterization or even as the sole experimental data in absence of a precise reaction force measurement (e.g., due to lack of the load sensor precision) in the inverse characterization of mechanical properties of the samples (Subheading 3.7).

25. If due to practical difficulties a lower imaging frequency than the recommended 20 Hz frequency is used, decreasing the loading rate in the proposed protocols of Subheading 3.5 (e.g., to 1%/s) can improve the quality of the collected data.
26. An image processing software is required to obtain the deformation of the lateral edge of the sample versus time using the obtained images from the optical edge detection method. We used a script developed in MATLAB R2018a (provided by the MathWorks Inc.) for image processing. The image processing script is provided in the Supplementary Materials section. The images include data regarding the deformation of the lateral sample edge during the unconfined compression test. These data can not only be used to increase the accuracy of the inverse mechanical characterization but also estimate the friction coefficient between the sample and loading platens. However, for simplicity, we assumed a frictionless contact in our mathematical models and used the maximum deformation of the lateral edge as the output of the optical edge detection method.

Acknowledgments

This work was supported by Marie Skłodowska-Curie Individual Fellowship (CREATION project: MSCA-IF-2019-893771) and KU Leuven/FWO Happy Joints project (C14/18/077 and G045320N). It is also part of the EOS excellence program Joint-Against OA (G0F8218N). The authors thank the KU Leuven core facility for biomechanical experimentation (FIBEr) and Mr. Wouter Willekens for their support in the experiments and analyzing the data.

Supplementary Materials

Supplementary material to this chapter contains the MATLAB scripts for image processing, and parameter fitting and can be found together with the guides to run the scripts on https://bme-soft-tissue.pages.gitlab.kuleuven.be/unconfined-compression/.

References

1. Cohen NP, Foster RJ, Mow VC (1998) Composition and dynamics of articular cartilage: structure, function, and maintaining healthy state. J Orthop Sports Phys Ther 28:203–215
2. Shefelbine SJ, Ma CB, Lee KY, Schrumpf MA, Patel P, Safran MR et al (2006) MRI analysis of in vivo meniscal and tibiofemoral kinematics in ACL- deficient and normal knees. J Orthop Res 24:1208–1217
3. Brittberg M, Lindahl A, Nilsson A, Ohlsson C, Isaksson O, Peterson L (1994) Treatment of deep cartilage defects in the knee with

autologous chondrocyte transplantation. N Engl J Med 331:889–895
4. Jacobi M, Villa V, Magnussen RA, Neyret P (2011) MACI-a new era? Sports Med Arthrosc Rehabil Ther Technol. https://doi.org/10.1186/1758-2555-3-10
5. Mirahmadi F, Tafazzoli-Shadpour M, Shokrgozar MA, Bonakdar S (2013) Enhanced mechanical properties of thermosensitive chitosan hydrogel by silk fibers for cartilage tissue engineering. Mater Sci Eng C 33:4786–4794
6. Ebrahimi M, Ojanen S, Mohammadi A, Finnilä MA, Joukainen A, Kröger H et al (2019) Elastic, viscoelastic and fibril-reinforced poroelastic material properties of healthy and osteoarthritic human tibial cartilage. Ann Biomed Eng 47:953–966
7. Koh YG, Lee JA, Kim YS, Lee HY, Kim HJ, Kang KT (2019) Optimal mechanical properties of a scaffold for cartilage regeneration using finite element analysis. J Tissue Eng. https://doi.org/10.1177/2041731419832133
8. Mohammadi H, Mequanint K, Herzog W (2013) Computational aspects in mechanical modeling of the articular cartilage tissue. Proc Inst Mech Eng H 227:402–420
9. Marchiori G, Berni M, Boi M, Filardo G (2019) Cartilage mechanical tests: evolution of current standards for cartilage repair and tissue engineering: a literature review. Clin Biomech 68:58–72
10. Evans S (2017) How can we measure the mechanical properties of soft tissues? In: Avril S, Evans S (eds) Material parameter identification and inverse problems in soft tissue biomechanics, CISM International Centre for Mechanical Sciences (Courses and Lectures). Springer, Cham
11. Nguyen VB, Wang CX, Thomas CR, Zhang Z (2009) Mechanical properties of single alginate microspheres determined by microcompression and finite element modelling. Chem Eng Sci 64:821–829
12. Danso EK, Julkunen P, Korhonen RK (2018) Poisson's ratio of bovine meniscus determined combining unconfined and confined compression. J Biomech 77:233–237
13. Olvera D, Daly A, Kelly DJ (2015) Mechanical testing of cartilage constructs. In: Doran P (ed) Cartilage tissue engineering. Humana Press, New York
14. DiSilvestro MR, Suh JK (2001) A cross-validation of the biphasic poroviscoelastic model of articular cartilage in unconfined compression, indentation, and confined compression. J Biomech 34:519–525
15. Wilson W, Van Donkelaar CC, Van Rietbergen B, Huiskes R (2005) A fibril-reinforced poroviscoelastic swelling model for articular cartilage. J Biomech 38:1195–1204
16. Ateshian GA, Hung CT (2003) Functional properties of native articular cartilage. In: Guilak F, Butler DL, Goldstein SA, Mooney DJ (eds) Functional tissue engineering. Springer, New York
17. Patel JM, Wise BC, Bonnevie ED, Mauck RL (2019) A systematic review and guide to mechanical testing for articular cartilage tissue engineering. Tissue Eng Part C Methods 25: 593–608
18. Qi J, Hu Z, Song H, Chen B, Xie D, Zhou L et al (2016) Cartilage storage at 4 C with regular culture medium replacement benefits chondrocyte viability of osteochondral grafts in vitro. Cell Tissue Bank 17:473–479
19. Armstrong CG, Lai WM, Mow VC (1984) An analysis of the unconfined compression of articular cartilage. J Biomech Eng 106:165–173
20. Elahi SA, Tanska P, Mukherjee S, Korhonen RK, Geris L, Jonkers I (2021) Guide to mechanical characterization of articular cartilage and hydrogel constructs based on a systematic in silico parameter sensitivity analysis. J Mech Behav Biomed Mater 124:104795
21. Shepherd DET, Seedhom BB (1999) Thickness of human articular cartilage in joints of the lower limb. Ann Rheum Dis 58:27–34

Chapter 20

Viral Gene Delivery in Chondrocytes

Christopher V. Nagelli, Christopher H. Evans, and Rodolfo E. De La Vega

Abstract

Viral gene transfer, known as transduction, is a powerful research tool for studying the biology of chondrocytes in novel ways and also a technology enabling the use of gene therapy for regenerating cartilage and treating diseases that affect cartilage, such as osteoarthritis. Adenovirus, retrovirus, lentivirus, and adeno-associated virus (AAV) are most commonly used to transduce chondrocytes. Although AAV is able to transduce chondrocytes in situ by intra-articular injection, chondrocytes are most commonly transduced in monolayer culture using the four vectors mentioned above. Protocols for achieving this are described, along with a discussion of the variables that can influence transduction efficiency.

Key words Adenovirus, Retrovirus, AAV, Lentivirus, Cartilage

1 Introduction

There are three main motives for introducing genes into chondrocytes: to study their biology, to promote cartilage regeneration, and to treat diseases affecting cartilage, such as osteoarthritis. In terms of scientific study, the introduction of novel genetic information allows sophisticated interrogation of chondrocytes in a manner that is otherwise hard to achieve. Examples include the transfer of reporter genes as well as genes that affect specific metabolic and molecular processes in a deliberate and controlled fashion. In terms of cartilage regeneration, gene transfer can be used to express one or more morphogens, growth factors, transcription factors, or other gene products to enhance chondrogenesis. These general strategies are also relevant to treating diseases affecting cartilage, such as osteoarthritis [1]. In all applications, the gene products of interest can be species of non-coding RNA as well as proteins, and their expression can be constitutive or inducible. It is also possible to deliver the machinery of gene editing by gene transfer [2].

Martin J. Stoddart et al. (eds.), *Cartilage Tissue Engineering*,
Methods in Molecular Biology, vol. 2598, https://doi.org/10.1007/978-1-0716-2839-3_20,

Although genes can be transferred to chondrocytes by nonviral vectors, viral gene delivery is much more efficient because of the innate ability of viruses to deliver their own genomes efficiently into the cells they infect. Nonviral gene delivery is discussed elsewhere in this volume. As described below, a variety of viral vectors are available for gene transfer to chondrocytes. The selection of which vector to use depends to a large degree on the purpose of the study. Moreover, if clinical use is envisaged, the choice is constrained by safety concerns. While there are many different types of viruses which have been modified to create viral vectors for gene delivery [3], the most commonly used viral vectors in both laboratory studies and human clinical trials are derived from adenovirus, adeno-associated virus (AAV), retrovirus, and lentivirus [4]; these also have the most relevance to orthopedic applications, including gene delivery to chondrocytes [5]. It is important to note that the design of viral vectors continues to evolve and several generations of such vectors are available, each with different properties to a lesser or greater degree. The nature of some of these variants is given in this article.

This chapter describes basic methods for using viral vectors to deliver genes to chondrocytes. Methods for the preparation of individual vectors are not given, as these are idiosyncratic. Moreover, most major medical centers have vector core facilities that produce these materials to order for both internal and external customers. Failing that, there are a variety of reliable commercial suppliers. Viral stocks should be aliquoted in small volumes according to envisaged need and stored at −80 °C. Freeze-thaw should be minimized.

1.1 Salient Properties of Common Viral Vectors (Table 1)

1.1.1 Adenovirus Vectors

Adenovirus vectors transduce many types of cells, both dividing and non-dividing, usually with high transduction efficacy. A major advantage is that these vectors are straightforward to produce at high titers in the laboratory using producer lines, such as HEK293 cells. Adenovirus is approximately 100 nm in size and has a double-stranded DNA genome of approximately 35 kb flanked by inverted terminal repeat at the 5′ and 3′ ends. Depending on the construct, adenovirus can accommodate inserts ranging from 8 to 30 kb in size [6].

Transduction with adenovirus activates MAP kinases and NF-κB, which could interfere with experimental data in vitro and cause inflammation in vivo. Moreover, of relevance to in vivo application, adenovirus vectors provoke both humoral and cell-mediated immune responses. Because cells infected with the commonly used "first-generation" adenoviruses, deleted in certain early genes, continue to express low levels of antigenic viral proteins, they are cleared by the immune system and in vivo gene expression is lost [7]. Later generations of adenoviral vectors with additional

Table 1
Salient properties of commonly used viral vectors

Viral vector	Advantages	Disadvantages	Other properties
Adenovirus	Easy to produce in high titers Transduces both dividing and non-dividing cells Relatively good freeze-thaw stability Easy to procure and produce (first- and second- generation vectors)	Immunogenic Difficult to procure and produce (third generation) Does not transduce chondrocytes in situ	~1 in every 50–100 viral particles is infectious Non-integrating Carrying capacity 8–30 kb Transient transduction of dividing cells
Adeno-associated virus (AAV)	Transduces both dividing and non-dividing cells Relatively good freeze-thaw and thermal stability Capable of transducing chondrocytes in vivo No human disease associated with AAV Multiple serotypes allow for directed tropism	Difficult to procure and produce Gene carrying capacity is small Large number of the human population have pre-existing neutralizing antibodies to certain serotypes Expensive	Depending on serotype ~1 in 50 particles is infectious Transducing capacity varies widely between serotypes, cells, species, and different preparations Non-integrating
Retrovirus (Moloney murine leukemia virus derived)	Easy to produce Selection of transduced cells straightforward	Modest titers Does not transduce non-dividing cells Risk of insertional mutagenesis Does not transduce chondrocytes in vivo	~1 in every 100–1000 viral particles is infectious ~8 kb of packaging capacity Integrating
Lentivirus	Transduces both dividing and non-dividing cells Selection of transduced cells straightforward	Risk of insertional mutagenesis Does not transduce chondrocytes in vivo	~1 in every 100–1000 viral particles is infectious ~8 kb of packaging capacity Integrating

deletions reduce this problem, culminating in “high capacity” or “gutted” adenoviruses that contain no viral coding sequences at all [8]. Whether the immune-privileged conditions within cartilage protect adenovirally transduced cells from destruction by the immune system is unknown.

Because adenovirus vectors are non-integrating, they do not cause insertional mutagenesis and, when used locally in vivo, are perceived to be safe. Although chondrocytes lack the coxsackievirus and adenovirus receptor (CAR), through which adenovirus typically infects cells [9], their use to transduce chondrocyte cultures has been reported in numerous publications [10–13]. It is likely that the virus uses secondary, low-affinity receptors such as integrins to gain entry to chondrocytes. It has been reported that transduction of chondrocytes with adenovirus is more efficient in three-dimensional alginate cultures than in monolayers [14].

1.1.2 Adeno-Associated Virus (AAV) Vectors

AAV is gaining favor as a viral vector for human gene therapy because it has been shown to be safe and to transduce non-dividing cells; it has demonstrated efficacy in multiple clinical trials which has led to approval by the American, European, and Asian drug regulating agencies. With a size of approximately 20 nm, AAV is much smaller than adenovirus and, unlike the latter, has reliably been shown to transduce chondrocytes in situ [15] (Fig. 1). This is a major advantage for both experimental research and the development of clinical gene therapies for diseases of cartilage. It is presently being delivered by intra-articular injection in a clinical trial of gene therapy for osteoarthritis (NCT 02790723).

Another advantage of AAV is its relative stability and its ability to provide extended periods of transgene expression if the host cell does not divide, despite the recombinant viral genome remaining episomal. The persistence of the viral genomes is thought to reflect the formation of large concatemeric episomes [16]. There are various serotypes of AAV with different tropisms [17]. For in vivo

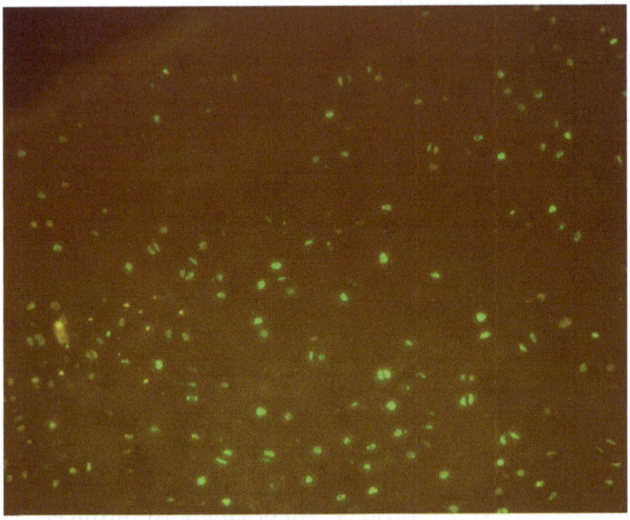

Fig. 1 Equine articular chondrocytes transduced in situ following intra-articular injection of AAV encoding GFP. (From ref. 15)

use, judicious selection of serotype allows AAV to target specific cell populations while avoiding the neutralizing humoral immune response to AAV2 present in a large portion of the population as a result of prior asymptomatic infection. AAV serotypes 2, 2.5, 5, 6, 7, and 8 have been reported to transduce chondrocytes with high efficiency [18, 15, 19]. Serotype tropisms and transduction efficiency can vary by species.

Wild-type AAV has a single-stranded DNA genome of approximately 4.7 kb. Early versions of AAV vectors suffered from the need for the host cell to synthesize the second, complementary DNA strand before transgene expression could occur. This was often slow and inefficient. The development of self-complementing, double-stranded AAV genomes [20] obviates the need for host cell second-strand synthesis and allows much more efficient transgene expression. However, the packaging capacity of the self-complementing viruses is reduced to 2–2.5 kb, which is problematic for large or multiple gene sets.

Producing AAV vectors can be challenging. Most investigators use a three-plasmid transfection of HEK293 cells to produce vector that then needs to be purified from the culture supernatant [21]. Manufacturing AAV under good manufacturing practice is still is a major problem and a bottleneck for clinical development [22].

1.1.3 Retrovirus Vectors

Retrovirus vectors derived from the Moloney murine leukemia virus, a gamma retrovirus, formed the basis of the first efficient vectors developed for gene therapy and were the first used in human clinical trials. These vectors are relatively straightforward to produce and manipulate, but their use is largely limited to ex vivo gene delivery because this type of retrovirus requires target cell division for efficient transduction. Retrovirus particles are approximately 100 nm in size, and the genome comprises two identical RNA molecules of about 8 kb. This sets an upper limit for their packaging capabilities. Retroviral titers also tend to be modest unless pseudotyped viruses are used.

Unlike adenovirus and AAV, retrovirus can be produced in a continuous fashion from producer cell lines. This greatly facilitates their use. Retroviral vectors are integrating; their genomes insert themselves into the host cell genome at unpredictable sites which introduces the risk of insertional mutagenesis. This is not an issue for in vitro and animal studies, but severely limits their clinical utility. Nevertheless, integration of retroviral genomes means that the transferred gene(s) stay with the transduced cells as they divide and after transplantation into cartilage [23]. The inclusion of a selectable marker gene allows cells to be selected to ensure the presence of viral genomes in all cells. Moloney-based retroviral vectors are reviewed in ref. 24.

1.1.4 Lentivirus Vector

Lentiviral vectors are also members of the retroviral vector family but have the advantage of transducing non-dividing cells. However, they still risk insertional mutagenesis which, in the context of chondrocytes, limits their use to in vitro experiments and studies with experimental animals. However, they can be produced in high titers and are highly infectious. Similar to the Moloney retrovirus, the genome comprises two identical RNA molecules of about 10 kb, which imposes an upper packaging capacity. The original lentivirus vectors were derived from HIV, but newer vectors have their origins in non-human primates and other sources (reviewed in ref. 25).

2 Materials

1. Low-protein binding 1.5 mL tubes (*see* **Note 1**).
2. Monolayer expanded chondrocytes (*see* **Note 2**).
3. Viral vector of choice (*see* Table 1).
4. Cell culture plastic vessel of suitable size for study design (*see* Table 2).
5. Growth medium:
   (a) Dulbecco's Modified Eagle Medium/Nutrient Mixture F-12 (DMEM/F12).
   (b) 10% Fetal bovine serum (FBS) (v/v ratio).
   (c) 100 U/mL penicillin.
   (d) 100 μg/mL streptomycin.
6. Phosphate-buffered saline (PBS).
7. Hexadimethrine bromide (Polybrene™).
8. Bleach.

Table 2
Transduction and culture media volume suggestions

Culture vessel	Transduction volume	Final culture medium volume
96-well plate	25–50 μL per well	60–160 μL
48-well plate	40–60 μL per well	0.15–0.38 μL
24-well plate	100–200 μL per well	400–1000 μL
12-well plate	300–500 μL per well	0.76–1.9 mL
6-well plate	0.5–1 mL per well	1.92–4.8 mL
T75 flask	1.5–3 mL per well	10 mL
T175 flask	4–5 mL per well	25 mL

3 Methods

General Considerations Because AAV is categorized as a Biosafety Level 1 agent and retrovirus, lentivirus, and adenovirus are Biosafety Level 2 agents, it is important to follow the appropriate biosafety recommendations when using viral vectors. While these may vary across institutions, general guidelines can be found on the websites of national and international institutes of health or disease control. At the very minimum, the use of a laboratory coat, nitrile gloves, eye protection, face mask, and a solution with a minimum bleach content of 10% is recommended. It is also a good idea to avoid the use of sharps while handling viral vectors.

As noted, direct in vivo gene delivery to chondrocytes within cartilage has only been confirmed robustly for AAV delivered by intra-articular injection (Fig. 1). For most other purposes, delivery occurs in cell culture, usually monolayer. Transducing chondrocytes in monolayer culture allows the use of different vectors, transduction in large numbers and in different fashions, the possibility of performing characterization after transduction, and the ability to use a known viral dose per cell, also known as multiplicity of infection (MOI). The MOI is used loosely in the literature to indicate the number of viral particles (vp) or viral genomes (vg) per cell, or the number of transducing or infectious units per cell. This distinction is important, because as indicated in Table 1, in most recombinant viral vector preparations less than 1 in 100 virions is infectious (*see* **Note 3**).

Of note, achieving a high percentage (>85%) of transduction may require considerable amounts of viral vector, the use of transduction enhancing techniques (described later), or both.

When performing transductions, we recommend including reporter gene constructs to serve both as a transduction efficiency marker and as a control for the experimental virus treatment. Using reporter genes, transduction efficiency can be tested with simple techniques. Green fluorescent protein (GFP) and its variants provide the most commonly used reporter genes. They allow quick and easy assessment of transduction by observation under fluorescence microscopy (Fig. 2). If more detailed information is required, flow cytometry of the transduced cells can be performed. Other convenient reporter genes encode various alternative fluorescent proteins (cyan, yellow, and red), the β-galactosidase gene (lacZ), and the luciferase gene (firefly, renilla, metridia).

In Vitro Transduction of Chondrocytes What follows is a generic method for transducing monolayer cultures of chondrocytes. When special steps for specific viruses are needed, these are indicated in the text (*see* **Notes 4** and **5**). As noted earlier, when testing for the appropriate MOI, it is always better to use the number of

Fig. 2 Human chondrocytes in monolayer culture transduced with AAV encoding GFP. Scale bar = 100 μm

transducing units rather than the physical titer. An initial test with MOIs of 10, 50, 100, 500, and 1000 serves as a good starting point for chondrocytes. In the interest of economy, the minimum amount of viral suspension should be used, consistent with cell coverage and survival. Table 2 lists some recommended volumes for transduction and subsequent culture.

Because our laboratory typically uses overnight transduction, this is the protocol described here.

3.1 Overnight Transduction

Day before transduction:

1. Review the complete experimental plan to determine the amount of cells, wells, dishes, and vectors needed. Refer to Table 2 for the suggested transduction volumes. At the end of the day, seed chondrocytes in the tissue culture vessel of choice at a density of 60–70%.

   Day of transduction, at the end of the day:

2. Thaw virus on ice.
3. Dilute virus to the required transduction concentration using growth medium—DMEM/F12 + FBS + penicillin/streptomycin (*see* **Note 6**).
4. If using retrovirus or lentivirus vectors, add hexadimethrine bromide (Polybrene™) to a final concentration of 5–10 μg/mL.
5. Aspirate the growth medium from the tissue culture vessel.

6. Wash the cells with PBS and aspirate.
7. Add the virus solution as calculated from Table 2 to each respective vessel following the established experimental plan, and then incubate the cells at 37 °C, 5% CO_2, and 95% humidity overnight (*see* **Note 7**).

   Day after transduction:

8. Next morning, aspirate the virus-containing media (*see* **Note 8**).
9. Perform two PBS washes to remove any remaining viral particles and toxic reagents, such as hexadimethrine bromide (Polybrene™).
10. Add fresh growth medium to the cells (*see* Table 2 for recommended volumes).
11. Allow the cells to incubate overnight at 37 °C, 5% CO2, and 95% humidity (*see* **Notes 9** and **10**).

3.2 Strategies to Enhance Transduction Efficiency

There is no single method that ensures 100% transduction of chondrocytes or any other cell type, although selection of transduced cells is possible using a selectable marker in conjunction with an integrating vector (*see* **Note 11**). Techniques such as flow cytometry can also be applied. There are, however, several strategies for enhancing transduction efficiency.

3.2.1 Spinoculation

Centrifugation has been shown to enhance transduction efficiency for adenovirus, retrovirus, and lentivirus vectors and is especially useful when short transduction periods (<4 h) are required. Its use should be limited to tissue culture plates as it minimizes the number of cells exposed without culture media because of the tilting process during centrifugation. Consider using 1000–2000 g for 20–30 min at 32–37 °C. It is performed only once, immediately after addition of the virus-containing media.

3.2.2 Charge Neutralization

The surfaces of viral vector particles and cells are negatively charged, which presents an electrostatic barrier to transduction. For this reason, polycations such as hexadimethrine bromide (polybrene™) are sometimes used to provide an electrostatic bridge. As indicated in the protocol described above, it is routinely used when transducing with retrovirus and lentivirus. However, it is toxic to cells and the ideal concentration should be determined prior to using it for the first time. Average concentrations range from 5 to 10 μg/mL of medium. Another polyamine known as "GeneJammer" has been reported to enhance adenoviral transduction very efficiently [26]. Charge neutralization using lanthanide ions has also been shown to improve adenoviral transduction [27].

3.2.3 Magnetic Transduction

In recent years, new commercial products have been developed to enhance in vitro transfection and transduction efficiency using magnetic forces. While most of the reports have focused on nucleic acids, there are kits to enhance adenovirus, AAV, and lentivirus transduction. The kits provide reagents for rendering the viruses magnetic, enabling them to be and thereby pulled onto the surface of the cell with a magnet [28].

3.2.4 Other

Flow-through has been reported to enhance the efficiency of retroviral transduction [29].

4 Notes

1. Use of low-protein binding tubes is recommended to minimize virus binding to plastic.
2. Prior to conducting experiments, articular chondrocytes should be isolated from cartilage, expanded as monolayers in chondrocyte growth medium to the number of cells that are required for the experiment. Do not allow cells to exceed 80% confluence (~15,000–20,000 cells/cm^2) to reduce the cells' tendency to dedifferentiate or to detach from the tissue culture surface. It is recommended to use cells at low passage (P0–P2).
3. Imprecise titering is a source of some confusion in the literature and frustration in the laboratory. It is also important to remember that viral preparations lose activity if subjected to excessive rounds of freeze-thaw.
4. The optimal transducing conditions will need to be determined empirically for each application, and these vary from laboratory to laboratory. This will usually require a series of pilot experiments to determine the optimal MOI, whether serum is necessary or inhibitory, the optimal time of transduction, etc.
5. Primary cells, such as chondrocytes, will require larger MOIs than those reported in the literature for established cell lines such as HEK293 and A549.
6. Prepare 10% more virus solution than actually needed, to account for pipetting losses.
7. Add everything that came in contact with virus to a container with a solution of 10% bleach in water. For bigger items that cannot be submerged, such as large conical tubes and dishes, add 10% bleach solution to these. Either way, bleach solution must remain in contact with the materials for more than 10 min before rinsing them with water and disposing off them as biohazard waste.
8. The liquid aspiration container must have enough bleach to account for a 10% concentration when full.

9. For adenovirus and AAV, transgene expression can be assessed starting the day after virus removal. For retrovirus and lentivirus, reverse transcription and integration into the cell's genome delays transgene expression, and this is best assessed 24–48 h after virus removal.
10. If required by the experimental aims, chondrocytes are suited for three-dimensional culture conditions as soon as 6 h after virus removal has been performed.
11. As indicated in this chapter, the mechanics of viral transduction of chondrocytes are straightforward—add the virus to the cells and transduction will occur spontaneously. However, transduction efficiency is not always high, which is why this chapter emphasizes the need for pilot experiments to optimize dosing, time of transduction, volume of the viral suspension, the question of serum, and other such variables. There is no algorithm for this, and all laboratories are different. However, the quality of the viral preparation is vital. No amount of transduction condition optimization will enable efficient transduction using a low-titer or low-quality preparations of vector. This is a particular risk when the virus is titered as viral particles or genomes rather than infectious particles. Moreover, even well-established vectors cores and commercial suppliers have batch-to-batch variations in the material they provide. And even a good preparation will not withstand multiple rounds of freeze-thaw.

References

1. Evans CH, Ghivizzani SC, Smith P et al (2000) Using gene therapy to protect and restore cartilage. Clin Orthop Relat Res (379 Suppl): S214–S219. https://doi.org/10.1097/00003086-200010001-00027
2. Kantor A, McClements ME, MacLaren RE (2020) CRISPR-Cas9 DNA base-editing and prime-editing. Int J Mol Sci 21(17). https://doi.org/10.3390/ijms21176240
3. Lundstrom K (2018) Viral vectors in gene therapy. Diseases 6(2). https://doi.org/10.3390/diseases6020042
4. Evans CH (2019) The vicissitudes of gene therapy. Bone Joint Res 8(10):469–471. https://doi.org/10.1302/2046-3758.810.BJR-2019-0265
5. Evans CH, Ghivizzani SC, Robbin PD (2021) Orthopaedic gene therapy: 25 years on. J Bone Joint Surg 9(8):e20.00220
6. Greber UF, Gomez-Gonzalez A (2021) Adenovirus – a blueprint for gene delivery. Curr Opin Virol 48:49–56. https://doi.org/10.1016/j.coviro.2021.03.006
7. Yang Y, Nunes FA, Berencsi K et al (1994) Cellular immunity to viral antigens limits E1-deleted adenoviruses for gene therapy. Proc Natl Acad Sci U S A 91(10):4407–4411. https://doi.org/10.1073/pnas.91.10.4407
8. Ricobaraza A, Gonzalez-Aparicio M, Mora-Jimenez L et al (2020) High-capacity adenoviral vectors: expanding the scope of gene therapy. Int J Mol Sci 21(10). https://doi.org/10.3390/ijms21103643
9. Ito T, Tokunaga K, Maruyama H et al (2003) Coxsackievirus and adenovirus receptor (CAR)-positive immature osteoblasts as targets of adenovirus-mediated gene transfer for fracture healing. Gene Ther 10(18):1623–1628. https://doi.org/10.1038/sj.gt.3302060
10. Brower-Toland BD, Saxer RA, Goodrich LR et al (2001) Direct adenovirus-mediated insulin-like growth factor I gene transfer enhances transplant chondrocyte function. Hum Gene Ther 12(2):117–129. https://doi.org/10.1089/104303401750061186

11. Palmer GD, Gouze E, Gouze JN et al (2003) Gene transfer to articular chondrocytes with recombinant adenovirus. Methods Mol Biol 215:235–246. https://doi.org/10.1385/1-59259-345-3:235
12. Shuler FD, Georgescu HI, Niyibizi C et al (2000) Increased matrix synthesis following adenoviral transfer of a transforming growth factor beta1 gene into articular chondrocytes. J Orthop Res 18(4):585–592. https://doi.org/10.1002/jor.1100180411
13. Smith P, Shuler FD, Georgescu HI et al (2000) Genetic enhancement of matrix synthesis by articular chondrocytes: comparison of different growth factor genes in the presence and absence of interleukin-1. Arthritis Rheum 43(5):1156–1164. https://doi.org/10.1002/1529-0131(200005)43:5<1156::AID-ANR26>3.0.CO;2-M
14. Oberholzer A, John T, Kohl B et al (2007) Adenoviral transduction is more efficient in alginate-derived chondrocytes than in monolayer chondrocytes. Cell Tissue Res 328(2): 383–390. https://doi.org/10.1007/s00441-006-0371-5
15. Watson Levings RS, Broome TA, Smith AD et al (2018) Gene therapy for osteoarthritis: pharmacokinetics of intra-articular self-complementary adeno-associated virus interleukin-1 receptor antagonist delivery in an equine model. Hum Gene Ther Clin Dev 29(2): 90–100. https://doi.org/10.1089/humc.2017.142
16. Yang J, Zhou W, Zhang Y et al (1999) Concatamerization of adeno-associated virus circular genomes occurs through intermolecular recombination. J Virol 73(11):9468–9477. https://doi.org/10.1128/JVI.73.11.9468-9477.1999
17. Lisowski L, Tay SS, Alexander IE (2015) Adeno-associated virus serotypes for gene therapeutics. Curr Opin Pharmacol 24:59–67. https://doi.org/10.1016/j.coph.2015.07.006
18. Chen Q, Luo H, Zhou C et al (2020) Comparative intra-articular gene transfer of seven adeno-associated virus serotypes reveals that AAV2 mediates the most efficient transduction to mouse arthritic chondrocytes. PLoS One 15(12):e0243359. https://doi.org/10.1371/journal.pone.0243359
19. Goodrich LR, Choi VW, Carbone BA et al (2009) Ex vivo serotype-specific transduction of equine joint tissue by self-complementary adeno-associated viral vectors. Hum Gene Ther 20(12):1697–1702. https://doi.org/10.1089/hum.2009.030
20. McCarty DM (2008) Self-complementary AAV vectors; advances and applications. Mol Ther 16(10):1648–1656. https://doi.org/10.1038/mt.2008.171
21. Grimm D (2002) Production methods for gene transfer vectors based on adeno-associated virus serotypes. Methods 28(2): 146–157. https://doi.org/10.1016/s1046-2023(02)00219-0
22. Srivastava A, Mallela KMG, Deorkar N et al (2021) Manufacturing challenges and rational formulation development for AAV viral vectors. J Pharm Sci. https://doi.org/10.1016/j.xphs.2021.03.024
23. Kang R, Marui T, Ghivizzani SC et al (1997) Ex vivo gene transfer to chondrocytes in full-thickness articular cartilage defects: a feasibility study. Osteoarthr Cartil 5(2):139–143. https://doi.org/10.1016/s1063-4584(97)80007-6
24. Maetzig T, Galla M, Baum C et al (2011) Gammaretroviral vectors: biology, technology and application. Viruses 3(6):677–713. https://doi.org/10.3390/v3060677
25. Milone MC, O'Doherty U (2018) Clinical use of lentiviral vectors. Leukemia 32(7): 1529–1541. https://doi.org/10.1038/s41375-018-0106-0
26. Fouletier-Dilling CM, Bosch P, Davis AR et al (2005) Novel compound enables high-level adenovirus transduction in the absence of an adenovirus-specific receptor. Hum Gene Ther 16(11):1287–1297. https://doi.org/10.1089/hum.2005.16.1287
27. Palmer GD, Stoddart MJ, Gouze E et al (2008) A simple, lanthanide-based method to enhance the transduction efficiency of adenovirus vectors. Gene Ther 15(5):357–363. https://doi.org/10.1038/sj.gt.3303092
28. Sapet C, Pellegrino C, Laurent N et al (2012) Magnetic nanoparticles enhance adenovirus transduction in vitro and in vivo. Pharm Res 29(5):1203–1218. https://doi.org/10.1007/s11095-011-0629-9
29. Del Vecchio MA, Georgescu HI, McCormack JE et al (2001) Approaches to enhancing the retroviral transduction of human synoviocytes. Arthritis Res 3(4):259–263. https://doi.org/10.1186/ar311

Chapter 21

Decellularization of Articular Cartilage: A Hydrochloric Acid–Based Strategy

Cornelia Schneider and Sylvia Nürnberger

Abstract

Removing cellular material from a tissue, a process called decellularization, reduces the risk of adverse host reactions, allows for efficient decontamination, and extends the shelf-life of the matrix. It facilitates the use of cartilage tissue as human-derived allograft, thus providing the field of cartilage regeneration with a biomaterial unmatched in its similarity to native cartilage in terms of structure, composition, and mechanical properties.

The dense extracellular matrix of articular cartilage requires a particularly thorough process to achieve the removal of cells, cell debris, and reagents used in the process. In our studies (Nürnberger et al., EBioMedicine 64:103196, 2021; Schneider et al., Tissue Eng Part C Methods 22(12):1095–1107, 2016), we have successfully developed a protocol for achieving decellularization via physical, chemical, and enzymatic steps. Combining freeze-thaw cycles for devitalization, hydrochloric acid as decellularization agent and the enzymatic removal of glycosaminoglycans, results in an acellular scaffold that is fully biocompatible and promotes cellular attachment. The structure and sophisticated architecture of collagen type II is left intact.

This chapter provides a comprehensive guide to the steps and reagents needed to decellularize articular cartilage. In addition to the standard decell-deGAG protocol, a fast option is given which is suitable for thin specimen. Histological evaluation is presented to illustrate treatment success.

Key words Cartilage, Decellularization, Glycosaminoglycans, Hydrochloric acid, Decontamination

1 Introduction

State-of-the-art regeneration of large cartilage defects by tissue engineering relies on two components: a biomaterial to provide structural support and guidance and cells to produce and maintain new functional tissue. In recent years, the role of the biomaterial has come into focus as critical factor for determining cellular behavior and the quality of the formed repair tissue. Biomechanics, chemical, and structural composition have been investigated, and properties resembling cartilage were found to be favorable to support chondrogenesis [1–3]. The majority of commercially available

Martin J. Stoddart et al. (eds.), *Cartilage Tissue Engineering*,
Methods in Molecular Biology, vol. 2598, https://doi.org/10.1007/978-1-0716-2839-3_21,

biomaterials only partly mimic the biomechanics, chemical, and/or structural composition found in articular cartilage [4–7].

The most natural habitat for chondrocytes and the most natural filling material for cartilage defects is cartilage itself. Autologous transplantations from non-load-bearing sites into the defect are in use but bear the disadvantages of donor site morbidity and limited availability.

In the search for alternative sources of cartilage tissue, allografts are a promising option. Human articular cartilage tissue is explanted in course of certain clinical procedures, such as total hip arthroplasty, and usually disposed of. Recently though, the value of this human tissue has been reevaluated.

Allografts contain cellular material from the donors that might be recognized as foreign by the host and trigger adverse immune responses. For cartilage, a discussion on potential immunoprivileged status is ongoing. Evidence suggests that immune reactions on allogenic tissue depend on donor source and defect location [8]. In addition, healthy cartilage tissue is limitedly available and elaborate to store. A solution for those shortcomings is the pretreatment of the tissue in a way that devitalizes donor cells and removes their residuals, a process that is called decellularization. This process also allows for full decontamination and prolongs the shelf-life of the thus created biomaterial.

There is a general consensus on minimal criteria to be matched, based on the findings of in vivo studies: a DNA content lower than 50 ng DNA per mg extracellular matrix (dry weight) and a lack of visible nuclear material on histological tissue sections. Decellularization can be achieved via physical, chemical, or enzymatic means, most effectively by a combination thereof [9–11].

Physical decellularization usually aims for cell burst through freeze/thawing, mechanical force, agitation, or sonication [12–16]. These techniques are most often used in combination with chemical or enzymatic steps to enhance their efficacy.

Some chemical agents, such as hypotonic or hypertonic buffers, utilize osmotic pressure to cause cytolysis [11, 17]. Most protocols however aim for decellularization via a degradation and solubilization of cellular components, which can then be washed out [18–22]. However, some detergents, such as the widely used sodium dodecyl sulfate, reportedly display cytotoxic effects on cells [23–26]; therefore, their complete removal from the final scaffold must be guaranteed. It needs to be kept in mind that chemical agents act in a non-targeted manner and may also cause unwanted harm to extracellular matrix components.

With their often specific substrates, enzymes allow for the degradation and removal of specific target molecules. Most noticeable for decellularization purpose are nucleases used to break down nucleic acids [17, 27, 28]. Other enzymes such as elastase or pepsin can aid decellularization indirectly, by depleting expandable matrix components and thus enhancing porosity, promoting better diffusion with decellularization reagents and removal of cellular components [10, 29, 30].

The exceptionally dense structure of articular cartilage poses a particular challenge for decellularization, full removal of reagents, and subsequent reseeding.

In our studies, widespread decellularization protocols were tested and adapted for the use on human articular cartilage, resulting in the hereby described protocol. Freeze/thaw cycles are used to devitalize the scaffold and enhance the efficacy of subsequent steps [31]. Hydrochloric acid degrades cells and cellular components and greatly reduces the DNA content. Pepsin is employed to selectively deplete glycosaminoglycans (GAG). To loosen the exceptionally dense matrix, the removal of GAG was accepted as a necessary compromise. Peracetic acid decontamination and extensive washing conclude the protocol.

The standard protocol takes a minimum of 6 days. For thin samples (<500 μm thickness), the protocol has been adapted to a faster version.

In the resulting scaffold the complex collagen type II matrix was preserved, while absence of glycosaminoglycans offered additional porosity to facilitate the removal of cellular debris as well as reagents.

2 Materials

1. Hypotonic buffer: 10 mM Tris base, pH 8.

   For 200 mL solution, dissolve 0.242 g Tris base (e.g., Promega, H5131) in about 180 mL water, and adjust the pH to 8.0 using HCl. Fill up the solution to 200 mL using a graduated cylinder. Store at 4 °C.

2. Decellularization reagent: 0.1 M hydrochloric acid (HCl).

   For example, 10 mL 1 M HCl, GMP grade (Puritan Products, 6384GMP), in 90 mL distilled water.

3. Phosphate-buffered saline (PBS), without Ca^{2+}/Mg^{2+}.

   For example, Lonza, BE17-512F, or with a self-mixed solution. Therefore, dissolve 200 mg potassium chloride (KCl), 200 mg potassium dihydrogen phosphate (KH_2PO_4), 8 g sodium chloride (NaCl), and 1.15 g anhydrous disodium hydrogen phosphate (Na_2HPO_4) in 1 L water.

4. Pepsin solution: 0.1% pepsin in 0.5 M acetic acid.

   Prepare the solution fresh before use. For 50 mL solution, add 1.43 mL glacial acetic acid (e.g., Merck, 8.18755) to 48.6 mL water to obtain 0.5 M acetic acid. Dissolve 50 mg pepsin (0.1%, Sigma, P7012) in the acetic acid solution.

5. Decontamination solution: 0.1% peracetic acid.
   For 100 mL solution, add 250 μL 40% peracetic acid (e.g., Sigma Aldrich, 1.07222) to 99.75 mL water.
6. 0.1 M sodium hydroxide (NaOH) (only for the quick decell-deGAG protocol).
   For example, 10 mL 1 M NaOH, cell culture grade (Sigma Aldrich, S2770), in 90 mL water.
7. Storage solution: PBS + 1% gentamycin +1% amphotericin B.
   For example, Gibco, 15710049 and 15290018.
8. Vials, 50 mL tubes, tweezers, pipetting device.
9. Shaking incubator, thermoshaker, or similar device.

3 Methods

3.1 Decellularization and GAG Depletion ("decell-deGAG" Procedure)

This protocol was developed for and optimized with macroscopically intact human articular cartilage, derived from femoral heads of patients undergoing total hip arthroplasty. The specimen featured a thickness of up to 1.5 mm and a diameter of 8 mm. The procedure takes a minimum of 6 days.

1. Cut the cartilage to the desired shape and thickness before starting the procedure. Keep it in storage solution (PBS + gentamycin + amphotericin B) at 4 °C until you are ready to start.
2. For performing the freeze/thaw cycles (*see* **Note 1**), transfer each biopsy into a vial or similar small vessel. Freeze the biopsies at −20 °C for at least 1 h (*see* **Note 2**), thaw for 1 h at room temperature, and then freeze again for at least 1 h at −20 °C.
3. Add hypotonic buffer (10 mM Tris base, pH 8) until the cartilage biopsy is fully submerged (*see* **Note 3**). Freeze the biopsies at −20 °C for at least 1 h, thaw for 1 h at room temperature, and then freeze again for at least 1 h at −20 °C.
4. Transfer the samples to 50 mL tubes (*see* **Note 4**), a maximum amount of three biopsies per tube. Add 5 mL hypertonic buffer per tube (*see* **Note 5**), and incubate at 37 °C under continuous shaking (*see* **Note 6**) overnight (*see* **Note 7**).
5. Discard the hypotonic buffer. Add 5 mL decellularization reagent (0.1 M HCl) (*see* **Note 8**) per tube, and incubate at 37 °C under continuous shaking overnight.
6. Discard the decellularization reagent. Add 5 mL PBS per tube and incubate at 37 °C under continuous shaking for 30 min. Discard the PBS and add 5 mL fresh PBS, and shake again at 37 °C for 30 min. Discard the PBS once more, add 5 mL fresh PBS, and incubate at 37 °C under continuous shaking overnight (*see* **Note 9**).

7. Discard the PBS. Add 10 mL of pepsin solution (0.1% pepsin in 0.5 M acetic acid) (*see* **Note 10**), and incubate at 37 °C under continuous shaking for 6 h. Discard the pepsin solution (*see* **Note 11**), add 10 mL fresh pepsin solution, and incubate at 37 °C under continuous shaking overnight.
8. Discard the pepsin solution. Add 5 mL PBS per tube, and incubate at 37 °C under continuous shaking for 30 min.
9. Discard the PBS. Add 5 mL decontamination solution (0.1% peracetic acid) (*see* **Note 12**) per tube, and incubate at room temperature under continuous shaking for 3 h.
10. Discard the decontamination solution. Add 5 mL PBS per tube, and incubate at room temperature under continuous shaking for 30 min. Discard the PBS and add 5 mL fresh PBS, and shake again at room temperature for 30 min. Discard the PBS once more, add 5 mL fresh PBS, and incubate at room temperature under continuous shaking overnight.
11. Discard the PBS. Store the samples at 4 °C, submerged in >1 mL storage solution per sample.

3.2 Alternative Protocol: Quick decell-deGAG

For thinner scaffolds (or thick laser engraved scaffolds [32]), the protocol was adapted toward a reduced duration, from 6 to 4 days. The specimen used to establish this protocol measured up to 500 μm in thickness (or 1 mm with laser engraving) and a diameter of 8 mm.

1. Cut the cartilage to the desired shape and thickness before starting the procedure. Keep it in storage solution (PBS + gentamycin + amphotericin B) until you are ready to start.
2. For performing the freeze/thaw cycles (*see* **Note 1**), transfer each biopsy into a vial or similar small vessel. Freeze the biopsies at −20 °C for at least 1 h (*see* **Note 2**), thaw for 1 h at room temperature, and then freeze again for at least 1 h at −20 °C.
3. Add hypotonic buffer (10 mM Tris base, pH 8) until the cartilage biopsy is fully submerged (*see* **Note 3**). Freeze the biopsies at −20 °C for at least 1 h, thaw for 1 h at room temperature, and then freeze again for at least 1 h at −20 °C.
4. Transfer the samples to 50 mL tubes (*see* **Note 4**), three biopsies per tube. Add 5 mL decellularization reagent (0.1 M HCl) per tube (*see* **Note 8**), and incubate at 37 °C under continuous shaking (*see* **Note 6**) overnight (*see* **Note** 7).
5. Discard the decellularization reagent. Add 10 mL of pepsin solution (0.1% pepsin in 0.5 M acetic acid) (*see* **Note 10**), and incubate at 37 °C under continuous shaking for 6 h. Discard the pepsin solution, add 10 mL fresh pepsin solution, and incubate at 37 °C under continuous shaking overnight.

6. Discard the pepsin solution. Add 5 mL PBS per tube and incubate at 37 °C under continuous shaking for 30 min.
7. Discard the PBS. Add 5 mL 0.1 M NaOH per tube and incubate at 37 °C under continuous shaking for 6 h.
8. Discard the NaOH. Add 5 mL PBS per tube and incubate at 37 °C under continuous shaking for 30 min. Discard the PBS and add 5 mL fresh PBS, and shake again at 37 °C for 30 min. Discard the PBS once more, add 5 mL fresh PBS, and incubate at 37 °C under continuous shaking overnight.
9. Discard the PBS. Store the samples at 4 °C, submerged in >1 mL storage solution per sample.

3.3 Evaluation of Treatment Success

Heidenhain's Azan staining is most efficient in visualizing cells and cellular residuals within the cartilage matrix (blue), since it stains the cytoplasm and especially the nuclei intensively red and the matrix blue (Fig. 1). No red remnants should be visible in successfully decellularized matrix (Fig. 2, a scaffold prepared from cartilage using our standard decell-deGAG protocol). Contrarily, when cartilage is treated with pepsin solution alone, the amount of cellular material can only be reduced, most probably due to the acetic acid within the solution but leaves cellular remnants within the lacunae (Fig. 3). Apart from histological assessment, there is a variety of other options to evaluate treatment success: DNA stainings using reagents such as DAPI or PI that bind to DNA of dead or dying cells, membrane lipid stainings such as DIL or DIO, or quantitative assays applied to papain-digested matrix material, such as CyQUANT GR or PicoGreen.

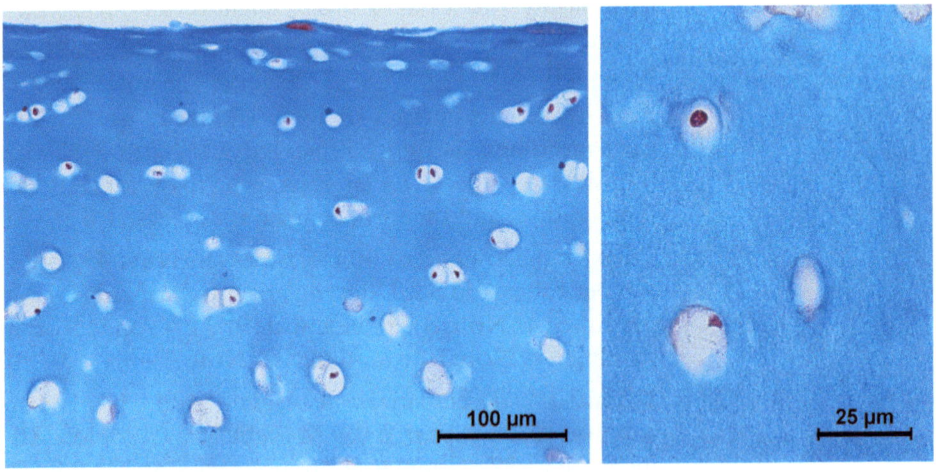

Fig. 1 Azan staining of native human articular cartilage. Vital cells (nuclei stained red) reside inside their lacunae in the dense collagen matrix (blue)

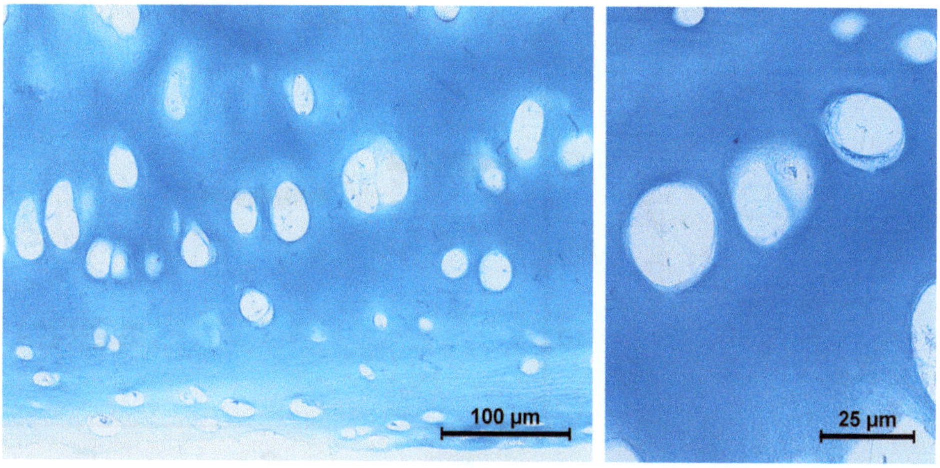

Fig. 2 Azan staining of human articular cartilage treated with the standard decell-deGAG protocol. No cell nuclei are visible after decell-deGAG treatment, and the lacunae are largely devoid of cellular remnants

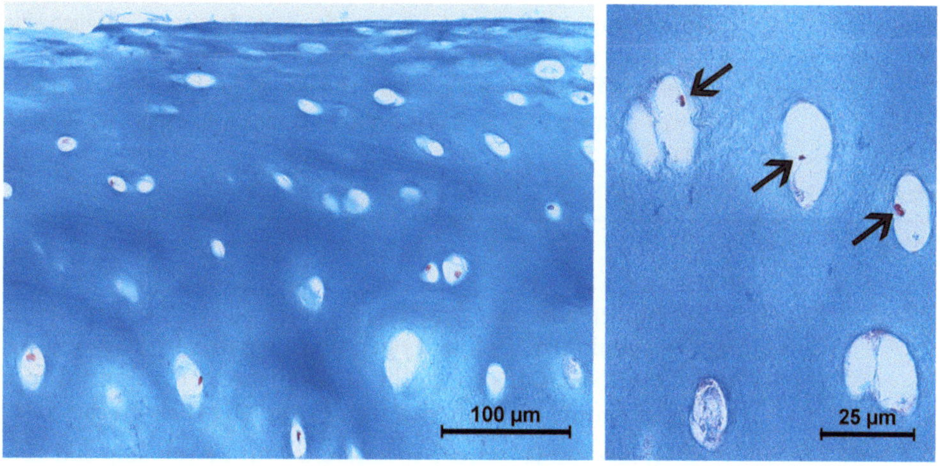

Fig. 3 Azan staining of human articular cartilage treated with pepsin solution alone. Matrix treated with pepsin in acetic acid contains less cellular material than matrix treated with the decell-deGAG protocols. Cellular remnants including some condensed nuclei (arrows) are, however, visible

The DNA content can be quantified following papain digestion of the samples [33], e.g., via CyQUANT™ GR (part of the CyQUANT® Cell Proliferation Kit, Life Technologies, C7026), a dye exhibiting fluorescence enhancement when bound to nucleic acids.

The DNA content is shown to be reduced from a mean 950 ± 160 ng/mg sample (dry) in untreated cartilage to 11 ± 3 ng/mg sample after the standard decell-deGAG protocol (Fig. 4). Pepsin in acetic acid reduces the content to 420 ± 49 ng/mg.

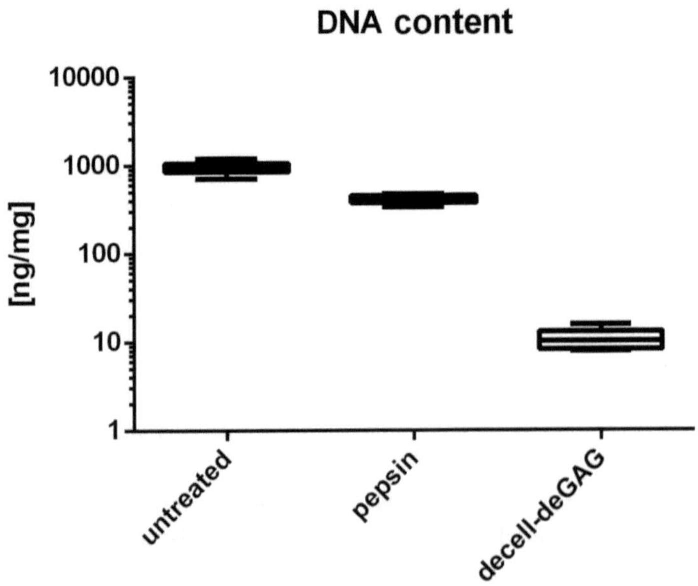

Fig. 4 DNA content per mg dry weight as an index for decellularization efficiency. The pepsin solution halves the amount of DNA per sample, while the standard decell-deGAG protocol reduces the content by about 99%. The final DNA content is well below 50 ng/mg (dry weight), a common threshold for decellularized tissues

4 Notes

1. Multiple freeze/thaw cycles cause cell burst by intracellular ice crystal formation and open up the matrix by extracellular crystal formation to facilitate diffusion of solutions used in subsequent treatments, thus enhancing their efficacy [31].
2. It is possible to store the material during this step and continue with the protocol 1 week before the intended use. Storage times of up to 3 months have been performed without any evidence of adverse effects.
3. 1 mL is sufficient in most vial types, given that scaffold dimensions allow the use of a vial. Take care not to fill the vial to the brim to avoid leakage when the buffer expands during freezing.
4. Alternatively, similar vessel with a sufficiently large base area can be used. 50 mL tubes were chosen for their base area diameter of about 3 cm, allowing 5 mL of liquid to generate a turbulent flow while shaking. The same amount of liquid in a narrow tube (e.g., 15 mL tube) will have a reduced treatment efficacy.
5. Hypotonic buffer is intended to cause residual cells and organelles to burst by osmotic swelling, the fragments of which can then be washed out by the buffer and subsequent steps.

6. The protocol was established using a GFL 3032 shaking incubator set to 180 rpm. Tubes were placed in the device slightly tilted to further increase turbulence, causing the scaffolds to move within the liquid rather than remaining at the bottom of the tube.
7. "Overnight" typically amounts to an incubation time between 22 and 26 h. No influence on the outcome has been detected when varying within this time frame.
8. Of all tested common decellularization agents, hydrochloric acid was found to be most efficient in reducing the scaffold's DNA content [34].
9. **Steps 4–6** (hypotonic buffer-decellularization reagent-PBS washing) can be performed multiple times to enhance the decellularization. This will increase the process duration from 6 to 12 days, with only a small improvement [34], but might be useful in samples where the DNA threshold of <50 ng/mg cannot be met with the "basic" protocol.
10. Pepsin diminishes the GAG content while leaving the collagen type II structure intact [34], leading to enhanced matrix porosity and cell attachment [32]. If the GAG content should remain unaltered, skip this step.
11. The pepsin solution is changed after 6 h in order to avoid inhibition by reaction products [35].
12. Peracetic acid has been shown to effectively inactivate viruses while having only a minimal effect on the extracellular matrix composition and structure [36, 37]. Additionally or alternatively, gamma irradiation is a widely used option to provide sterility including virus inactivation. Its use, however, still needs to be validated in order to exclude interactions with the scaffold matrix.

Acknowledgments

This work was supported by funding from the Austrian Research Promotion Agency FFG (project CartiScaff, #842455), as well as the Lorenz Böhler Fonds (16/13). The authors wish to thank all cooperation partners who were involved in those projects and coauthors of the publication Schneider and coworkers [34].

References

1. Chang CH, Chen CC, Liao CH et al (2014) Human acellular cartilage matrix powders as a biological scaffold for cartilage tissue engineering with synovium-derived mesenchymal stem cells. J Biomed Mater Res A 102(7): 2248–2257. https://doi.org/10.1002/jbm.a.34897
2. Chen CC, Liao CH, Wang YH et al (2012) Cartilage fragments from osteoarthritic knee promote chondrogenesis of mesenchymal stem cells without exogenous growth factor induction. J Orthop Res 30(3):393–400. https://doi.org/10.1002/jor.21541

3. Chiu LH, Chen SC, Wu KC et al (2011) Differential effect of ECM molecules on re-expression of cartilaginous markers in near quiescent human chondrocytes. J Cell Physiol 226(8):1981–1988. https://doi.org/10.1002/jcp.22530
4. Getgood A, Brooks R, Fortier L et al (2009) Articular cartilage tissue engineering: today's research, tomorrow's practice? J Bone Joint Surg Br 91(5):565–576. https://doi.org/10.1302/0301-620X.91B5.21832
5. Jeuken RM, Roth AK, Peters R et al (2016) Polymers in cartilage defect repair of the knee: current status and future prospects. Polymers (Basel) 8(6). https://doi.org/10.3390/polym8060219
6. Nürnberger S, Cyran N, Albrecht C et al (2011) The influence of scaffold architecture on chondrocyte distribution and behavior in matrix-associated chondrocyte transplantation grafts. Biomaterials 32(4):1032–1040. https://doi.org/10.1016/j.biomaterials.2010.08.100
7. Rai V, Dilisio MF, Dietz NE et al (2017) Recent strategies in cartilage repair: a systemic review of the scaffold development and tissue engineering. J Biomed Mater Res A 105(8): 2343–2354. https://doi.org/10.1002/jbm.a.36087
8. Arzi B, DuRaine GD, Lee CA et al (2015) Cartilage immunoprivilege depends on donor source and lesion location. Acta Biomater 23: 72–81. https://doi.org/10.1016/j.actbio.2015.05.025
9. Badylak SF, Freytes DO, Gilbert TW (2009) Extracellular matrix as a biological scaffold material: structure and function. Acta Biomater 5(1):1–13. https://doi.org/10.1016/j.actbio.2008.09.013
10. Crapo PM, Gilbert TW, Badylak SF (2011) An overview of tissue and whole organ decellularization processes. Biomaterials 32(12): 3233–3243. https://doi.org/10.1016/j.biomaterials.2011.01.057
11. Gilbert TW, Sellaro TL, Badylak SF (2006) Decellularization of tissues and organs. Biomaterials 27(19):3675–3683. https://doi.org/10.1016/j.biomaterials.2006.02.014
12. Funamoto S, Nam K, Kimura T et al (2010) The use of high-hydrostatic pressure treatment to decellularize blood vessels. Biomaterials 31(13):3590–3595. https://doi.org/10.1016/j.biomaterials.2010.01.073
13. Oliveira AC, Garzon I, Ionescu AM et al (2013) Evaluation of small intestine grafts decellularization methods for corneal tissue engineering. PLoS One 8(6):e66538. https://doi.org/10.1371/journal.pone.0066538
14. Pegg DE (1987) Mechanisms of freezing damage. Symp Soc Exp Biol 41:363–378
15. Pegg DE (2020) The relevance of ice crystal formation for the cryopreservation of tissues and organs. Cryobiology 93:3–11. https://doi.org/10.1016/j.cryobiol.2020.01.005
16. Sasaki S, Funamoto S, Hashimoto Y et al (2009) In vivo evaluation of a novel scaffold for artificial corneas prepared by using ultrahigh hydrostatic pressure to decellularize porcine corneas. Mol Vis 15:2022–2028
17. Kheir E, Stapleton T, Shaw D et al (2011) Development and characterization of an acellular porcine cartilage bone matrix for use in tissue engineering. J Biomed Mater Res A 99(2):283–294. https://doi.org/10.1002/jbm.a.33171
18. Beck EC, Barragan M, Libeer TB et al (2016) Chondroinduction from naturally derived cartilage matrix: a comparison between devitalized and decellularized cartilage encapsulated in hydrogel pastes. Tissue Eng Part A 22(7–8): 665–679. https://doi.org/10.1089/ten.TEA.2015.0546
19. Elder BD, Eleswarapu SV, Athanasiou KA (2009) Extraction techniques for the decellularization of tissue engineered articular cartilage constructs. Biomaterials 30(22): 3749–3756. https://doi.org/10.1016/j.biomaterials.2009.03.050
20. Ghassemi T, Saghatoleslami N, Mahdavi-Shahri N et al (2019) A comparison study of different decellularization treatments on bovine articular cartilage. J Tissue Eng Regen Med 13(10):1861–1871. https://doi.org/10.1002/term.2936
21. Schwarz S, Koerber L, Elsaesser AF et al (2012) Decellularized cartilage matrix as a novel biomatrix for cartilage tissue-engineering applications. Tissue Eng Part A 18(21–22): 2195–2209. https://doi.org/10.1089/ten.TEA.2011.0705
22. Seddon AM, Curnow P, Booth PJ (2004) Membrane proteins, lipids and detergents: not just a soap opera. Biochim Biophys Acta 1666(1–2):105–117. https://doi.org/10.1016/j.bbamem.2004.04.011
23. Bleeg HS (1990) Non-specific cleavage of collagen by proteinases in the presence of sodium dodecyl sulfate. Scand J Dent Res 98(3): 235–241. https://doi.org/10.1111/j.1600-0722.1990.tb00967.x
24. Cartmell JS, Dunn MG (2004) Development of cell-seeded patellar tendon allografts for anterior cruciate ligament reconstruction. Tissue Eng 10(7–8):1065–1075. https://doi.org/10.1089/ten.2004.10.1065

25. Rieder E, Kasimir MT, Silberhumer G et al (2004) Decellularization protocols of porcine heart valves differ importantly in efficiency of cell removal and susceptibility of the matrix to recellularization with human vascular cells. J Thorac Cardiovasc Surg 127(2):399–405. https://doi.org/10.1016/j.jtcvs.2003.06.017
26. Wilson GJ, Courtman DW, Klement P et al (1995) Acellular matrix: a biomaterials approach for coronary artery bypass and heart valve replacement. Ann Thorac Surg 60(2 Suppl):S353–S358. https://doi.org/10.1016/0003-4975(95)98967-y
27. McFetridge PS, Daniel JW, Bodamyali T et al (2004) Preparation of porcine carotid arteries for vascular tissue engineering applications. J Biomed Mater Res A 70(2):224–234. https://doi.org/10.1002/jbm.a.30060
28. Mendoza-Novelo B, Avila EE, Cauich-Rodriguez JV et al (2011) Decellularization of pericardial tissue and its impact on tensile viscoelasticity and glycosaminoglycan content. Acta Biomater 7(3):1241–1248. https://doi.org/10.1016/j.actbio.2010.11.017
29. Nimeskern L, Utomo L, Lehtoviita I et al (2016) Tissue composition regulates distinct viscoelastic responses in auricular and articular cartilage. J Biomech 49(3):344–352. https://doi.org/10.1016/j.jbiomech.2015.12.032
30. Porzionato A, Sfriso MM, Macchi V et al (2013) Decellularized omentum as novel biologic scaffold for reconstructive surgery and regenerative medicine. Eur J Histochem 57(1):e4. https://doi.org/10.4081/ejh.2013.e4
31. Burk J, Erbe I, Berner D et al (2014) Freeze-thaw cycles enhance decellularization of large tendons. Tissue Eng Part C Methods 20(4):276–284. https://doi.org/10.1089/ten.TEC.2012.0760
32. Nürnberger S, Schneider C, Keibl C et al (2021) Repopulation of decellularised articular cartilage by laser-based matrix engraving. EBioMedicine 64:103196. https://doi.org/10.1016/j.ebiom.2020.103196
33. Kim YJ, Sah RL, Doong JY et al (1988) Fluorometric assay of DNA in cartilage explants using Hoechst 33258. Anal Biochem 174(1):168–176. https://doi.org/10.1016/0003-2697(88)90532-5
34. Schneider C, Lehmann J, van Osch GJ et al (2016) Systematic comparison of protocols for the preparation of human articular cartilage for use as scaffold material in cartilage tissue engineering. Tissue Eng Part C Methods 22(12):1095–1107. https://doi.org/10.1089/ten.TEC.2016.0380
35. Greenwell P, Knowles JR, Sharp H (1969) The inhibition of pepsin-catalysed reactions by products and product analogues. Kinetic evidence for ordered release of products. Biochem J 113(2):363–368. https://doi.org/10.1042/bj1130363
36. Hodde J, Hiles M (2002) Virus safety of a porcine-derived medical device: evaluation of a viral inactivation method. Biotechnol Bioeng 79(2):211–216. https://doi.org/10.1002/bit.10281
37. Hodde J, Janis A, Ernst D et al (2007) Effects of sterilization on an extracellular matrix scaffold: part I. Composition and matrix architecture. J Mater Sci Mater Med 18(4):537–543. https://doi.org/10.1007/s10856-007-2300-x

Chapter 22

Self-Assembly Culture Model for Engineering Musculoskeletal Tissues

Nicholas N. DePhillipo, Jerahme Martinez, and George R. Dodge

Abstract

The goal of a self-assembly tissue engineering is to create functional tissue following a natural cell-driven process that mirrors natural development. This approach to tissue engineering has tremendous potential for the development of reparative strategies to treat musculoskeletal injuries and diseases, especially for articular cartilage which has poor regenerative capacity. Additionally, many bioengineering and culture methods fail to maintain the chondrocyte phenotype and contain the correct matrix composition in the long term. Existing cartilage-engineering approaches have been developed, but many approaches involve complicated culture techniques and require foreign substances and biomaterials as scaffolds. While these scaffold-based approaches have numerous advantages, such as an instant or rapid creation of biomechanical properties, they frequently result in dedifferentiation of cells in part, due to the adherence to foreign scaffold materials. In this chapter, we describe a novel approach of developing a scaffold-less cartilage-like biomaterial, using the simple principle that cells at high density bear a capacity to coalesce when they cannot attach to any culture substrate. We refer to the biomaterial formed as a cartilage tissue equivalent or CTA and have published to describe their characteristics and utility in high-throughput drug screening. The method is described to generate reproducible cartilage analogs using a specialized high-density suspension culture technique using a hydrogel poly-2-hydroxyethyl methacrylate (polyHEMA) coating of a culture dish. We have demonstrated that this approach can rapidly form biomass of chondrocytes that over time becomes very synthetically active producing a cartilage-like extracellular matrix that closely mimics the biochemical and biomechanical characteristics of native articular cartilage. The culture approach can also be used to form CTA from other than articular cartilage-derived chondrocytes as well as mesenchymal stem cells (MSCs) (while differentiating MSCs into chondrocytes). Some of the advantages are phenotype stability, reproducible CTA size, and biomechanical and biochemical characteristics similar to natural cartilage.

Key words Self-assembly culture, Scaffold-free, Cartilage tissue analog, Bioengineering materials, Regenerative medicine, Cartilage tissue engineering

1 Introduction

The self-assembly method is a practical approach for growing various cell types, including differentiated chondrocytes and precursor cells. This method works particularly well for creating soft tissues such as cartilage, which have poor regenerative capacity on their

Martin J. Stoddart et al. (eds.), *Cartilage Tissue Engineering*,
Methods in Molecular Biology, vol. 2598, https://doi.org/10.1007/978-1-0716-2839-3_22,

own, and circumvents issues dealing with limited availability of viable donor cells and lack of culture methods to maintain the chondrocyte phenotype [1, 2]. The process for generating engineered cartilage constructs is rather straightforward, whereby simply controlling the initial cell seeding density, substratum, and culture media is sufficient to promote the chondrocytes to coalesce and form a natural-like cartilaginous tissue. This approach or slight variations have been utilized and proven as a reliable method to create various types of cartilaginous tissue for the purpose of different outcomes [1–5]. This is routinely achieved using cells at a high density and low media volume and pioneered by a number of earlier studies [6–8]. By selecting the appropriate cell source, or even a combination of cells, one can design constructs with desirable properties. Constructs created by the self-assembly process have a striking similarity to that native cartilage, in terms of biochemical and matrix composition, cellularity, and mechanical properties.

Self-assembly in vitro cultures take advantage of chondrocytes' natural ability to coalesce, differentiate, and synthesize an abundance of extracellular matrix, as occurs during chondrogenesis during normal development [5]. The advantage over scaffold-based engineering models is that self-assembly models typically result in a more stable phenotype and can result in natural surrogates of cartilage without non-cartilaginous-derived materials such as substances like agarose or synthetic hydrogels. One could essentially consider that self-assembly platforms allow the cells themselves to produce the scaffold and environment to support cartilage formation. In vitro models have long been able to recapitulate aspects of this phenomenon for some time now and have provided us with key insights to understanding the signaling mechanisms that drive chondrocyte differentiation and cartilage development [6–8]. Early in vitro models, consisting of monolayer cultures of chondrocytes or precursor cells stimulated with various factors (e.g., ITS, ascorbic acid, and beta-glycerophosphate), achieved chondrocyte differentiation, but did not maintain the chondrocyte phenotype nor create 3D microtissues. The 2D culture models are often used to explore the effect of exogenous cues on cell differentiation, in which gene expression and immunostaining are often outcome measures for these studies. However, monolayer cultures lack the ability to retain a desired cartilage-specific phenotype, particularly when it comes to primary cell cultures, and have no functional application. From 2D cultures came the development of pelleted micromass cultures that took on a cartilage nodule-like formation and often including exogenous bioactive materials or artificial scaffolds [3].

Researchers learned to leverage chondrocytes' intrinsic properties to engineer native 3D tissue structures under minimal manipulated conditions and without the need of an exogenous material. The term "self-assemble" is used to describe tissue engineering

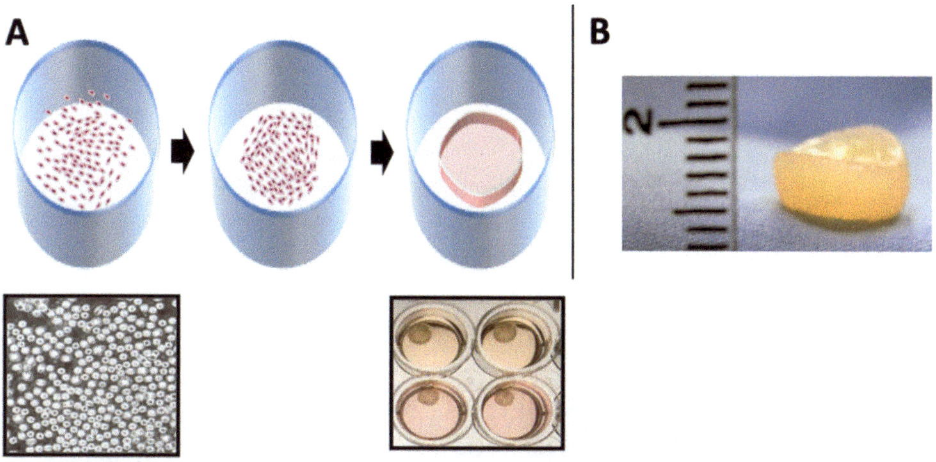

Fig. 1 Cartilage tissue analog (CTA) formation over 24–48 h. (**a**) Schematic representation of cell coalescence over time generated by high-density suspension culture above a polyHEMA hydrogel coating. (**b**) Image of a mature CTA at 14 weeks. (Modified from Mohanraj et al., J Biomech (2014) [9])

strategies that simply rely on the intrinsic ability of cells to coalesce without the requirement of external manipulation, such as centrifugation, reshaping, or use of exogenous material. This technique allows cells to self-assemble in a suspension culture and form three-dimensional spheroid constructs that resemble native tissues in terms of biochemical and biomechanical properties. The self-assembly approach we describe here is simple to establish and reproducible. The cartilage-like mass forms quickly (Fig. 1) and maintains the cartilage-like phenotype and bears resemblances in biomechanical properties (Fig. 2). Here we describe our approach to "engineer" cartilage-like tissue analogs by self-assembly and without the use of foreign scaffolds. There has been good success in generating cartilage surrogates, and these straightforward approaches hold great potential for creative cartilage tissue engineering with new regenerative orthopedic approaches, as well as in applications where large numbers of identical cartilage-like biomasses are desired [4, 11–13].

2 Materials

In this chapter, we describe our self-assembly method for engineering cartilage-like tissue that bears a high level of mimicry with native cartilage. We point out key steps that are most crucial to the process. The goal is to provide readers with a detailed experimental protocol to create or enhance their cartilage engineering needs. As much of the initial CTA characterization was done using bovine articular cartilage as the source of chondrocytes, we describe that approach. The protocol works successfully with all mammalian

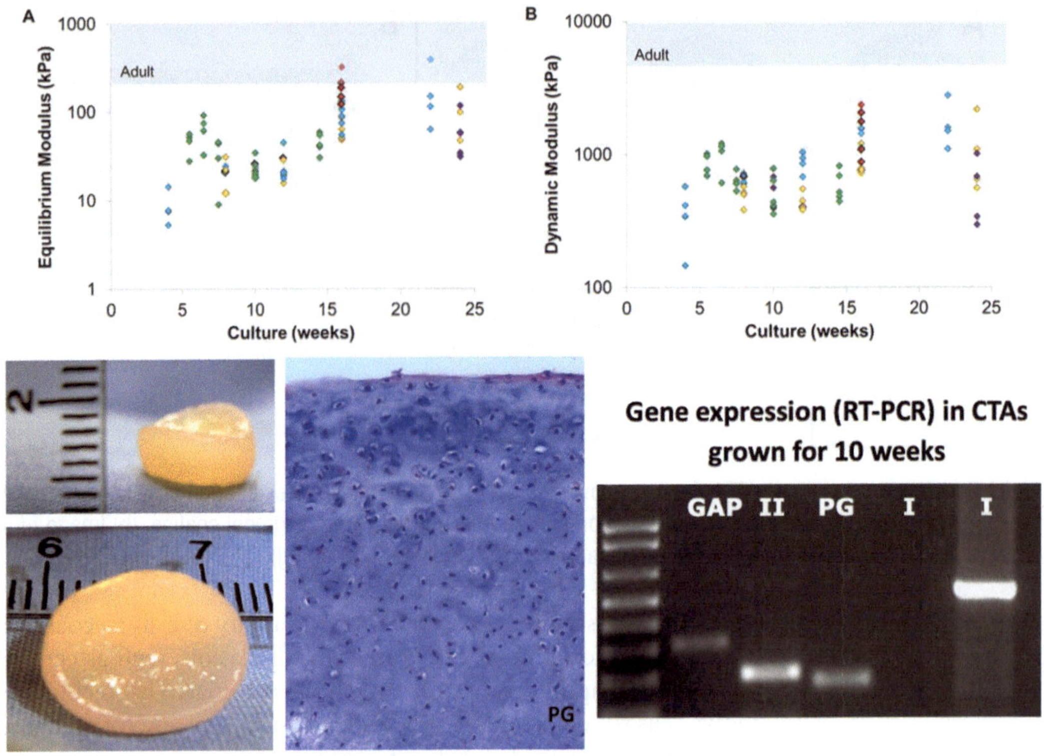

Fig. 2 CTA biomechanics over maturation time, CTA morphology, histological and basic gene expression characteristics. (**a**) Equilibrium modulus and (**b**) dynamic modulus of CTAs increased with culture duration, and approached native adult bovine cartilage properties by 16 weeks. Bovine articular cartilage chondrocytes (1×10^6/well) were grown in polyHEMA-coated plates for 10 weeks. Histological image stained with alcian blue shows the typical cell to matrix distribution of mature articular cartilage. RT-PCR demonstrates the expected constitutive cartilage genes, type II collagen (II) and aggrecan (PG). Lack of type I collagen (I) is shown alongside of synovial fibroblasts, as a type I collagen control. (Modified from Mohanraj et al., J Biomech (2014) [9] and Novotny et al., Tissue Eng (2006) [10])

cell sources explored (including pig, horse, cow, human). We describe first materials needed for harvesting chondrocytes from juvenile bovine knee joints, followed by the supplies needed to prepare the cartilage tissue analog (CTA), and their fabrication.

2.1 Harvest Supplies

1. Dissection tools.
2. Betadine and 70% ethanol (to clean the outer surface of joint prior to opening and cartilage dissection).
3. Sterile specimen cups.
4. Phosphate-buffered saline (PBS).
5. Penicillin-streptomycin-fungizone (PSF).
6. PBS with 2× PSF (Superwash).
7. Amphotericin B (to add to the Superwash at final conc 5 μg/mL).

8. Blade holder + #22 scalpel blades.
9. Petri dish (to dissect and chop cartilage).
10. Single edged "razor" blades (autoclaved).
11. Weigh boat (to weigh cleaned and chopped cartilage).
12. Collagenase type II, Worthington.
13. 50 mL filter tubes or 250–500 mL media filters (to filter media and collagenase).

2.2 Chondrocyte Isolation and CTA Generation Supplies

1. 50 mL conical bottom centrifuge tubes.
2. Mesh filters 70–100 μm.
3. Trypan blue.
4. Dulbecco's minimum essential medium with high glucose (DMEM).
5. Fetal bovine serum, final conc 10%.
6. Penicillin, final conc 100 U/mL.
7. Streptomycin, final conc 100 μg/mL.
8. Amphotericin B, final conc 2.5 μg/mL.
9. L-Glutamine, final conc 2 mM.
10. MEM vitamin solution, final conc 1%.
11. HEPES buffer, final conc 25 mM.
12. Ascorbic acid, final conc 50 μg/mL.
13. Ultra-Low Adhesion Well Plates (Corning) (available in 6-, 24-, 96-well plate formats). *Alternatively, culture vessel of any shape can be coated directly with poly-2-hydroxyethyl methacrylate (polyHEMA) a10% (volume/volume) in 90% ethanol, air-dried in a laminar flow hood and sterilized with UV light.*

3 Methods

Self-assembly protocols for developing neocartilage tissue constructs generally require three conditions: (1) an initial high seeding density in the range of several million cells per construct, (2) an optimal growth media supplemented with common chondrogenic factors, and (3) a culture substrate that facilitates cell-cell adhesion over cell-substratum attachment. Of course, one should determine appropriate conditions best suited for their needs, especially when working with a new cell source. Construct size, matrix content, and tissue mechanical properties are all affected by these three key conditions. Following these guidelines, one can achieve uniform and mature functional tissue-like structures within a matter of weeks. The first step of the process begins with obtaining a suitable cell source. Generally, the most strenuous part of the protocol will be isolating cells, if indeed working with a primary tissue source. There are common approaches to isolating various musculoskeletal cell populations, but typically require a lengthy and tedious extraction protocol with various enzymatic digestion steps [14, 15]. Ultimately the cell source will depend on the desired application or experimental use of the CTA. Well-established methods for creating hyaline cartilage use the self-assembly culture method [6, 16, 17].

Several methods are used to induce cell aggregation, including round-bottom or V-shaped culture dishes, hanging drop suspension, rotation cultures, and pellet centrifugation. The self-assembly approach described here differs from these approaches in that it does not require external input or energy. The cell-driven process replicates the natural mechanisms of tissue formation. Simply seeding dissociated cells above a non-adherent substrate, such as tissue culture plates coated with the hydrogel poly(2-hydroxyethyl methacrylate), will drive cell-cell aggregation over cell-substratum attachment [6, 10, 14].

Within hours of seeding, cells will coalesce into a tight compact structure, with integrins functioning as primary mediators of this activity [17]. N-Cadherin activity occurs shortly after and further drives neocartilage formation. A single spheroid will take shape and be visible by the naked eye within 24 h of the initial seeding (Fig. 1). From here, it's just a matter of allowing sufficient time for the tissue to mature. Cells remain proliferative for approximately the initial 3 weeks followed by a biosynthetic phase that continues where the CTA expands and occurs radially and in thickness as matrix production increases. The endpoint depends largely on the quality of tissues desired, with natural-like tissue properties developing over the course of weeks. Below are protocols for chondrocyte harvest and CTA preparation.

3.1 Chondrocyte Harvest from Bovine Femoral Condyles

1. Prepare 1 specimen cup per joint with 2× (2%v/v) penicillin-streptomycin-fungizone (PSF) in PBS (filter 2× PSF/PBS).
2. Clean the joint with Betadine followed by 70% ethanol prior to opening the joint. Take care to maintain as sterile a field as possible.
3. Typical success (lack of contamination) is from carrying out the initial dissection in a dissection with laminar flow hood. Once the cartilage is dissected from the joint surfaces (as described below) and in the boosted antibiotic antimycotic solution, it is safe to take the tissue to a tissue culture facility for further processing.
4. Once the joint surfaces are exposed of the joint of choice, score the femoral cartilage surface using sterile blade (22 blade w/ #4 holder) to score cartilage surface into grids (make squares as small as possible) and transfer pieces to the specimen cup.

5. Shake specimen cups to wash cartilage, aspirate, and repeat 2× with Superwash in PBS.
6. Transfer cartilage to petri dishes; remove cartilage that looks like it has bone marrow, vessels, or soft-fibrous tissue; and chop up cartilage into *very* (<1 mm) small pieces (need greatest surface area for digestion).
7. Weigh chopped cartilage using a specimen cup (previously weighed emptied).
8. Remove all PSF/PBS and split all chopped cartilage into specimen cups containing no more than 5 g of tissue in each. Prepare the collagenase solution.
   (a) Collagenase solution:
      (i) Dulbecco's minimum essential medium (DMEM with high glucose), with the components describe above as Complete Media.
      (ii) Dissolve collagenase type II in basal medium for a final concentration of 2 mg/mL collagenase solution.
      (iii) Sterile filter collagenase solution using a 0.22 or 0.45 μm syringe filter or bottle-top style.
      (iv) Need 40 mL (equivalent to 80 mg) collagenase solution per cup. This is appropriate for ~5 g tissue per specimen cup.
      (v) Weigh approx. 80 mg collagenase and adjust volume of Complete Media to achieve 2 mg/mL. Exact volume is not critical.
9. Put specimen cups in incubator at 37 °C on a slow-moving gyrating platform or spin plate overnight (digest for at least 14 but no more than 18 h for best CTA formation).
   (a) Mix cartilage/collagenase solution with 25 mL pipette after a few hours.
   (b) If digestion does not look complete (matrix should look amorphous), add more collagenase and digest for a few more hours if needed.

3.2 CTA Preparation

1. After the overnight digestion period (12–18 h), mix the cartilage digest with a pipette to further break up cartilage. If needed continue to incubate for an additional period of 1–2 h to ensure complete digestion before starting CTA preparation. (Note: additional collagenase can be added; however, it is not very effective unless the digest is split 1:2 with fresh media and new collagenase. This will reduce the viscosity and improve enzymic activity on remaining cartilage pieces).

2. Filter cartilage-media suspension through 70/100 μm mesh filter into a 50 mL tube (change filters out when they get blocked). Rinse the filters to ensure collecting as many cells as possible.
   (a) Split each digest into two 50 mL tubes, and dilute the digest at least 1:2 with fresh PBS to reduce the viscosity. Failure to do this dilution will result in loss of cells not pelleting due to the viscous proteoglycan dense solution.
3. Centrifuge for 20 min at 1750 RPM at 4–10 °C.
4. Remove supernatant and save temporarily to check that most cells have been spun down. If there are many cells present, split the supernatant with fresh media and spin down again.
   (a) Use a slide or hemocytometer to check for cells under microscope. With so many washes, loss of cells is typical in the wash solutions unless care is taken to ensure proper centrifugation. This is successful when the digest is dilute enough to reduce viscosity and proteoglycan content.
5. Start to combine tubes when resuspending pellets in fresh PBS with 2× PSF, centrifuge again (wash process), and then repeat this process 3× until there is only one tube after the last wash.
   (a) Suspend cells by gentle pipetting in 20 mL of Complete Medium for counting.
6. Count cells using hemocytometer or automatic cell counter.
7. Adjust volume based on cell count and plate format being used. If necessary, centrifuge again to reduce volume, and remove supernatant and resuspend in desired volume of Complete Medium.
8. Choose the appropriate culture plate and wet the hydrogel surface with enough Complete Media to cover the well, and wait 10 min in 37 °C incubator.
9. Remove the wetting solution with care not to scratch the surface of the plate. Scratching the plate surface will result in cell attachment and compromise the CTA formation.
10. Seeding densities per plate (Corning Ultra-Low Adhesion Well Plates):
    (a) 6-well plate: 100 million cells/well in 5 mL of media.
    (b) 24-well plate: 10 million cells/well in 2 mL of media.
    (c) 96-well plate: 1 million cells/well in 200uL of media (note: stock cell suspension should be 5 million cells/mL before seeding).

11. 24 h later (next day), do a "half-media change" (CTAs should be mostly formed—if not formed, wait another 24 h before doing a full-media change— remove all media and replace with fresh media).
    (a) Half-media change: carefully remove half the media in the well and replace with fresh Complete Media.
12. Feed two to three times weekly taking care to not disturb the CTA. The volume is not critical once the cells have coalesced. Since CTA form rather quickly, after 1–2 weeks, one can remove nearly all the media and it can be replaced during the media changes.
13. After 2–4 weeks, using the 96-well format and fabricating 100s of CTA, there is a logistical benefit to moving the CTAs to a 6-well plate (10–20 CTA/well). This enables easier media removal and greater volumes of media used for longer periods.
14. CTAs can be grown for many weeks in culture, and additional manipulations such as mechanical loading and growth factors including TGFb or FGFs will enhance the extracellular matrix formed and the speed in which a more native cartilage-like biomaterial is formed.
15. Using the CTA culture approach described here, one can generate hundreds of identical cartilage surrogates for experimental procedures such as drug screening and cartilage defect repair approaches.

4 Notes

1. The initial seeding density is critical for the formation of tightly compact and uniform tissue structures. Too few cells will create irregular, non-spheroid shapes, where cells in excess may not enhance construct formation and will deplete nutrition and increase risk of contamination. Typically, one to five million cells are needed to create cartilage tissue construct that is millimeters in dimension [9]. The literature cited in this text will serve as reference point to guide initial cell seeding density; however, according to the goal of the CTA or its use, it is recommended to test a range of seeding densities. This is particularly critical when using a new cell source such as different species or mesenchymal stem cells instead of differentiated chondrocytes. Our experience has been that even if the cells are not in a solid mass as shown in Figs. 1 and 2, the clusters of cells still remain phenotypically correct in terms of their constitutive cartilage gene expression. We suggest finding the minimum number of cells required to create uniform spheroid structure. Once that information is known, slightly altering cell number can further enhance construct size or shape, as well as the biochemical properties.

2. If long-term culture and/or more mature CTA is the goal (>4 weeks), then after CTA are formed, they should be "scooped up" with a sterile spatula and placed in a new low adhesion plate to (1) enable greater volumes of media to be used and reduce tedious media changes, (2) prevent contamination, and (3) decrease probable scratching of the surface of the plate and having adherent cells in the culture.
3. Our experience has been human adult (normal and OA) cartilage-derived chondrocytes form CTA, however, typically smaller, and grow more slowly (OA tissue derived) [10].
4. Mesenchymal stem cells can also be used and will form CTA when chondrogenic factors are added to the process; however, their formation and characteristics typically require 3 or more weeks to show the same growth profile as differentiated cells [9].

References

1. Huey DJ, Hu JC, Athanasiou KA (2012) Unlike bone, cartilage regeneration remains elusive. Science 338(6109):917–921. https://doi.org/10.1126/science.1222454
2. Athanasiou KA, Shah AR, Hernandez RJ et al (2001) Basic science of articular cartilage repair. Clin Sports Med 20(2):223–247. https://doi.org/10.1016/s0278-5919(05)70304-5
3. Elder BD, Athanasiou KA (2008) Effects of confinement on the mechanical properties of self-assembled articular cartilage constructs in the direction orthogonal to the confinement surface. J Orthop Res 26(2):238–246. https://doi.org/10.1002/jor.20480
4. Mohanraj B, Meloni GR, Mauck RL et al (2014) A high-throughput model of post-traumatic osteoarthritis using engineered cartilage tissue analogs. Osteoarthritis Cartilage/OARS, Osteoarthritis Res Soc 22(9):1282–1290. https://doi.org/10.1016/j.joca.2014.06.032
5. Ofek G, Revell CM, Hu JC et al (2008) Matrix development in self-assembly of articular cartilage. PLoS One 3(7):e2795. https://doi.org/10.1371/journal.pone.0002795
6. Estrada LE, Dodge GR, Richardson DW et al (2001) Characterization of a biomaterial with cartilage-like properties expressing type X collagen generated in vitro using neonatal porcine articular and growth plate chondrocytes. Osteoarthritis Cartilage/OARS, Osteoarthritis Res Soc 9(2):169–177. https://doi.org/10.1053/joca.2000.0373
7. Dodge GR, Diaz A, Sanz-Rodriguez C et al (1998) Effects of interferon-gamma and tumor necrosis factor alpha on the expression of the genes encoding aggrecan, biglycan, and decorin core proteins in cultured human chondrocytes. Arthritis Rheum 41(2):274–283. https://doi.org/10.1002/1529-0131(199802)41:2<274::AID-ART11>3.0.CO;2-Z
8. Reginato AM, Sanz-Rodriguez C, Diaz A et al (1993) Transcriptional modulation of cartilage-specific collagen gene expression by interferon gamma and tumour necrosis factor alpha in cultured human chondrocytes. Biochem J 294(Pt 3):761–769. https://doi.org/10.1042/bj2940761
9. Mohanraj B, Farran AJ, Mauck RL et al (2014) Time-dependent functional maturation of scaffold-free cartilage tissue analogs. J Biomech 47(9):2137–2142. https://doi.org/10.1016/j.jbiomech.2013.10.022
10. Novotny JE, Turka CM, Jeong C et al (2006) Biomechanical and magnetic resonance characteristics of a cartilage-like equivalent generated in a suspension culture. Tissue Eng 12(10):2755–2764. https://doi.org/10.1089/ten.2006.12.2755
11. Johnstone B, Alini M, Cucchiarini M et al (2013) Tissue engineering for articular cartilage repair--the state of the art. Eur Cell Mater 25:248–267. https://doi.org/10.22203/ecm.v025a18
12. Zhang L, Hu J, Athanasiou KA (2009) The role of tissue engineering in articular cartilage repair and regeneration. Crit Rev Biomed Eng 37(1–2):1–57. https://doi.org/10.1615/critrevbiomedeng.v37.i1-2.10
13. Mohanraj B, Hou C, Meloni GR et al (2014) A high throughput mechanical screening device for cartilage tissue engineering. J Biomech

47(9):2130–2136. https://doi.org/10.1016/j.jbiomech.2013.10.043

14. Revell CM, Reynolds CE, Athanasiou KA (2008) Effects of initial cell seeding in self assembly of articular cartilage. Ann Biomed Eng 36(9):1441–1448. https://doi.org/10.1007/s10439-008-9524-x
15. Hu JC, Athanasiou KA (2006) The effects of intermittent hydrostatic pressure on self-assembled articular cartilage constructs. Tissue Eng 12(5):1337–1344. https://doi.org/10.1089/ten.2006.12.1337
16. Kraft JJ, Jeong C, Novotny JE et al (2011) Effects of hydrostatic loading on a self-aggregating, suspension culture-derived cartilage tissue analog. Cartilage 2(3):254–264. https://doi.org/10.1177/1947603510383686
17. Kim M, Kraft JJ, Volk AC et al (2011) Characterization of a cartilage-like engineered biomass using a self-aggregating suspension culture model: molecular composition using FT-IRIS. J Orthop Res 29(12):1881–1887. https://doi.org/10.1002/jor.21467

Chapter 23

Osteochondral Explant Isolation and Culture Under a Compression and Shear Bioreactor

M. Letizia Vainieri and Sibylle Grad

Abstract

Osteochondral explants harvested from different species are valuable preclinical ex vivo models for tissue engineering research. In this chapter, we describe the isolation of osteochondral plugs from bovine stifle joints, followed by defect creation, and plug preparation in a straightforward manner before mechanical loading using a compression and shear bioreactor. The method can be adapted to isolate osteochondral plugs from any animal species and to load explants in any type of bioreactor.

Key words Articular cartilage, Osteochondral defects, Ex vivo model, Bioreactor

1 Introduction

Articular joint injury poses a significant clinical challenge in orthopedics. Advances in recent decades are placing cartilage regeneration in the spotlight, paving the potential to overcome limitations of current treatments. Osteochondral ex vivo explants, in which chondral or osteochondral defects can be created, are valuable models for translational research [1]. As defects in the osteochondral unit often occur in the weight-bearing region, it appears inevitable that repair of focal traumatic lesions should take into account mechanical loading as an essential factor influencing osteochondral tissue regeneration [2]. In contrast to two- and three-dimensional cell and biomaterial culture models [3, 4], the osteochondral explant culture model allows investigations of the interplay between cellular and extracellular signals involved in cartilage repair [5, 6]. Explant models are also invaluable for assessing integration of a tissue-engineered graft with the surrounding cartilage, which is critical for its function and presents a significant challenge in the field. Although tremendous progress has recently been made to generate functional tissue and several cell-seeded and acellular biomaterials have been investigated [7–10], there is still much to be

Martin J. Stoddart et al. (eds.), *Cartilage Tissue Engineering*,
Methods in Molecular Biology, vol. 2598, https://doi.org/10.1007/978-1-0716-2839-3_23,

learned. To date inadequate biomechanical stability of the graft has frequently been observed [11], and the mismatch between repair and native tissue following surgery still remains one of the major clinical challenges contributing to a continued disruption in joint biomechanics and repair failure. This unsatisfactory repair scenario demonstrates the need for improved treatments. Since translation toward clinics needs proof of functionality, the development of ex vivo models under mechanical stimuli, closely representing an articulating joint in vivo, would allow an accurate screening of implants and factors in order to preselect promising conditions for in vivo studies.

Here we describe the protocol for the generation and preparation of an ex vivo osteochondral defect model under mechanical compression and shear that provides a representative physiological joint-like environment, which can allow reproducible prediction of biomaterial graft performance and of biomolecule treatment efficacy [12]. Osteochondral explants were isolated from the patellar groove of bovine stifle joints to obtain plugs with flat articular cartilage surface, and cartilage defects of 4 mm diameter were created. The explants were prepared to be used as an interface against a ceramic ball applying dynamic compressive and shear loading at the desired magnitude. Osteochondral defects can be filled with biomaterial and biomolecules and then subjected to mechanical stimulation. The implementation of mechanical forces is essential for preclinical testing of novel technologies in the field of cartilage repair, biomaterial engineering, and regenerative medicine [13].

2 Materials

Osteochondral explants were harvested from bovine stifle joints of 3- to 5-month-old calves, obtained from a local abattoir (Angst AG, Zurich, CH) within 48 h of slaughter. All solutions are prepared using Milli-Q water, phosphate-buffered saline (PBS), and tissue culture medium in sterile hood (unless indicated otherwise). Diligently follow all waste disposal regulations when disposing waste materials.

1. Prepare 50 mL conical centrifuge tubes containing 10 mL of sterile PBS supplemented with either 1% or 10% penicillin-streptomycin (Pen-Strep, Gibco) to be used for washing steps: (1) Wash in 1% Pen-Strep/PBS; (2) wash in 10% Pen-Strep/PBS; (3) wash in 1% Pen-Strep/PBS. The number of tubes depends on the experiment, usually using one tube per five plugs harvested from the same joint.
2. Prepare high-glucose Dulbecco's Modified Eagle Medium (DMEM-HG) supplemented with 1% Pen-Strep and 10% fetal bovine serum (FBS) for overnight culture.

3. Fermacidal™.
4. 70% ethanol.
5. Two scalpel holders (#4) and several blades (#20).
6. Two surgical tissue forceps (containing "teeth" on the tip to grasp a tissue).
7. Surgical drapes (Foliodrape, Hartmann).
8. Saline solution (4 °C).
9. Infusion pump (Intrafix SafeSet).
10. Bosch compact drill press (PBD 40).
11. Plastic box underneath the Bosch press.
12. Small circular saw (e.g., Proxxon bench circular saw KS 230).
13. Tools needed for osteochondral plugs harvesting and defect creation (Figs. 1 and 2):
    1. Figure 1a.1 Adjustable table.
    2. Figure 1a.2 PEEK plate placed on top of the adjustable table to protect it from drilling as shown in Fig. 1b.
    3. Figure 1a.3 Trephine (Ø 8.35 mm) for plug harvesting.
    4. Figure 1a.4 Steel tube to remove plug harvested and stuck within the trephine as shown in Fig. 1b.
    5. Figure 1c Steel plug holder for bone horizontal cutting using the small circular saw.
    6. Figure 2a.1 Trephine (Ø 4 mm) for defect creation.
    7. Figure 2a.2 Steel plug holder for defect creation.
    8. Figure 2a.3 Steel block to lift the steel plug holder as shown in Fig. 2b.

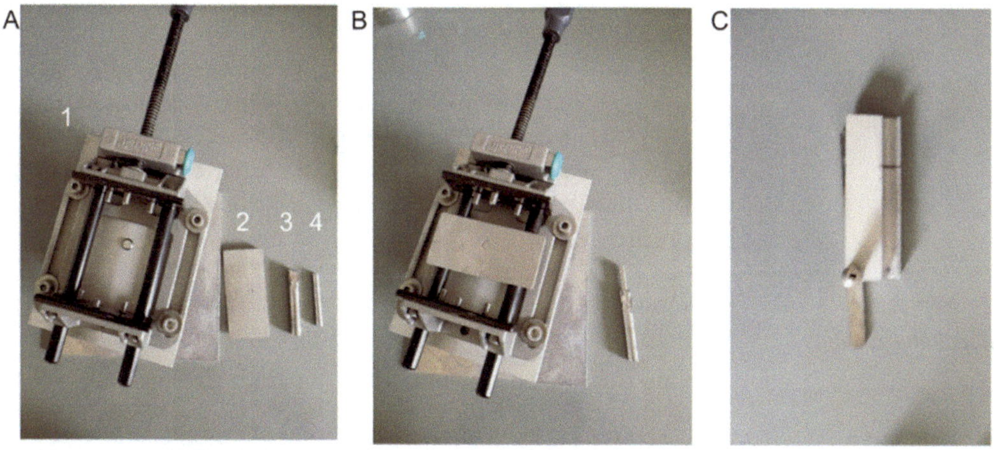

Fig. 1 Tools for osteochondral plug harvesting. (**a**) (1) Adjustable table for clamping the femur to firmly hold the joint. (2) PEEK plate placed on top of the adjustable table to protect it from drilling as shown in **b**. (3) Trephine (Ø 8.35 mm) for plug harvesting. (4) Steel tube to remove harvested plug stuck within the trephine as shown in **b**. (**c**) Steel plug holder for bone horizontal cutting using the small circular saw

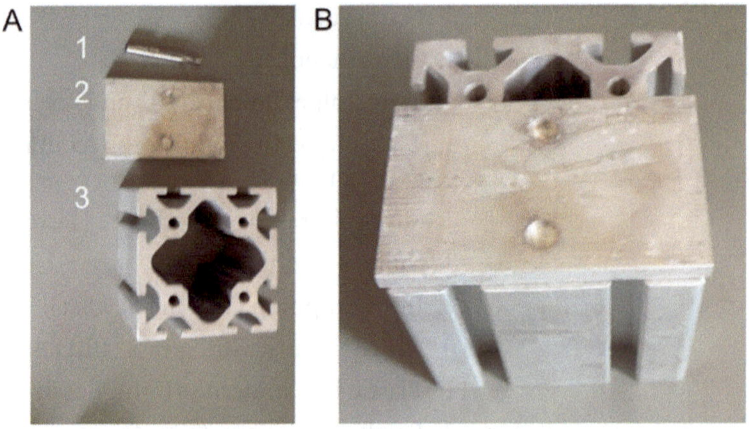

Fig. 2 Tools used to create chondral/osteochondral defects. (**a**) (1) Trephine (Ø 4 mm) for defect creation. (2) Steel plug holder for plug defect creation. (3) Steel block to lift the steel plug holder as shown in **b**

14. Tools needed to load plugs in the bioreactor (Fig. 3):
    1. Figure 3.1 Black blocks (custom made from PEEK).
    2. Figure 3.2 White polyethylene rings (different Ø, ranging from 7.3 to 7.9 mm).
    3. Figure 3.3 PEEK holders for bioreactors (top, middle) and for unloaded control (bottom).

3 Methods

3.1 Preparation of Working Space

1. In the dissection room, clean working surfaces first with water, then Fermacidal™, and 70% ethanol (EtOH).
2. Cover the space aside the drilling machine, the bench, and the hood with sterile surgical drapes.
3. Pipe the roll controller infusion pump (Intrafix SafeSet) through the port of entry of the saline solution. Hook the saline solution up to the hanger-like tool (S shape above the drilling machine), and connect the other end of the pump to the steel tube of the drilling machine to help deliver a precise dose.

3.2 Harvesting Osteochondral Plugs

1. On the bench: use scalpel holder and blade to cut the synovial membrane around the joint, so that the joint will open.
2. Cut the cruciate ligaments to separate the menisci and tibia from the femur condyles.
3. Keep cutting the muscles and ligaments around to separate the patella and tibial side from the condyles and femoral side.

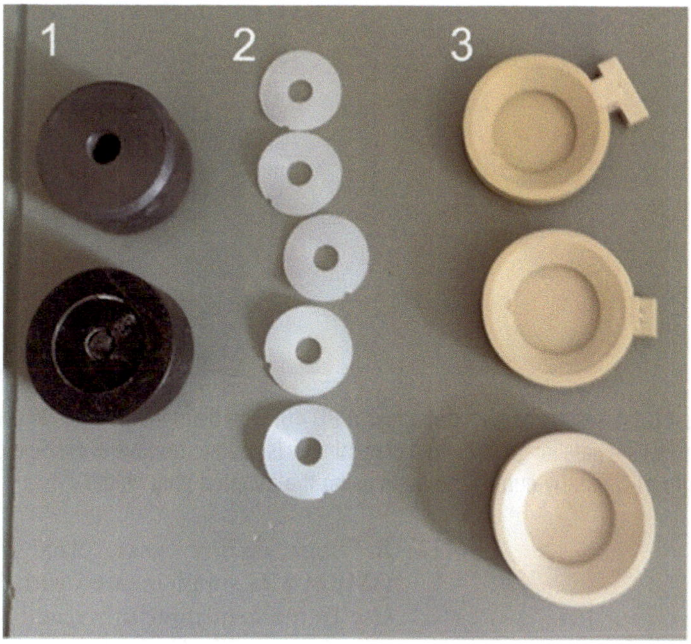

Fig. 3 Tools used for culture in the bioreactor. (1) Black PEEK blocks used for holding the plugs and press-fitting the plugs into the white rings. (2) White polyethylene rings (different Ø, ranging from 7.3 to 7.9 mm) to be pressed onto the bony part of the plug and afterward press-fitted into the PEEK holders for bioreactors/unloaded controls. (3) PEEK holders for bioreactors (top, middle) and for unloaded control (bottom) holding the white ring with the press-fitted plug

4. Place a piece of surgical drape above the adjustable table to better hold the femur with the cartilage surface of the patellar groove facing upward.
5. Clamp the femur on the adjustable table to firmly hold the joint.
6. Open the tube of saline solution so that the cartilage is kept wet throughout the whole procedure.
7. Use trephine diamond drill (Ø 8.35 mm) to drill osteochondral plugs in the cartilage of the patellar groove (Fig. 4a). Stop drilling when the whole subchondral bone is pierced (*see* **Note 1**).
8. Remove plugs from the trephine by pushing the steel tube (Fig. 1a.4) into the trephine (Fig. 1b), and rinse the plugs in PBS solution containing 1% Pen-Strep. Keep the plugs in this solution until all the plugs are harvested.
9. Shorten the plugs to a length of 5 mm by cutting the bone with small circular saw and using the adapted plug holder (Fig. 4b).
10. If defect creation is not needed, skip Subheading 3.3 and then proceed with culture of the plugs as detailed below.

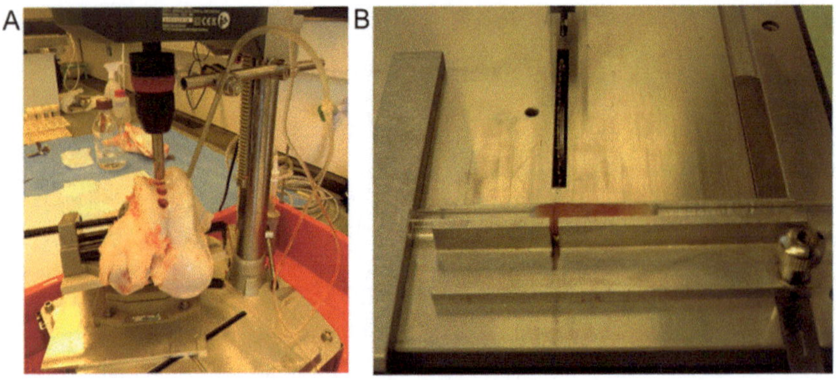

Fig. 4 (**a**) A Bosch compact drill press (PBD 40) with diamond trephine is used to harvest osteochondral plugs from bovine patellar groove. (**b**) A circular saw is used to remove exceeding bone. The plugs are shortened to a length of 5 mm by cutting the bone using the adapted plug holder

11. *In tissue culture room*, place the osteochondral plugs in DMEM-HG supplemented with 1% Pen-Strep and 10% FBS. Use 15 mL centrifuge tubes, add 3 mL of medium, and culture overnight at 37 °C, 5% CO_2, 95% humidity in the incubator (*see* **Note 2**).
12. The following day check for sterility, dispose of the plugs showing sign of contamination, and start with the specific experiment.

3.3 Chondral or Osteochondral Defect Creation

1. Once all plugs have been harvested, change trephine (Ø 4 mm) and place the steel block with plug holder.
2. Place the osteochondral plug in the plug holder, and switch on the laser integrated in the Bosch compact drill press (red circle button) to identify the center of the plug (Fig. 5a; *see* **Note 3**).
3. To create a defect (chondral/osteochondral) with a specific depth, press the black round button (Bosch drilling machine) so that the display shows the height (mm) instead of speed (rpm) (Fig. 5b).
4. Lower the trephine until it reaches and slightly touches the articular surface of the plug.
5. While the trephine is touching the articular surface (this will be the 0 point), press the white circular button (Bosch drilling machine, Fig. 5b), so that the digital drilling depth will show 0 mm.
6. Go up with the trephine, switch on the drill, and go down again to create the desired defect depth (*see* **Note 4**).
7. Rinse the plug in 10% Pen-Strep solution in PBS. Leave the plugs in solution until all the defects are created (rinse for 10 min at least).
8. Rinse plugs in 1% Pen-Strep solution in PBS for 10 min.

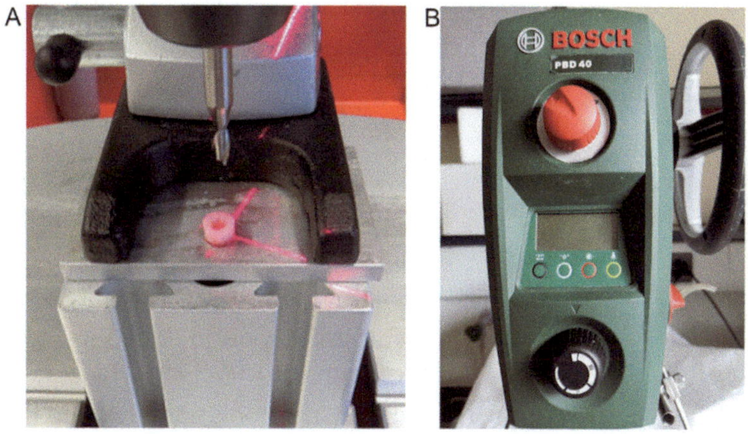

Fig. 5 (**a**, **b**) How to create chondral/osteochondral defects in the center of the plug. (**a**) The osteochondral plug is placed in the steel plug holder, and the laser (integrated in the Bosch compact drill press; red circle button) is switched on to identify the center of the plug. (**b**) To create a defect (chondral/osteochondral) with a specific depth, the black round button (Bosch drilling machine) is pressed so that the display shows the height (mm) instead of speed (rpm)

9. *In the tissue culture room*, place osteochondral plugs in DMEM-HG supplemented with 1% Pen-Strep and 10% FBS. Use 15 mL centrifuge tubes, add 3 mL of medium, and culture overnight at 37 °C, 5% CO_2, 95% humidity in the incubator (*see* **Note 2**).
10. The following day check for sterility, discard the plugs that show sign of contamination, and proceed according to the specific experiment.

3.4 Clean and Sterilize Tools

1. Remove all the tools used for plug harvesting and place them in a plastic beaker with water and soap.
2. Discard blades in the blade disposal.
3. Clean the drilling machine with soap and water, also underneath.
4. Use a toothbrush with soap and water to remove blood and bone pieces from the Bosch machine and from the trephine holder of the Bosch compact drill press (very important, since the machine can be sterilized only once per year).
5. Rinse the Bosch machine with water.
6. Remove the Bosch machine from the box and clean it with 70% EtOH. Let it air-dry.
7. Remove the saline solution and watery soap drained into the plastic box. Make sure the bony pieces are not discarded in the sink.

8. Rinse the box with Fermacidal and 70% EtOH. Let it air-dry, place the Bosch machine back into the box, and cover everything with plastic sheet.
9. Clean the surface of the small circular saw, then carefully spray 70% EtOH on the saw, and wipe everything with tissues. On the right side of the circular saw, there is a screw with a skew tool insert, unscrew it, and clean all the blood and bony pieces trapped inside.
10. Clean all the surfaces with water, Fermacidal, and 70% EtOH.
11. Sterilize the tools indicated in the following figures by using the steam autoclave: Figs. 1a.2–4, c, 2a.2, 3, and 3.1–3.
12. Sterilize the tools indicated in the following figures using ethylene oxide: Figs. 1a.1, and 2a.1. In addition, sterilize the circular saw machine if you have drilled human samples.
13. Bosch machine can be sterilized once per year in ethylene oxide. If human material is used, please be extra careful in cleaning and disinfecting the machine (especially inside the trephine machine holder).
14. Store autoclaved instruments at a dry place.

3.5 Fixation of Osteochondral Plugs in Bioreactor Loading Holders

1. After overnight incubation and selection of sterile plugs, prepare 2% low-melting agarose (Lonza). Dissolve it in sterile PBS by boiling using a microwave. Allow to cool in the sterile hood (to avoid heat-induced cell damage), but make sure it does not solidify.
2. Use black holder to insert the osteochondral plug with cartilage side facing down (Fig. 6a). Screw it from the bottom to accommodate the whole plug until it reaches the exact length (about 6 mm). Follow this procedure only for plugs undergoing load (*see* Fig. 3.1, top and middle PEEK holders). For unloaded plugs go to **step 9** of this section.
3. Use white polyethylene rings to hold the plugs in the bioreactor holder (*see* Fig. 3.2 and **Note 5**).
4. Press-fit the white ring (in Fig. 6a is a gray ring made of PEEK, please use the white polyethylene one) onto the bony part of the plug in the black holder.
5. Unscrew from the bottom of the black holder to have the plug aligned with the polyethylene ring (Fig. 6b).
6. Take the press-fit polyethylene ring and the plug from the black holder, and further press-fit it into the PEEK holder (Fig. 6c).
7. Repeat this procedure with all the plugs that undergo load. Be sure that the polyethylene ring is placed firmly on the bottom of the PEEK holder. While repeating this procedure with all the plugs, make sure that the other explants are always covered with medium.

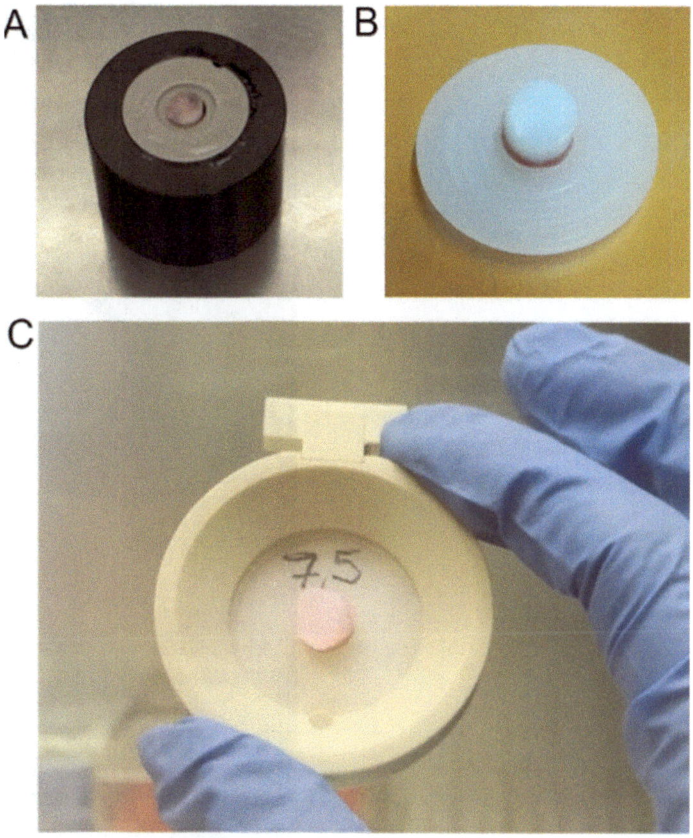

Fig. 6 (**a**–**c**) How to hold the plug in the PEEK holder. (**a**) The white polyethylene ring (in the fig. it is a gray ring made of PEEK) is pressed onto the bony part of the plug in the black holder. (**b**) The plug is aligned with the polyethylene ring and removed from the black holder. (**c**) The ring with the plug is press-fitted into the PEEK holder

8. Remove medium.
9. Slightly cut the tip of the P1000 pipette tip and add 1–2 mL of agarose to the holders until the bony part is fully covered (*see* **Note 6**).
10. Wait until the agarose solidifies.
11. Repeat this step for all the plugs harvested (*see* **Note** 7).
12. Add 3 mL of medium (according to the specific experiment) to the PEEK holders, and culture in the incubator or load in the bioreactor (depending on the experiment performed, Fig. 7).
13. After loading, place the plugs + PEEK holders in 10 mm Petri dish and culture them in the incubator (*see* Subheading 3.2).

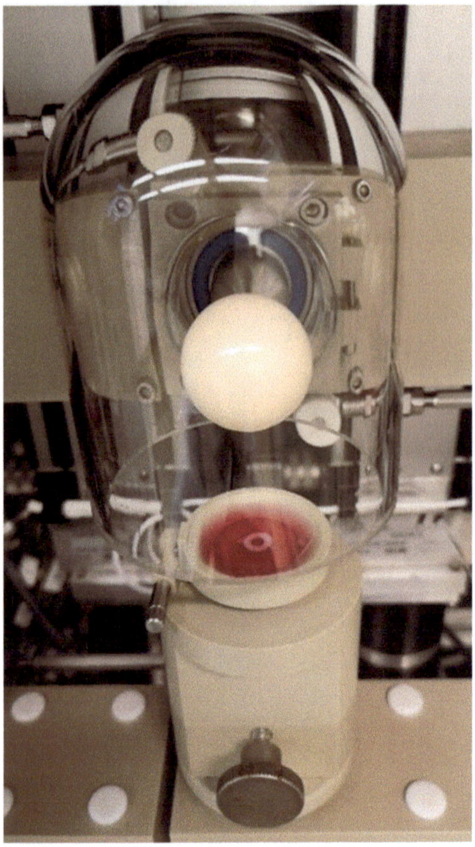

Fig. 7 Osteochondral defect explant cultured under mechanical stimulation in compression and shear load bioreactor

4 Notes

1. Care should be taken if the plugs are used for loading within the bioreactor; cut in such a way that the articular cartilage surface is as flat as possible. This is to provide a constant displacement and full contact between the articulating interface and the cartilage surface. Depending on the size of the joint, five to six flat plugs are harvested per patellar groove.
2. Culture osteochondral plugs in conical centrifuge tubes leaving the lids slightly open in the incubator to permit gas exchange if you use tubes without filter caps.
3. Care should be exercised about the position of the plug while placing the steel plug holder for defect creation, since the center of the laser could not necessarily reflect the center of the plug. The downside of creating non-centric defects is that the articulating interface pressed onto the osteochondral explant results in an unequal distribution of load, since then the center of the plug will not correspond to the center of the defect.

4. Pay attention to the millimeters on the digital screen of the Bosch compact drill press. Once the 0 point is reached, the articular surface will be touched, and from this point onward, the more you drill, the deeper will be the defect.
5. Engrave on each polyethylene ring the diameter of the inner circle created, in order to exactly press-fit the osteochondral explants to fully support the loading. Since biological samples are used and they tend to have slightly different diameters, a selection of polyethylene rings ranging from 7.3 to 7.9 mm should be available.
6. Low-melting gelling agarose is added to tissue culture plate or bioreactor holders to cover the subchondral bone and prevent cell outgrowth during culture.
7. Do not keep the osteochondral explants too long without medium, in order to avoid samples drying out and in turn cell death.

Acknowledgments

This project has received funding from the European Union's Horizon 2020 research and innovation programme under Marie Sklodowska-Curie Grant Agreement No. 642414.

References

1. de Vries-van Melle ML, Mandl EW, Kops N et al (2012) An osteochondral culture model to study mechanisms involved in articular cartilage repair. Tissue Eng Part C Methods 18(1):45–53. https://doi.org/10.1089/ten.TEC.2011.0339
2. Peroglio M, Gaspar D, Zeugolis DI et al (2018) Relevance of bioreactors and whole tissue cultures for the translation of new therapies to humans. J Orthop Res 36(1):10–21. https://doi.org/10.1002/jor.23655
3. Grad S, Eglin D, Alini M et al (2011) Physical stimulation of chondrogenic cells in vitro: a review. Clin Orthop Relat Res 469(10): 2764–2772. https://doi.org/10.1007/s11999-011-1819-9
4. Stoddart MJ, Ettinger L, Hauselmann HJ (2006) Enhanced matrix synthesis in de novo, scaffold free cartilage-like tissue subjected to compression and shear. Biotechnol Bioeng 95(6):1043–1051. https://doi.org/10.1002/bit.21052
5. de Vries-van Melle ML, Narcisi R, Kops N et al (2014) Chondrogenesis of mesenchymal stem cells in an osteochondral environment is mediated by the subchondral bone. Tissue Eng A 20(1–2):23–33. https://doi.org/10.1089/ten.TEA.2013.0080
6. Vainieri ML, Lolli A, Kops N et al (2020) Evaluation of biomimetic hyaluronic-based hydrogels with enhanced endogenous cell recruitment and cartilage matrix formation. Acta Biomater 101:293–303. https://doi.org/10.1016/j.actbio.2019.11.015
7. Tan AR, Hung CT (2017) Concise review: mesenchymal stem cells for functional cartilage tissue engineering: taking cues from chondrocyte-based constructs. Stem Cells Transl Med 6(4):1295–1303. https://doi.org/10.1002/sctm.16-0271
8. Taraballi F, Bauza G, McCulloch P et al (2017) Concise review: biomimetic functionalization of biomaterials to stimulate the endogenous healing process of cartilage and bone tissue. Stem Cells Transl Med 6(12):2186–2196. https://doi.org/10.1002/sctm.17-0181
9. Antunes BP, Vainieri ML, Alini M et al (2020) Enhanced chondrogenic phenotype of primary

bovine articular chondrocytes in Fibrin-Hyaluronan hydrogel by multi-axial mechanical loading and FGF18. Acta Biomater 105: 170–179. https://doi.org/10.1016/j.actbio.2020.01.032

10. Lolli A, Sivasubramaniyan K, Vainieri ML et al (2019) Hydrogel-based delivery of antimiR-221 enhances cartilage regeneration by endogenous cells. J Control Release 309:220–230. https://doi.org/10.1016/j.jconrel.2019.07.040
11. Khan IM, Gilbert SJ, Singhrao SK et al (2008) Cartilage integration: evaluation of the reasons for failure of integration during cartilage repair. A review. Eur Cell Mater 16:26–39. https://doi.org/10.22203/ecm.v016a04
12. Vainieri ML, Wahl D, Alini M et al (2018) Mechanically stimulated osteochondral organ culture for evaluation of biomaterials in cartilage repair studies. Acta Biomater 81:256–266. https://doi.org/10.1016/j.actbio.2018.09.058
13. Vainieri ML, Alini M, Yayon A et al (2020) Mechanical stress inhibits early stages of endogenous cell migration: a pilot study in an ex vivo osteochondral model. Polymers (Basel) 12(8):doi:10.3390/polym12081754

Chapter 24

A Mouse Model of Acute Cartilage Injury and Repair

Anne-Sophie Thorup, Francesco Dell'Accio, and Suzanne E. Eldridge

Abstract

Chondral defects are common and disabling. The development of pharmacological approaches for cartilage repair requires the availability of in vivo models which are amenable for gain and loss of function and ideally to genetic modification. In this chapter, we describe a method to induce full-thickness cartilage defects which, in young DBA/1 mice, heal spontaneously, but fail to heal in C57BL/6 mice of the same age or in aged DBA/1 mice. This model (or variants) has been used for genetic screenings to identify genes associated to repair capacity, to study stem cells involved in cartilage repair, and to study the function of molecules involved in repair mechanisms.

Key words Cartilage, Joint injury, Repair, Mouse models, Osteoarthritis

1 Introduction

Cartilage defects in the joints are reported in 61% of all arthroscopies [1, 2]. In young and healthy individuals, relatively small acute cartilage defects can heal, but with aging, increased bone mass, and comorbidities, this capacity to heal is progressively reduced [3]. Chronic lesions can drive further cartilage loss [3] and therefore will require treatment. Available treatments include microfracture, which however leads to relatively short-lived results [4]; autologous chondrocyte implantation, which however is too costly and cumbersome for most centers; and unloading strategies [3].

There is a clinical need for a pharmacological approach to promote cartilage regeneration. Cartilage defects can evolve to osteoarthritis, in which abnormal load results in cartilage breakdown, joint pain, and reduced mobility. Osteoarthritis is the leading cause of permanent disability and absenteeism and affects up to 1/3 of the people over 60 [5]. To identify molecular targets to achieve cartilage healing, a mouse model is an essential tool because it enables gain- and loss-of-function experiments using mouse genetics and traditional pharmacological approaches.

Martin J. Stoddart et al. (eds.), *Cartilage Tissue Engineering*,
Methods in Molecular Biology, vol. 2598, https://doi.org/10.1007/978-1-0716-2839-3_24,

This protocol describes a method to study the healing of full-thickness cartilage defects, in which the outcome is age and strain dependent: young DBA/1 mice will heal and therefore can be used for loss-of-function approaches, whereas aged DBA/1 or C57BL/6 mice of any age will not heal and can be used for gain-of-function approaches.

2 Materials

1. Mice (*see* **Notes 1** and **2**).
2. Weighing scales.
3. Anesthetic (example can include isoflurane for inhalation or ketamine and xylazine for parenteral administration).
4. Electric razor.
5. 70% ethanol.
6. Dissection microscope and appropriate lighting.
7. Extra Narrow Scissors (FST 14088-10).
8. Bonn Micro Forceps (FST 11084-07).
9. Blade Holder & Breaker—Flat Jaws (FST 10052-11).
10. Scalpel Blades—Breakable (FST 10050-00).
11. Mathieu needle holders (FST).
12. 26G needle.
13. Glass beads (4 mm diameter)—optional (*see* **Note 3**).
14. Vicryl sutures 6.0 absorbable 13 mm 1/2 circle taper point plus needle.
15. Ethilon sutures 5.0 atraumatic needle.
16. Recovery cage.
17. Heat pad.
18. Bone scissors.

3 Methods

Medial parapatellar arthrotomy:

1. Preheat a heat pad at approximately 22–25 °C.
2. Place a sterile drape over the heat pad.
3. Prepare a clean recovery cage with plenty of bedding and hiding place and a small towel over the bedding, and place it over another heat pad set at 22–25 °C.
4. Prepare a drape with a 2 cm cleft for putting through the operated leg. The opening should be approximately 3–4 cm from one margin and perpendicular to it, so that the drape will

not reach the head and the breathing can be constantly observed during the operation. It is useful if the drape is made of waterproof material to protect the body from becoming wet when sterilizing the limb with ethanol spray. Typically, we use the sleeve of a glove that has been cut out to size.

5. Prepare a tray to keep the surgical tools clean.
6. Weigh the mouse (*see* **Note 2**).
7. Give mouse health score (*see* **Note 4**) [6].
8. Anesthetize using appropriate anesthetic and route of delivery (*see* **Note 5**).
9. Once anesthetized (confirmed by pinching of foot/tail), use an electric razor to shave the mouse knee.
10. Place the anesthetized mouse on its back over the drape and the heating pad with its snout into the isoflurane mask (if using the inhalation method).
11. Cover the mouse with the surgical drape, with the limb to be operated passed through the cleft. Make sure that the head is not covered so that the operator can monitor that the snout is constantly in the mask and the breathing rate.
12. Spray the limb with 70% ethanol or wipe it with chlorhexidine.
13. Open joint skin by holding taut with the micro forceps and make incision using the extra narrow scissors. Separate the skin from the muscle by blunt dissection.
14. Place the mouse under the dissection microscope. Fully extend the knee. Open the joint capsule by inserting the scalpel at the medial side of the insertion of the patellar tendon on the tibial tuberosity with the cutting edge oriented toward yourself, and cut alongside the medial margin of the patellar tendon and, further, along the medial margin of the vastus medialis.
15. Observe by eye and confirm the joint structures are visible and appear normal (Fig. 1a).
16. With the joint still fully extended, dislocate the patella laterally using the micro forceps (Fig. 1b). Flex the knee so that the patella stays in place over the lateral side of the knee.
17. To make the longitudinal full-thickness injury in the patellar groove using a 26G needle, place tip of the needle anteriorly to the intercondylar notch, and gently move it upward along the entire length of the patellar groove (Fig. 1c) (*see* **Note 3**). Repeat this until the bone is reached. This can be ascertained by the presence of bleeding from the subchondral bone.
18. Reposition the patella within the groove (Fig. 1d).
19. Suture the joint capsule and secure the patella in place with Vicryl 6.0 resorbable sutures and Mathieu needle holder.

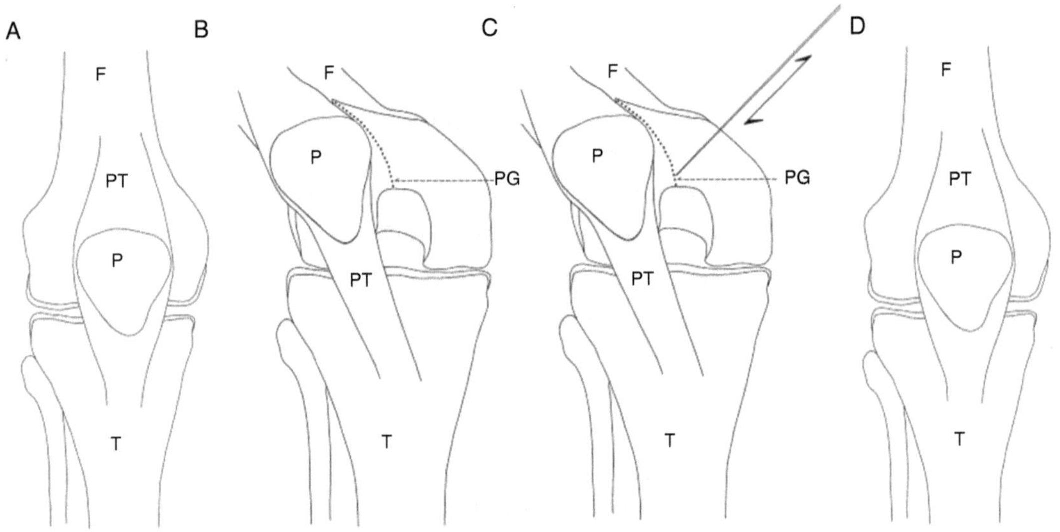

Fig. 1 Graphical representation of the procedure. (**a**) **Step 13**: the knee joint is opened to expose the structures. (**b**) **Steps 14–16**: the patella is dislocated to lateral side of the knee and the joint is flexed to expose the patella groove. (**c**) **Step 17**: the needle is inserted into the patella groove and moved up and down to create the injury. (**d**) **Step 18**: the knee is extended, and the patella is replaced into the patella groove. F, femur; T, tibia; P, patella; PT, patella tendon; PG, patella groove

20. Suture the skin using Ethilon 5.0 sutures using the Mathieu needle holder.
21. The contralateral knee can either be left non-operated or subjected to arthrotomy and patellar dislocation without cartilage injury, to produce sham-operated controls depending on the experimental design. Optionally, a treatment can be delivered (*see* **Note 6**).
22. Transfer the mouse to a clean recovery cage that has been pre-warmed to 22–25 °C by placing a heat pad underneath. The recovery time will vary from mouse to mouse and depending on anesthetic and route of administration. Keep recovering mice under observation.
23. Once fully recovered, keep the mice as specified by local guidance (standard housing conditions) for 8 weeks (*see* **Notes 7** and **8**).
24. Euthanize the mouse.
25. Remove the skin, and, using the bone scissors, cut the middle of the femur and the middle of the tibia to enable removal of the intact knee joint.
26. Process the knee joints for histology and section the joints 5 μm thick coronally on the femur (*see* **Note 9**).
27. Perform histological staining with Safranin O according to standard protocols (*see* Fig. 2 and **Note 10**).

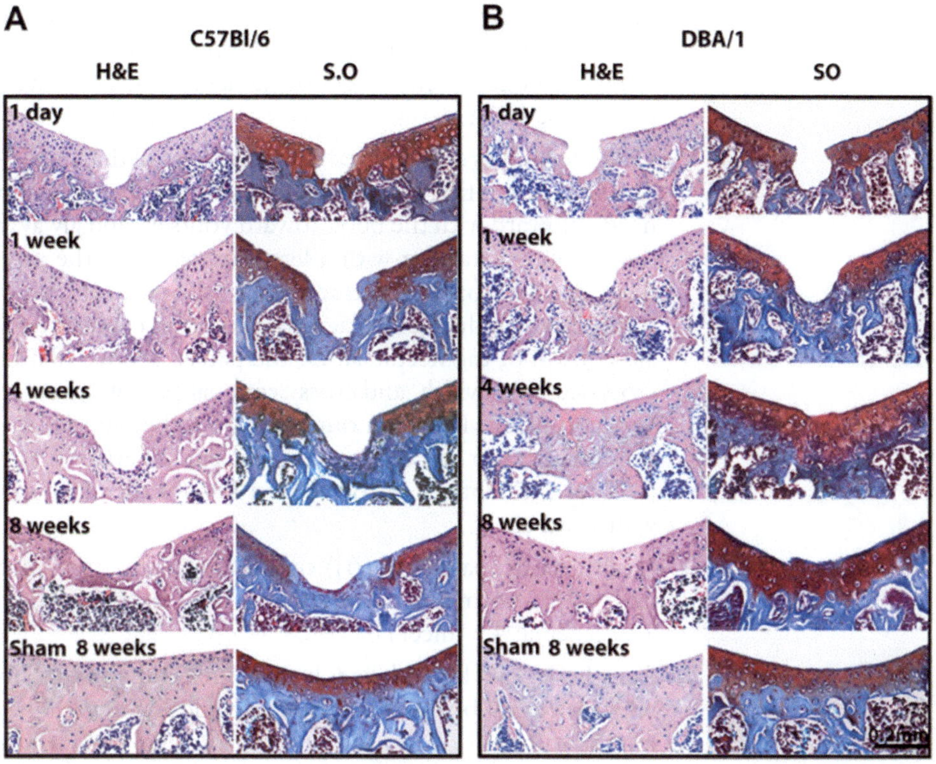

Fig. 2 Outcome of the procedure in different mouse strains. (**a**) Hematoxylin and eosin (H&E) (left) and Safranin O (SO) (right) staining of sections from sham operated and injured joints of 8-week-old C57BL/6 and (**b**) DBA/1 mice at the indicated time points. (Reproduced from Eltawil et al. [8] with permission (License number 5073180270918))

28. Score the level of repair using the Pineda scale [7] (*see* **Note 11**).
29. Assess other outcomes as desired, bearing in mind limitations (*see* **Notes 12** and **13**).

4 Notes

1. Power calculation: based on our study [8], 16 mice per group are required to detect a 50% difference of the means with a power of 0.8 and with a signal level of 0.05. Considering the rate of patellar dislocations, it is recommended to start from 20 animals per group.
2. Mice (strain/age/sex) will determine the outcome and the size of a critical defect (*see* Fig. 2). For instance, young (10-week-old) DBA/1 mice heal, but C57BL/6 mice of the same age won't; aged (>6 months old) DBA/1 mice, however, won't heal either [8]. Similarly, transgenic mice [9] and lineage tracking mice [10, 11] can be used; however, background strain

differences may induce variation. Check the weights are consistent throughout the groups. Furthermore, the model can be used in genetic screens to identify genes associated with a repair phenotype [12].

3. Multiple strokes are required to generate a full thickness defect. To avoid generating multiple defects or irregular ones, slide the needle upward with the bevel toward yourself, and always move the needle up starting with a light stroke. Once the groove is formed, more force can be applied. Variation to the site of the injury can be reduced by placing a glass bead over the needle to aid control of the needle along the patella groove [8]. Depth, cross-sectional width, and cross-sectional area of the defects are very consistent. Based on our study published in [8], an experienced operator can achieve a coefficient of variation as low as 5.3% for the depth of the defect and 19% for the cross-sectional width and area.
4. Health score (based on [6]) overall visual observation of the limb (0 = normal; 1 = erythema; 2 = erythema+ edema; 3 = eczema; 4 = ulcer).
5. Anesthesia can be induced by any method licensed by the relevant ethical board. This could include isoflurane inhalation for the duration of the surgery; or ketamine and xylazine by parenteral administration which will result in approximately 30 min of deep anesthesia.
6. Pharmacological/bioactive molecules cannot be tested in this model as the cartilage defect is too small to hold biomaterials loaded or not with bioactive molecules. However, one could perform an intra-articular injection to test molecules delivered to the entire joint [6]. It is possible to do this at any stage of the model. Alternatively, gain-of-function or loss-of-function assays can be performed using genetic tools (knockout or overexpression) or adenoviral delivery.
7. Postsurgical analgesia may be administered if pain is not an outcome measure.
8. Variations to the model include changing the time frame, diet [13], and circadian rhythm.
9. When the sections intercept the growth plate in four points, the correct level has been reached to assess outcomes.
10. Other histological stains can be used including hematoxylin and eosin (Fig. 2), toluidine blue, and alcian blue.
11. Other cartilage scoring systems can be used such as the Wakitani and Mankin [14, 15].
12. Potential outcomes can be assessed in terms of cartilage, bone, synovium, secondary osteoarthritis [16], expression of RNA or proteins by in situ hybridization or immunofluorescence staining, or cell death by TUNEL assay [8].

13. A limitation of this model is that it requires dislocation of the patella with subsequent reduction. Unfortunately, this results in secondary patellar dislocations (usually in 20% of cases but can be in up to 50% when fights occur within the cage). Mice with patellar dislocation have poor repair and severe secondary osteoarthritis and need to be removed from the analysis. This should be taken into account when planning the experiment in order to have sufficient statistical power (*see* power calculation in **Note 1**).

Acknowledgments

This work was supported by Grants from FOREUM, Grant/Award Number: 1016807; Versus Arthritis, Grant/Award Numbers: 21515 and 22628; and Rosetrees Trust, Grant/Award Number: A2575. The authors thank Mr Jacob Wilson for contributing to the design of the figures.

References

1. Curl WW, Krome J, Gordon ES et al (1997) Cartilage injuries: a review of 31,516 knee arthroscopies. Arthroscopy 13:456–460
2. Hjelle K, Solheim E, Strand T et al (2002) Articular cartilage defects in 1,000 knee arthroscopies. Arthroscopy 18:730–734
3. Dell'Accio F, Vincent TL (2010) Joint surface defects: clinical course and cellular response in spontaneous and experimental lesions. Eur Cell Mater 20:210–217
4. Kreuz PC, Steinwachs MR, Erggelet C et al (2006) Results after microfracture of full-thickness chondral defects in different compartments in the knee. Osteoarthr Cartil 14: 1119–1125. https://doi.org/10.1016/j.joca.2006.05.003
5. Arthritis UKR (2013) Osteoarthritis in general practice. Med Press 222:253–258
6. Thorup AS, Strachan D, Caxaria S et al (2020) ROR2 blockade as a therapy for osteoarthritis. Sci Transl Med 12:3063. https://doi.org/10.1126/SCITRANSLMED.AAX3063
7. Pineda S, Pollack A, Stevenson S et al (1992) A semiquantitative scale for histologic grading of articular cartilage repair. Acta Anat (Basel) 143: 335–340. https://doi.org/10.1159/000147272
8. Eltawil NM, De Bari C, Achan P et al (2009) A novel in vivo murine model of cartilage regeneration. Age and strain-dependent outcome after joint surface injury. Osteoarthr Cartil 17: 695–704. https://doi.org/10.1016/j.joca.2008.11.003
9. Zhu L, Donhou S, Burleigh A et al (2020) TSG-6 is weakly chondroprotective in murine OA but does not account for FGF2-mediated joint protection. ACR Open Rheumatol 2: 605–615. https://doi.org/10.1002/acr2.11176
10. Decker RS, Um H-B, Dyment NA et al (2017) Cell origin, volume and arrangement are drivers of articular cartilage formation, morphogenesis and response to injury in mouse limbs. Dev Biol 426:56–68. https://doi.org/10.1016/j.ydbio.2017.04.006
11. Roelofs AJ, Zupan J, Riemen AHK et al (2017) Joint morphogenetic cells in the adult mammalian synovium. Nat Commun 8:15040. https://doi.org/10.1038/ncomms15040
12. Cheverud JM, Lawson HA, Bouckaert K et al (2014) Fine-mapping quantitative trait loci affecting murine external ear tissue regeneration in the LG/J by SM/J advanced intercross line. Heredity (Edinb) 112:508–518. https://doi.org/10.1038/hdy.2013.133
13. Wei W, Bastiaansen-Jenniskens YM, Suijkerbuijk M et al (2017) High fat diet accelerates cartilage repair in DBA/1 mice. J Orthop Res 35:1258–1264. https://doi.org/10.1002/jor.23280

14. Wakitani S, Goto T, Pineda SJ, Young RG, Mansour JM, Caplan AI, Goldberg VM (1994) Mesenchymal cell-based repair of large, full-thickness defects of articular cartilage. J Bone Joint Surg Am 76:579–592
15. Mankin HJ, Lippiello L (1971) The glycosaminoglycans of normal and arthritic cartilage. J Clin Invest 50:1712–1719. https://doi.org/10.1172/JCI106660
16. Eldridge SE, Barawi A, Wang H et al (2020) Agrin induces long-term osteochondral regeneration by supporting repair morphogenesis. Sci Transl Med 12:eaax9086

Chapter 25

Small-Animal Compression Models of Osteoarthritis

Blaine A. Christiansen, Deva D. Chan, Marjolein C. H. van der Meulen, and Tristan Maerz

Abstract

The utility of nonsurgical, mechanical compression-based joint injury models to study osteoarthritis pathogenesis and treatments is increasing. Joint injury may be induced via cyclic compression loading or acute overloading to induce anterior cruciate ligament rupture. Models utilizing mechanical testing systems are highly repeatable, require little expertise, and result in a predictable onset of osteoarthritis-like pathology on a rapidly progressing timeline. In this chapter, we describe the procedures and equipment needed to perform mechanical compression-induced initiation of osteoarthritis in mice and rats.

Key words Osteoarthritis, Post-traumatic, Animal models, Cartilage, Anterior cruciate ligament, Compression, Injury

1 Introduction

Rodent models of osteoarthritis (OA) are essential tools for studying the pathogenesis and treatment of OA on an accelerated timeline (from 10–20 years in humans to 4–12 weeks in rodent models). All animal models of OA initiate cartilage degradation, but many currently used models do not reproduce clinically relevant injury conditions, due to invasive and nonphysiologic injury methods most commonly involving surgery. In addition, these methods often cause an injury response due to invasive surgical procedures, complicating studies of the inflammatory response associated with the injury itself. Due to these limitations, there is emerging interest in compression models of OA that can noninvasively induce joint injury and resultant degeneration in mice and rats. These procedures use externally applied mechanical loads to overload the articular cartilage and/or injure the anterior cruciate ligament (ACL), more closely mimicking injury conditions relevant to humans. Noninvasive models do not break the skin or disrupt the joint capsule and are therefore completely aseptic, which avoids potential

Martin J. Stoddart et al. (eds.), *Cartilage Tissue Engineering*,
Methods in Molecular Biology, vol. 2598, https://doi.org/10.1007/978-1-0716-2839-3_25,

confounding effects caused by the surgical/invasive injury procedure. Furthermore, in models utilizing mechanical testing systems, injury loading is highly repeatable and less prone to operator variability, potentially affecting disease onset and progression in surgical OA models.

Established noninvasive compression models of OA generally apply controlled loads to the rodent limb and include the cyclic compression model in mice [1–5] and the overload ACL rupture (ACL-R) model in mice [6–10] and rats [11–15]. These models can be implemented using a variety of methods, each with its own advantages, limitations, and technical considerations (*see* **Notes 1–10**). The objectives of this chapter are to discuss the implementation of compression models of OA, including cyclic compression in mice (*see* **Notes 11–19**), ACL rupture in mice (*see* **Notes 20–33**), and rat ACL rupture in rats (*see* **Notes 34–44**) and to share best practices to facilitate repeatable injury and disease induction when employing these methods in preclinical studies of OA.

2 Materials

1. Animals—mice or rats. These procedures can be applied for any genetic strain, age, or sex (*see* **Notes 11**, **12**, **20**, and **35**).
2. Commercially available materials testing systems such as the Electroforce 3200 (TA Instruments, New Castle, DE) or equivalent (*see* **Note 6**).
3. Load cells for mice require approximately 20 N capacity; load cells for rats require approximately 200 N capacity (*see* **Notes 33** and **44**).
4. Appropriately sized fixtures that restrict medial-lateral motion of the knee and foot (*see* **Note** 7).
5. Anesthetic during loading: ketamine/xylazine or, preferably, inhaled isoflurane (2–4%) (*see* **Note 4**).
6. Analgesic post-injury for ACL-R models: e.g., 5 mg/kg SC carprofen and/or 0.05 mg/kg SC buprenorphine.

3 Methods

3.1 Mouse Cyclic Compression (See Notes 11–19)

1. Anesthetize mice and ensure adequate anesthesia with toe pinch and/or tail pinch.
2. Place the mouse supine, with the lower leg positioned horizontally (parallel to the body) between two loading platens (*see* Fig. 1).

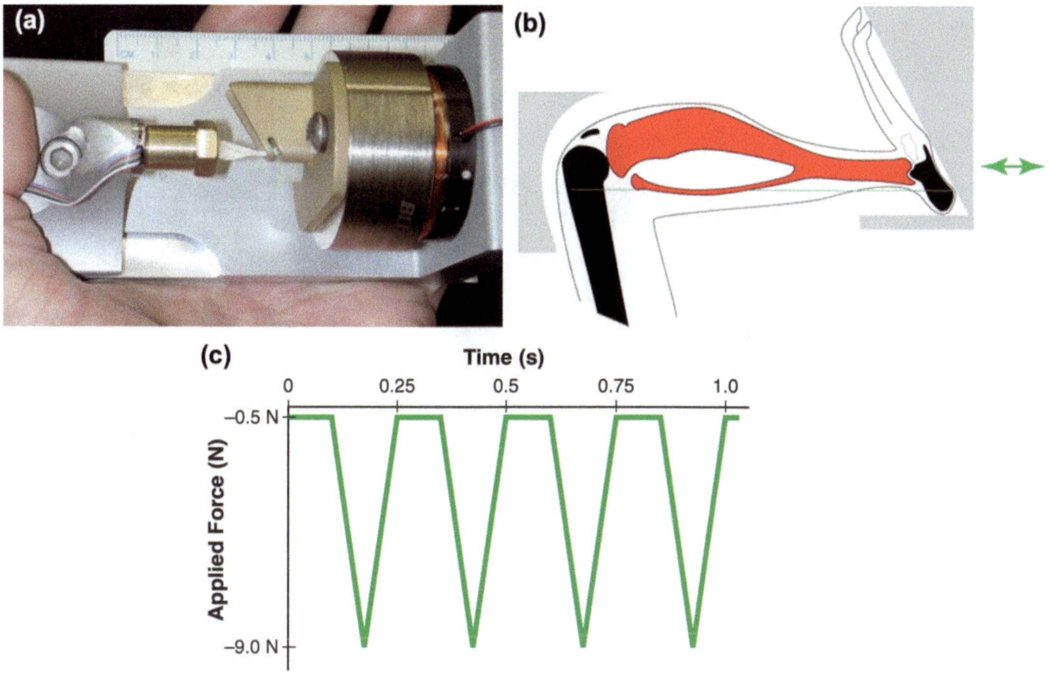

Fig. 1 (**a**, **b**) Handheld loading system used for cyclic compression loading of mice. (**c**) Commonly used waveform for the cyclic compression model

3. Apply a preload of 0.5–1 N to position the lower leg.
4. Apply cyclic loading for the designated number of cycles (*see* **Notes 11–13**).
5. Repeat procedures for each daily loading bout for the designated number of days and weeks to induce OA (*see* **Note 13**).

3.2 Mouse ACL-R (See Notes 20–33)

1. Anesthetize animals and ensure adequate anesthesia with toe pinch and/or tail pinch.
2. Place the mouse prone, with the lower leg positioned vertically (perpendicular to the body) between two loading platens with the hip in full extension (*see* Fig. 2).
3. Apply a preload of 1–2 N to hold the lower leg in place.
4. Apply a single compressive load with a loading rate of 1–10 mm/s to a target force (~12–15 N) or until ACL injury.
5. Injury is typically accompanied by an auditory cue ("click" or "crunch") and force release on the force-displacement plot, after which the load can be manually stopped (*see* Fig. 2).
6. Alternatively, the ACL can be injured by applying a considerably faster load (~200 mm/s) to a target displacement (1.7 mm) (*see* **Note 21**).

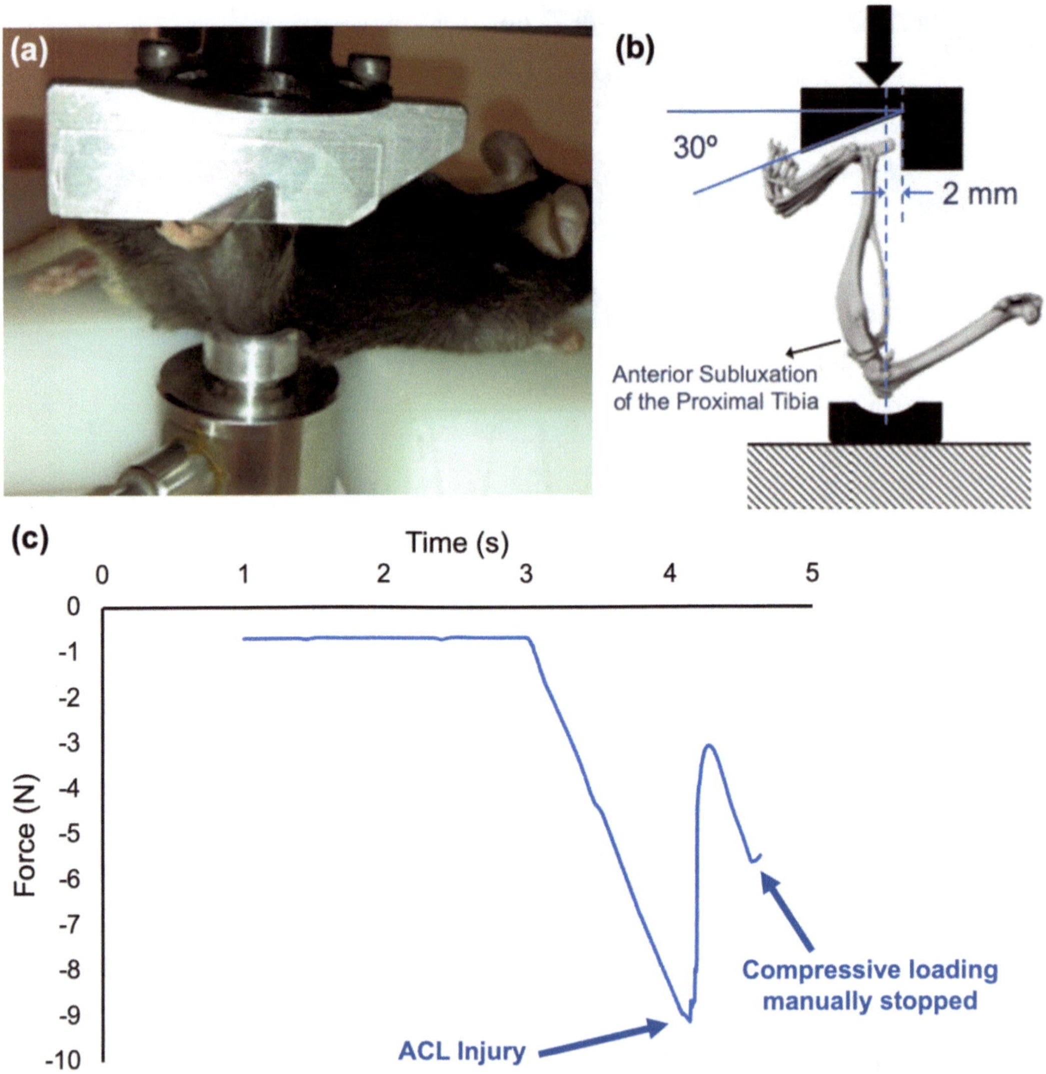

Fig. 2 (**a**, **b**) Configuration for inducing ACL-R in mice. (**c**) Representative compressive load that resulted in ACL rupture in a mouse knee

7. ACL injury can be confirmed using an anterior drawer test or with standard radiography (e.g., Faxitron) following injury loading to rule out fractures.
8. Provide at least one dose of analgesic post-injury and additional doses as needed.

3.3 Rat ACL-R (See Notes 34–44)

1. Anesthetize animals and ensure adequate anesthesia with toe pinch and/or tail pinch.
2. Place the rat prone, with the lower leg positioned vertically in the system (*see* Fig. 3).
3. Apply a 3 N preload for 10 s.

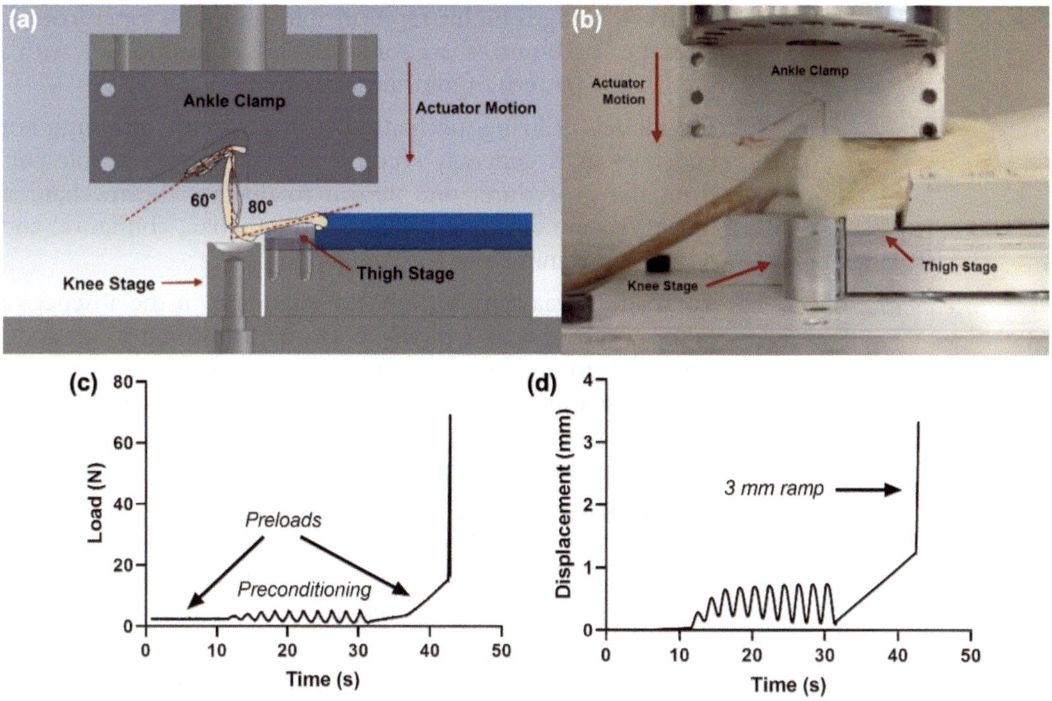

Fig. 3 (**a**, **b**) Configuration for inducing ACL-R in rats. (**c**, **d**) Load and displacement plots for preconditioning and injury load in a rat ACL-R model

4. Optional: apply ten preconditioning cycles, cycling from 1 to 5 N at 0.5 Hz. Preconditioning cycles can aid in seating the joint in the fixtures but may not be necessary.
5. Apply a preload ramp at 0.1 mm/s from 1 to 15 N.
6. Apply a failure ramp at 8 mm/s to a target displacement of 3 mm.
7. Injury is typically accompanied by an auditory cue ("click" or "crunch") and force release on the force-displacement plot (*see* **Note 35**).
8. ACL injury can be confirmed using an anterior drawer test or with standard radiography (e.g., Faxitron) following injury loading to rule out fractures (*see* **Note 36**).
9. Provide at least one dose of analgesic post-injury and additional doses as needed.

4 Notes

1. Once successfully set up, these methods are easy to learn with not much expertise required, which likely reduces inter- and intra-user variability compared to surgical or even injection models.

2. Injury/loading is highly reproducible (although faster loading rates, i.e., >10 mm/s, are somewhat less reproducible), with a predictable pattern of joint degeneration.
3. For models utilizing mechanical testing systems, although not necessary, it is generally recommended that two people perform the procedure: one person to operate the mechanical testing system and one person to anesthetize, configure, and monitor the animals.
4. Total anesthesia time for inhaled isoflurane, in the absence of other procedures, is typically 5–10 min, followed by an additional ~5 min of recovery monitoring after loading. Short procedure time and use of isoflurane for anesthesia enable rapid recovery and minimal anesthesia-dependent effects and allow these methods to be high throughput.
5. Due to being noninvasive and completely aseptic, all of these methods allow for the study of inflammation, ectopic intra-articular bone formation, joint biomechanics, or other processes at early time points, even only a few minutes after injury.
6. Implementation typically involves a commercially available materials testing system, but can also be achieved using a custom-built system or a small handheld device and laptop computer. However, these custom systems often have limited capabilities for high velocities and real-time measurement and control of force and displacement.
7. These procedures require custom-made fixtures that often vary between research groups. Knee fixtures must be able to hold the lower leg stationary during loading without the knee "slipping out." A shallow knee cup that does not constrain translation of the tibia will be more likely to result in ACL rupture. A deeper knee cup will allow for higher magnitude loading without acute ACL injury [16]. As an alternative to a cup, a channel/trough may be milled with a ball-end mill (facilitating rounded edges), which restricts medial-lateral motion while allowing for anterior translation of the tibia. Quadriceps tendon injury during loading can be avoided/mitigated with padding and/or rounding of edges on the knee fixture.
8. Degree of knee flexion may influence joint damage and likelihood of ACL injury. Tension in the quadriceps during hip extension makes ACL rupture more likely by pulling the tibia anteriorly, while tension in the hamstrings during flexion makes ACL rupture less likely by pulling the tibia posteriorly [16]. For both mouse and rat ACL-R, the knee joint should be in ~90–100° of flexion.
9. The flexion angle of the ankle may also influence likelihood of ACL injury. Greater ankle flexion creates tension in the gastrocnemius and soleus and causes anterior tibial translation,

making ACL injury more likely [16]. For both mouse and rat ACL-R procedures, the ankle joint should be in 30° of flexion (*see* Figs. 2 and 3).

10. Rotation of the lower limb also plays a role in the likelihood of inducing an ACL rupture during tibial compression [12]. Similarly, body and contralateral limb position are important, as pelvic rotation induces ipsilateral femoral internal/external rotation. Elevating the contralateral limb to the height of the loaded limb ensures a neutral pelvic rotation.

4.1 Cyclic Compression in Mice

11. The applied loading waveform and loading rate for mouse cyclic compression are based on osteogenic loading of the metaphysis [17, 18]. A common loading frequency is 4 Hz, corresponding to the walking frequency of the mouse at ~4% strain per second, with a dwell between each load-unload cycle (*see* Fig. 1c).
12. Load magnitude varies with animal age and strain. Generally, loads are in the 7–10 N range. Altering the load magnitude and number of loading bouts modulates the amount of joint damage generated.
13. For mouse cyclic compression, 1200 cycles are typically applied in each daily loading bout, 5 days per week for multiple weeks, although a single bout of loading produces similar effects to daily loading after 1 and 2 weeks [19].
14. Cyclic compression loading of mice induces compression and shear at the joint through AP translation of the tibia and rotation of the femur [20].
15. Daily in vivo loading induces cartilage damage in both male and female mice; degenerative changes are mouse-strain specific and progress over time [4].
16. Bone mass changes occur in the subchondral, epiphyseal, and metaphyseal bone. Over time, daily loading enhances bone mass at the metaphysis. Changes in bone mass closer to the joint surface are more variable and may initially decrease at early time points, followed by increased bone mass with time.
17. Osteophytes/enthesophytes form on the medial aspect of the joint at the MCL insertion. These structures are cartilaginous after 1 and 2 weeks of loading and are mineralized at 6 weeks of loading.
18. Joint fibrosis is present, but not classical inflammation. The degree of fibrosis differs substantially across different mouse strains.
19. Gait analysis shows no major differences in loaded animals.

4.2 ACL-R in Mice

20. ACL injury of mice typically occurs between 8 and 15 N; injury force correlates positively with body mass (but not sex) [10].
21. Using a higher speed loading rate is more likely to induce midsubstance tear of the ACL, rather than avulsion from the distal femur [7]. Real-time monitoring of injury on the force-displacement plot may be more difficult with high-speed loading.
22. Following ACL-R in mice, joint degeneration progresses to severe OA by 4–6 weeks; the medial side of the joint is more affected than the lateral side.
23. OA in older mice typically progresses slightly faster than in young mice [21]; OA progression is typically similar in male and female mice [22].
24. ACL deficiency causes the articulation of the femur to move to a more posterior position on the proximal tibia; this can lead to dislocation of the posterior horn of the medial meniscus.
25. ACL-R results in joint instability, and OA progression may be largely driven by mechanical factors. This model typically cannot be adjusted to produce a milder injury or slower PTOA progression.
26. Joint inflammation, synovitis, and protease activity are observed primarily during the first 2–3 weeks, peaking around 1–14 days post-injury [23].
27. Considerable synovial fibrosis and chondrophyte formation occur within the first 1–2 weeks; mineralized osteophytes are present by 4 weeks post-injury, which will continue to progressively grow and mineralize [24].
28. Osteophytes are commonly observed on the medial side of the distal femur, the posterior-medial proximal tibia, and the anterior horn of the medial meniscus [24].
29. Epiphyseal trabecular bone loss (~20–30% decrease in BV/TV) occurs during the first 2 weeks post-injury, primarily due to trabecular thinning; this bone loss is partially recovered at later time points.
30. Thinning of the subchondral bone plate also occurs during the first 2 weeks post-injury; at later time points, this is reversed and subchondral bone plate thickness is increased in injured joints.
31. No significant differences in overall activity level of mice following injury, though gait analysis indicates some gait asymmetry [25].

32. If possible, the mechanical testing system can be programmed to return to the "zero-displacement" position immediately after a sudden drop in load is detected to reduce the joint from the subluxed position.
33. It is recommended to program load and displacement limits to avoid overloading or damage to load cells, fixtures, or the mechanical testing system. For mice, a load limit of 20–25 N and a displacement limit of ~2.0 mm are recommended. These are dependent on strain and skeletal size, and careful preliminary testing is recommended to determine these limits.

4.3 ACL-R in Rats

34. ACL-R creates joint instability, and OA progression may be largely driven by mechanical factors; this model typically can't be adjusted to produce a milder injury or slower PTOA progression.
35. For a 200–250 g Lewis rat, the failure load is typically 60–70 N, at a displacement of 2.2–2.5 mm. Force release associated with ACL rupture may not be easily detectable on force-displacement plot for rat injuries.
36. *Important*: distal femoral physeal dislocations/fractures can give a false-positive anterior drawer test. These are most common in the rat ACL-R model. Cadaveric animals should be used to practice the procedure and confirm ACL injury by dissection to avoid physeal injuries.
37. Early articular cartilage damage is observed by 1–2 weeks post-injury, including surface fibrillation and acute necrosis [26].
38. Moderate OA is observed by 4–6 weeks post-injury, and severe OA is observed by 10 weeks.
39. Medial compartment damage is more severe than lateral compartment, with up to 2× OARSI grade on the medial femoral condyle compared to the lateral femoral condyle at 4 weeks [27].
40. Synovitis is observed as early as 3 days post-injury, maximal between 3 and 14 days, and synovial fibrosis persists up to 16 weeks post-injury [28, 14].
41. Osteophyte formation is commonly observed in the medial compartment, notably at the anterior medial condyle, posterior medial tibial plateau, and anterior medial meniscus. In general, osteophytes are larger in mice than in rats relative to the size of the joint.
42. Acute epiphyseal trabecular bone loss (~20–30% decrease in BV/TV) and subchondral bone thinning (~15–20%) are observed within the first 2 weeks. Persistent bone volume deficit of 5–10% is observable for at least 10 weeks post-injury

[29, 26]. Bone loss has been associated with both reduced bone formation rates and increased bone resorption and osteoclast numbers acutely after injury [26].

43. Subchondral bone thinning and volumetric loss are observed within first 2 weeks of injury, recovering up to 4 weeks post-injury. Subchondral bone thickening is observed during chronic disease [29, 26].
44. It is recommended to program load and displacement limits to avoid overloading or damage to load cells, fixtures, or the mechanical testing system. For rats, a load limit of 90–150 N and a displacement limit of 3.5–4.5 mm are recommended. These are dependent on strain and skeletal size, and careful preliminary testing is recommended to determine these limits.

Acknowledgments

Dr. Christiansen is supported by the National Institute of Arthritis and Musculoskeletal and Skin Diseases under Award Numbers R01 AR071459, AR075013, and AR075013-02S1 and by the Department of Defense under Award Number PR180268P1. Dr. Chan is supported by the National Science Foundation (CAREER Award 1944394) and the Indiana Clinical and Translational Sciences Institute, which is funded in part by Award Number UL1TR002529 from the National Institutes of Health, National Center for Advancing Translational Sciences, Clinical and Translational Sciences Award). Dr. van der Meulen is supported by the National Institute of Arthritis and Musculoskeletal and Skin Diseases under Award Numbers R21 AR071587.
R01 AR071459, AR075013, and AR075013-02S1 and by the Department of Defense under Award Number PR180268P1. Dr. Maerz is supported by the Congressionally Directed Medical Research Program (CDMRP, Award W81XWH-15-1-0186, PI: Kevin Baker) and the Michigan Integrative Musculoskeletal Health Core Center (P30 AR069620, National Institute of Arthritis and Musculoskeletal and Skin Diseases). The content is solely the responsibility of the authors; the funding bodies were not involved with the writing of this chapter. The authors have no conflicts of interest to disclose.

References

1. Poulet B, Hamilton RW, Shefelbine S et al (2011) Characterizing a novel and adjustable noninvasive murine joint loading model. Arthritis Rheum 63(1):137–147. https://doi.org/10.1002/art.27765
2. Poulet B, Westerhof TA, Hamilton RW et al (2013) Spontaneous osteoarthritis in Str/ort mice is unlikely due to greater vulnerability to mechanical trauma. Osteoarthr Cartil 21(5): 756–763. https://doi.org/10.1016/j.joca.2013.02.652
3. Poulet B, de Souza R, Kent AV et al (2015) Intermittent applied mechanical loading induces subchondral bone thickening that

may be intensified locally by contiguous articular cartilage lesions. Osteoarthr Cartil 23(6): 940–948. https://doi.org/10.1016/j.joca.2015.01.012

4. Ko FC, Dragomir C, Plumb DA et al (2013) In vivo cyclic compression causes cartilage degeneration and subchondral bone changes in mouse tibiae. Arthritis Rheum 65(6): 1569–1578. https://doi.org/10.1002/art.37906
5. Wu P, Holguin N, Silva MJ et al (2014) Early response of mouse joint tissue to noninvasive knee injury suggests treatment targets. Arthritis Rheumatol 66(5):1256–1265. https://doi.org/10.1002/art.38375
6. Christiansen BA, Anderson MJ, Lee CA et al (2012) Musculoskeletal changes following non-invasive knee injury using a novel mouse model of post-traumatic osteoarthritis. Osteoarthr Cartil 20(7):773–782. https://doi.org/10.1016/j.joca.2012.04.014
7. Lockwood KA, Chu BT, Anderson MJ et al (2014) Comparison of loading rate-dependent injury modes in a murine model of post-traumatic osteoarthritis. J Orthop Res 32(1): 79–88. https://doi.org/10.1002/jor.22480
8. Onur TS, Wu R, Chu S et al (2014) Joint instability and cartilage compression in a mouse model of posttraumatic osteoarthritis. J Orthop Res 32(2):318–323. https://doi.org/10.1002/jor.22509
9. Rai MF, Duan X, Quirk JD et al (2017) Post-traumatic osteoarthritis in mice following mechanical injury to the synovial joint. Sci Rep 7:45223. https://doi.org/10.1038/srep45223
10. Blaker CL, Little CB, Clarke EC (2017) Joint loads resulting in ACL rupture: effects of age, sex, and body mass on injury load and mode of failure in a mouse model. J Orthop Res 35(8): 1754–1763. https://doi.org/10.1002/jor.23418
11. Maerz T, Kurdziel MD, Davidson AA et al (2015) Biomechanical characterization of a model of noninvasive, traumatic anterior cruciate ligament injury in the rat. Ann Biomed Eng 43(10):2467–2476. https://doi.org/10.1007/s10439-015-1292-9
12. Ramme AJ, Lendhey M, Raya JG et al (2016) A novel rat model for subchondral microdamage in acute knee injury: a potential mechanism in post-traumatic osteoarthritis. Osteoarthr Cartil 24(10):1776–1785. https://doi.org/10.1016/j.joca.2016.05.017
13. Ramme AJ, Lendhey MS, Strauss EJ et al (2018) A biomechanical study of two distinct methods of anterior cruciate ligament rupture, and a novel surgical reconstruction technique, in a small animal model of posttraumatic osteoarthritis. J Knee Surg 31(1):43–49. https://doi.org/10.1055/s-0037-1600088
14. Brown SB, Hornyak JA, Jungels RR et al (2020) Characterization of post-traumatic osteoarthritis in rats following anterior cruciate ligament rupture by non-invasive knee injury (NIKI). J Orthop Res 38(2):356–367. https://doi.org/10.1002/jor.24470
15. Lepley LK, Davi SM, Butterfield TA et al (2020) Visualization of knee joint degeneration after non-invasive ACL injury in rats, JoVE Science Education Database Biomedical Engineering. JoVE, Cambridge
16. Hsia AW, Tarke FD, Shelton TJ et al (2018) Comparison of knee injury threshold during tibial compression based on limb orientation in mice. J Biomech 74:220–224. https://doi.org/10.1016/j.jbiomech.2018.04.014
17. Lynch ME, Main RP, Xu Q et al (2010) Cancellous bone adaptation to tibial compression is not sex dependent in growing mice. J Appl Physiol (1985) 109(3):685–691. https://doi.org/10.1152/japplphysiol.00210.2010
18. Holguin N, Brodt MD, Sanchez ME et al (2013) Adaptation of tibial structure and strength to axial compression depends on loading history in both C57BL/6 and BALB/c mice. Calcif Tissue Int 93(3):211–221. https://doi.org/10.1007/s00223-013-9744-4
19. Ko FC, Dragomir CL, Plumb DA et al (2016) Progressive cell-mediated changes in articular cartilage and bone in mice are initiated by a single session of controlled cyclic compressive loading. J Orthop Res 34(11):1941–1949. https://doi.org/10.1002/jor.23204
20. Adebayo OO, Ko FC, Goldring SR et al (2017) Kinematics of meniscal- and ACL-transected mouse knees during controlled tibial compressive loading captured using roentgen stereophotogrammetry. J Orthop Res 35(2): 353–360. https://doi.org/10.1002/jor.23285
21. Sebastian A, Murugesh DK, Mendez ME et al (2020) Global gene expression analysis identifies age-related differences in knee joint transcriptome during the development of post-traumatic osteoarthritis in mice. Int J Mol Sci 21(1). https://doi.org/10.3390/ijms21010364
22. Satkunananthan PB, Anderson MJ, De Jesus NM et al (2014) In vivo fluorescence reflectance imaging of protease activity in a mouse model of post-traumatic osteoarthritis. Osteoarthr Cartil 22(10):1461–1469. https://doi.org/10.1016/j.joca.2014.07.011

23. Gilbert SJ, Bonnet CS, Stadnik P et al (2018) Inflammatory and degenerative phases resulting from anterior cruciate rupture in a non-invasive murine model of post-traumatic osteoarthritis. J Orthop Res. https://doi.org/10.1002/jor.23872
24. Hsia AW, Anderson MJ, Heffner MA et al (2017) Osteophyte formation after ACL rupture in mice is associated with joint restabilization and loss of range of motion. J Orthop Res 35(3):466–473. https://doi.org/10.1002/jor.23252
25. Anderson MJ, Diko S, Baehr LM et al (2016) Contribution of mechanical unloading to trabecular bone loss following non-invasive knee injury in mice. J Orthop Res 34(10): 1680–1687. https://doi.org/10.1002/jor.23178
26. Maerz T, Newton MD, Fleischer M et al (2021) Traumatic joint injury induces acute catabolic bone turnover concurrent with articular cartilage damage in a rat model of post-traumatic osteoarthritis. J Orthop Res 39(9): 1965–1976. https://doi.org/10.1002/jor.24903
27. Maerz T, Newton MD, Kurdziel MD et al (2016) Articular cartilage degeneration following anterior cruciate ligament injury: a comparison of surgical transection and noninvasive rupture as preclinical models of post-traumatic osteoarthritis. Osteoarthr Cartil 24(11): 1918–1927. https://doi.org/10.1016/j.joca.2016.06.013
28. Maerz T, Fleischer M, Newton MD et al (2017) Acute mobilization and migration of bone marrow-derived stem cells following anterior cruciate ligament rupture. Osteoarthr Cartil 25(8):1335–1344. https://doi.org/10.1016/j.joca.2017.03.004
29. Maerz T, Kurdziel M, Newton MD et al (2016) Subchondral and epiphyseal bone remodeling following surgical transection and noninvasive rupture of the anterior cruciate ligament as models of post-traumatic osteoarthritis. Osteoarthr Cartil 24(4):698–708. https://doi.org/10.1016/j.joca.2015.11.005

Chapter 26

Modeling and Assessing Osteoarthritis in Mice by Destabilization of the Medial Meniscus (DMM)

Jadwiga Miotla-Zarebska, Ida Parisi, Vicky Batchelor, Tonia L. Vincent, and Paul K. Potter

Abstract

In this chapter, we describe an induced model of osteoarthritis in mice, frequently employed in the study of this disease. We outline in detail the surgical induction of disease and preparation of samples for histological assessment of disease.

Key words Osteoarthritis, Mouse model, Meniscus, Cartilage, Bone, Pain

1 Introduction

Osteoarthritis represents an increasing clinical burden with a paucity of interventions beyond joint replacement surgery [1]. While postoperative human tissue is generally available, and useful, the ability to study disease from its inception in a model organism has advanced our understanding of disease susceptibility, modification, and pathogenesis [2].

Destabilization of the medial meniscus (DMM) has now become a standardized methodology [3, 4] for the induction and analysis of osteoarthritis pathogenesis and associated phenotypes such as pain, a clinically relevant and important outcome measure [5]. Various animal models [6, 7] and in vitro systems [8] are also employed in studies, but the ability to induce disease at a defined point and track disease to different endpoints is extremely useful and when combined with genetic trackability and convenience of mice makes a formidable system for the study of this disease [2].

Pain is a very useful in vivo assessment of the development of OA symptoms and minimizes the number of animals required, as longitudinal studies can be carried out. It is also of clear clinical relevance. While there are a range of ways of assessing pain in mice,

Martin J. Stoddart et al. (eds.), *Cartilage Tissue Engineering*,
Methods in Molecular Biology, vol. 2598, https://doi.org/10.1007/978-1-0716-2839-3_26,

weight-bearing measurements are clinically relevant, and we have found them to be robust and reproducible after training. We employ the Linton incapacitance tester to assess spontaneous painful behavior in mice with murine OA and have also tested analgesia using this setup [9].

While there are an increasing number of noninvasive and molecular techniques to identify and analyze OA [4, 10, 11], histological analysis is a key methodology to assess the impact and stage of disease [12]. It allows for a thorough and detailed analysis of bone and soft tissue across the joint and is still used to validate new techniques.

This is a surgical procedure on live animals to induce disease and should be carried out with appropriate ethical permissions according to local regulations, under sterile conditions and in an appropriate surgical suite. The minimization of pain, discomfort, and suffering to animals is an ethical imperative. To achieve this detailed planning of experiments should be undertaken, including:

Choice of strain As with most phenotypes in mice, there are differences between strains in the response to DMM surgery [13], and therefore experiments should be carried out on an inbred background, as a mixed background will result in increased variation in results. The appropriate control for a congenic genetically modified mouse line is an appropriate congenic line without the genetic modification [14]. Differences in genetic background should also be taken into account in the determination of endpoints and power calculations (*see* below).

Determination of endpoints The age of mice at the induction of OA is an important consideration as the progression and severity of disease in old mice is accelerated [15]. Furthermore, the consideration of the endpoint should include consideration; is the aim to study early disease, pain responses, or late-stage histology? All of these may be influenced by the time between induction of disease, strain, sex, and genetic modification. Hence the time between induction of disease and analysis is an important factor in experimental design.

Power calculations To avoid using too many animals, or an underpowered experiment resulting in a wasted experiment, power calculations should be carried out during the planning of the experiment. The influence of strain, age, and sex should be incorporated into these calculations as they may affect the variability of results.

Appropriate controls To minimize the use of animals, the contralateral limb should be used as a negative/normal control. This has the advantage of controlling any treatments or interventions within the same animal. However, these do not recapitulate any nonspecific effects of the surgery, and as such sham operated mice, where all the surgical procedures are undertaken except the severing of the medial meniscotibial ligament (MMTL), should be included in the experimental plan. Age-matched and sex-matched littermate controls should be employed, with individual mice randomly allocated to DMM or sham operated groups. The individual mice should also be randomly distributed in cages and do not have cages of DMM operated mice and cages of sham operated mice housed separately. Caging density should be maintained across through the experiment as the number of mice within a cage can affect the activity of individual mice and hence may increase variation in the severity of disease.

Good scientific method Those undertaking the surgery should be blinded to treatment groups/genotypes, and those undertaking pain or histological assessment should be blinded as to those mice that underwent sham or DMM surgery. Histological scoring should be scored independently, and blinded, by two individuals and an average score used.

These considerations should be part of the careful planning of experiments, and we recommend using the ARRIVE guidelines and planning tools developed by the NC3Rs [16]. Both sexes should be used in any experiment to account for sex-specific differences [17].

This protocol covers the procedure for inducing OA by destabilization of the medial meniscus, assessing pain through hind paw weight distribution, and preparation of histology slides for assessment of damage.

2 Materials

All equipment must be sterilized or autoclaved and appropriate to the procedure. Ensure there are sufficient reagents and equipment to complete all planned surgeries before commencing. Appropriate controls should be used and power calculations used as described above to determine the experimental cohort size. Appropriate pre- and postsurgical care is essential to maximize the welfare of the mice and ensure appropriate experimental outcomes. Appropriate equipment substitutions can be used. Surgery should be carried out using sterile procedures (*see* **Note 1**).

2.1 Surgical Equipment

1. 26-gauge sterile needles.
2. Surgical needle holders.
3. Surgical #5 Dumont jeweler's forceps.
4. Surgical forceps.
5. 3 mm micro surgical knife with no. 15 blade.
6. 15-blade disposable scalpel.

7. Vicryl 6-0 (0.7 Ph.EuR) round bodied needle suture kit.
8. Ethilon 5-0 (1 Ph.EuR) reverse cutting needle suture kit.
9. Scissors.
10. Iodine/chlorhexidine.
11. 1 mL syringe.
12. Sterile cotton buds or surgical spears.
13. Swabs.
14. Micropore tape.
15. Surgical drapes.
16. Hair trimmer.
17. Viscotears.

2.2 Surgical Suite

1. A dissecting microscope and cold light source.
2. An oxygen concentrator or oxygen cylinder (*see* **Note 2**).
3. A clean Perspex surgery plate (~18 cm × 14 cm).
4. A small animal anesthesia workstation, including mask and box (*see* **Note 3**).

2.3 Anesthesia and Analgesia

1. Anesthetic: 1 mL of 0.3 mg/mL Vetergesic diluted with 19 mL of sterile water for injection at final concentration of 0.015 mg/mL.
2. Analgesic: 10 mg/mL of Torbugesic.
3. 100% isoflurane canister.

2.4 Post-surgery Recovery

1. Recovery chamber/hot box.

2.5 Pain Measurement

1. The Linton incapacitance tester (*see* **Note 4**).
2. Specimen tweezers.
3. Using the small sturdy sheet of paper.

2.6 Histology Sample Preparation

1. 10% v/v neutral buffered formalin.
2. Tissue cassettes.
3. 20% (v/v) formic acid.
4. 5% formalin in distilled water.
5. 0.5 M ethylenediaminetetraacetic acid (EDTA) pH 7.4 in distilled water.
6. Tissue processor.
7. 70% (v/v) ethanol.
8. 100% (v/v) ethanol.
9. Xylene.

10. Molten paraffin.
11. Forceps.
12. Embedding metallic molds.
13. Tissue embedding station.
14. Microtome.
15. Blades.
16. Water bath.
17. Adhesive microscope glass slides.
18. Drying oven.
19. 0.1% safranin O dye in 0.5 M acetic acid in distilled water.
20. 0.1% fast green dye in distilled water.
21. Harris hematoxylin.
22. Acid alcohol: 70% ethanol and 1% hydrochloric acid.
23. Blueing solution: 0.2% ammonium hydroxide in distilled water, pH 10.0.
24. 1% acetic acid in distilled water.
25. Glass coverslips (24 × 50 mm).
26. DPX mounting medium.
27. Ice box.

3 Methods

3.1 Surgery

To ensure sterile working, all surfaces and equipment (except sterilized surgical equipment) should be cleaned with TriGene (or another suitable disinfectant). Surgery should be carried out while wearing sterile gloves, and put on with appropriate sterile technique, a lab coat, and a surgical mask. The surgical suite should be set up so everything is accessible and within reach, and surgery can be carried out with the minimum of disruption. The surgery area may be set also up in LAF cabinet (*see* Fig. 1).

1. Turn on recovery box and ensure it has reached 29 °C before commencing surgery. The recovery box should be placed away from the surgery suite in a quiet area of the room to allow the mice to recover with the procedures recorded against each animal (*see* **Note 5**).
2. In the surgery area, set up the Perspex surgery plate, dissecting microscope, cold light source, oxygen concentrator, and small-animal anesthesia workstation, including mask and box.
3. Lay out a sterile drape beside the Perspex plate and placed within easy reach (*see* Fig. 2).

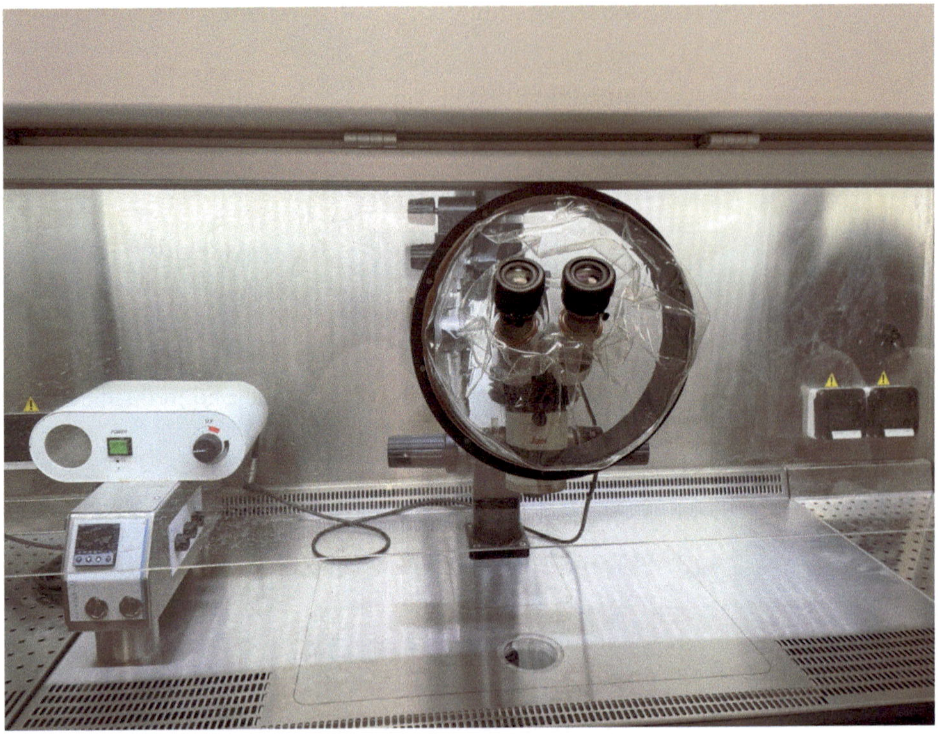

Fig. 1 Surgery may be carried out in a laminar airflow cabinet for increased biological security. Here is an example of the setup in a germ-free facility. The setup must allow a comfortable position for the operator

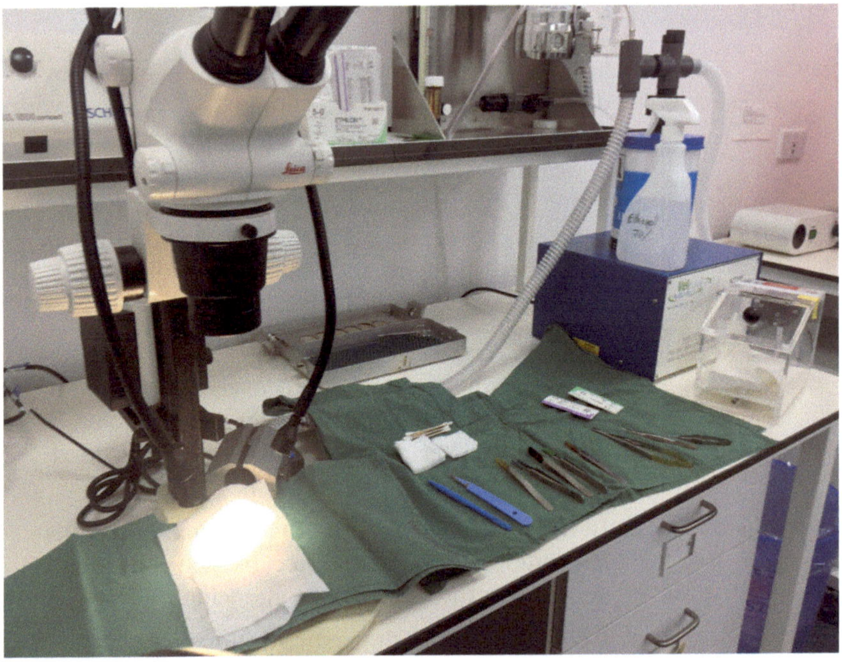

Fig. 2 A typical surgical setup, with equipment ready and placed conveniently, ready for use

4. Tip the following surgical equipment onto the drape: the surgical kit, microsurgical knife, disposable scalpel, Vicryl suture kit, and Ethilon suture kit. The equipment should be tipped onto the sterile drape from their wrapping rather than placed, to maintain sterility.
5. A clean hair trimmer should be placed so it is easily accessible.
6. Viscotears should also be placed so it is easily accessible.
7. Identify appropriate mouse and weigh it.
8. Place the mouse in anesthetic box, and turn on the isoflurane flow to induce anesthesia (*see* **Note 5**). Turn on oxygen cylinder or oxygen concentrator (allow 5 min for the correct concentration of oxygen, if using the concentrator and set the flow rate of oxygen to 2 L/min).
9. When the mouse is sedated sufficiently (This is characterized by a lack of response to sound and touch stimuli such as clicking fingers, brushing whiskers, and an absence of a pedal reflex), transfer it to preparation area, and fit the anesthetic mask.
10. Shave the anterior right knee using the hair trimmer (*see* **Note 6**).
11. Using micropore tape or suction, collect loose hair and discard.
12. Administer 0.1 mL of Vetergestic (0.00125 mg at 0.015 mg/mL) and 0.1 mL of Torbugesic (0.025 mg at 0.25 mg/mL) via subcutaneous injection (*see* **Note 7**).
13. Swab the shaved area with iodine/chlorhexidine solution.
14. Administer Viscotears to prevent eyes from drying out.
15. Turn on the isoflurane flow to the mask to maintain appropriate level of anesthesia (up to 3% isoflurane) and reduce concentration for maintenance (~0.5% isoflurane), and check pedal reflex is absent (*see* **Note 8**).
16. Transfer the mouse to the surgery area set up within the scope of the microscope with the mouse on its back and its nose in the mask (*see* **Note 9**).
17. Cover the mouse with a small sterile drape with the right leg accessible through a hole (Fig. 3).
18. Put on a new pair of sterile surgical gloves.
19. Ensure the mouse is positioned such that the right knee can be observed through the microscope.
20. Check the level of anesthesia by ensuring the mouse has no corneal or pedal reflex before starting surgery. Check periodically throughout the surgery to ensure the mouse is still anesthetized.
21. Make initial incision medially to the knee with the no. 15 blade, approximately 8–12 mm long, exposing the white parapatellar ligament.

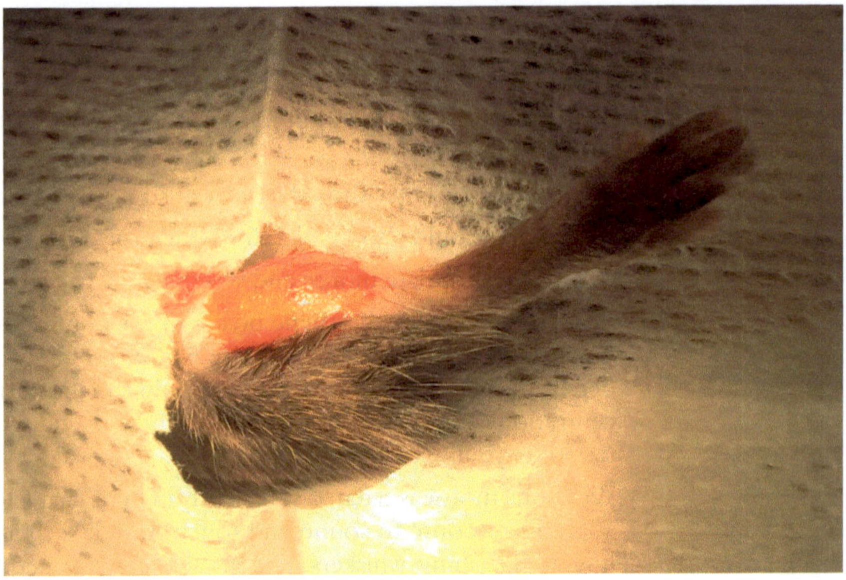

Fig. 3 Position the leg for surgery through the surgical drape ensuring the mouse is in the appropriate position for the operator and the mouse is protected

22. Make the secondary incision medially to the white parapatellar ligament.
23. Staunch any bleeding with a sterile cotton bud/surgical spear.
24. Open the incision with the jeweler's forceps (*see* **Note 10**).
25. Identify the MMTL, using the jeweler's forceps, blunt dissecting the fat pads as necessary.
26. Insert the micro surgical knife into the incision below where the MMTL was identified, with the blade facing up.
27. Sham operated controls should include everything up to this point, but the MMTL should be left intact (skip **step 28**).
28. Apply gentle but steady pressure upward, until you feel the ligament snap.
29. Staunch any bleeding with a sterile cotton bud/surgical spear.
30. Identify the cut end of the MMTL, the cut should have been close to the tibia, and the loose MMTL running from the MM should be long enough to just protrude from the incision (Fig. 4).
31. Further blunt dissection of the fat pads may be necessary to identify the cut end of the MMTL.
32. If the cut end is not visible, the MMTL may not have been completely severed (Fig. 4). In which case re-insert the jeweler's forceps where the MMTL was initially identified and see if you can locate the intact (or partially intact) ligament.

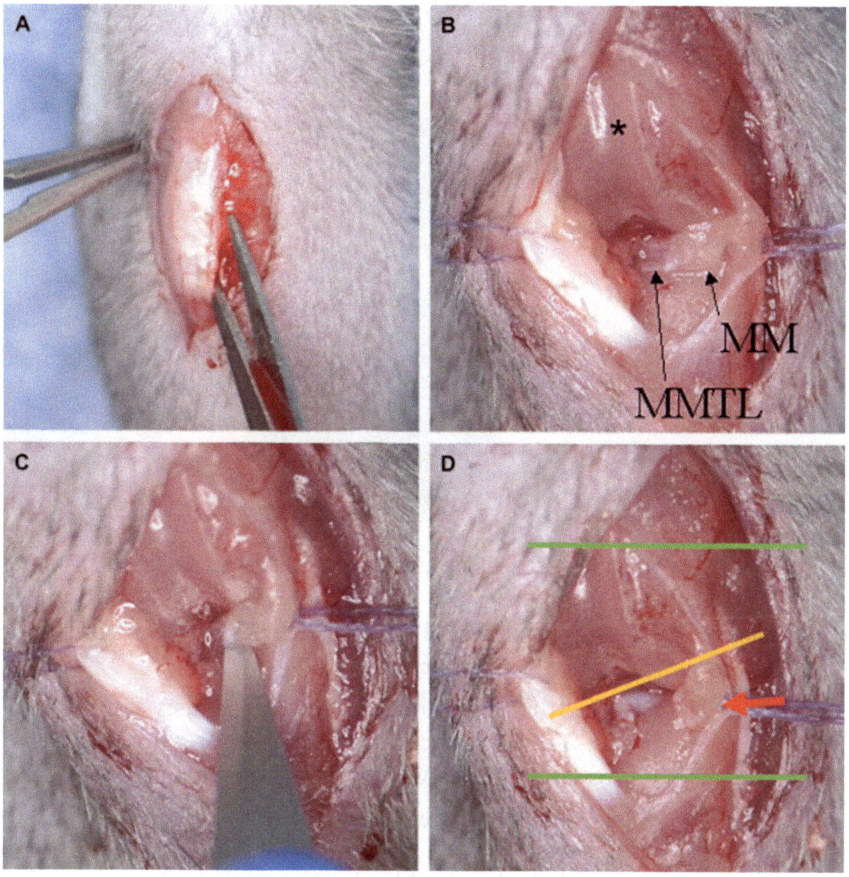

Fig. 4 Stages of the DMM surgery. (**a**) This incision in the skin can be a bit longer than that of the capsule. Use jeweler's forceps to open up the cavity behind the patella. (**b**) The incision shown is larger than usual to enable features to be demonstrated clearly. The femoral head (*) is visible and the medial meniscus (MM) and the meniscotibial ligament (MMTL) are shown. (**c**) The tip of the scalpel is inserted under the MMTL. (**d**) After cutting the MMTL, the cut end should be visible (red arrow). The appropriate position for internal (yellow line) and external (green lines) sutures are shown

33. When the MMTL is definitely severed, close the capsule with a surgeon's knot using Vicryl suture material (one suture should be sufficient).
34. Then close the dermal layer, again using a surgeon's knot with Ethilon suture material (depending on the length of the incision, two to three sutures should be sufficient). Ensure the loose ends of the sutures are cut fairly close to the skin to prevent the animal pulling the sutures (Fig. 5).
35. Refresh Viscotears and place the mouse in hot box or recovery chamber. Monitor the mice until they are fully recovered (*see* **Note 11**).

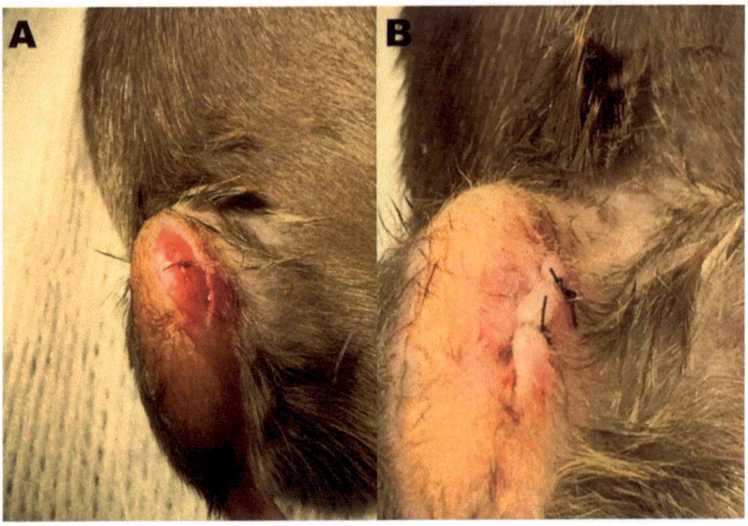

Fig. 5 Correct suturing after surgery of the capsule (**a**) and dermal layer (**b**)

3.2 Pain Assessment

1. Bring the animal cages to the testing room with as little disturbance as possible.
2. Let them acclimatize in their cages for a minimum of 15–30 min to reduce the impact of a novel environment and resulting stress, which may affect the outcome of the testing.
3. During this time, calibrate both platforms of the Linton incapacitance tester (Fig. 6) using the supplied 50 g weight. If needed, also double-check the tare and ensure it is set to zero between both platforms as described in the manual.
4. Set the recording time to 1 s interval (*see* **Note 12**).
5. One by one, move the animal from the cage into the Linton incapacitance box as gently as possible, ideally letting the animals walk off the operator's arm into the box by themselves. Again, this helps prevent a stress response.
6. Wait until the animal spontaneously turns into the correct position with the forelimbs resting on the curved side of the box (*see* Fig. 6).
7. Lightly extract the tail using the specimen tweezers, taking care not to disturb or startle the animals to minimize stress. Hold the tail lightly to ensure the animal stays in the correct position.
8. Using the small sturdy sheet of paper, reposition the hindlimbs if necessary, to encourage a symmetric stance (*see* Fig. 6), and when this is achieved, press "enter" to assess the weight difference.
9. Record three consecutive readings with the animal in the correct position. If the animal moves during this time, wait until it has settled again before recording.

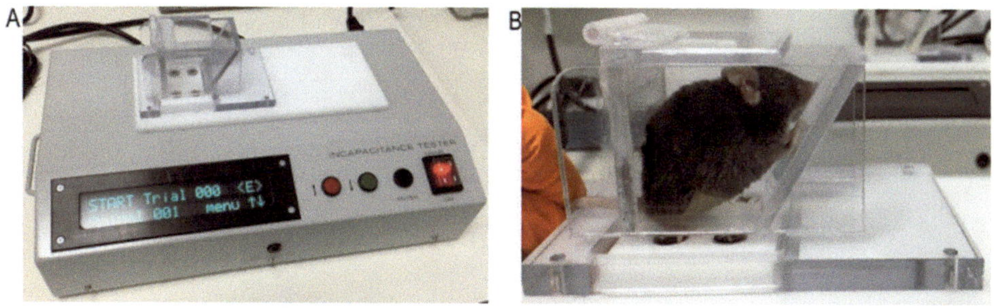

Fig. 6 The Linton incapacitance setup (**a**) with a mouse in situ in the correct position (**b**) (note the tail is being held gently to keep the mouse in position)

10. After obtaining all required readings, return the animal to its cage.
11. Clean the platform and box with TriGene (or another suitable disinfectant) before starting with the next animal.
12. After collection of all technical replicates, calculate the average percentage weight difference between the legs per animal.
13. Take the average of all animals within groups (biological replicates) to receive a final reading for that group (*see* **Note 13**).

3.3 Histology Sample Preparation

1. The mice are euthanized using the inhalant anesthetic CO_2. A secondary physical method of euthanasia such as cervical dislocation is used to confirm death.
2. Dissect the knee joints carefully to remove skin and excess muscles (Fig. 7).
3. Fix the mouse knees in 10% neutral buffered formalin, with a volume of at least four times that of the specimen at room temperature, ensuring during this time that the knee is maintained fully extended (*see* **Note 14**).
4. The samples can be fixed for up to 24 h (*see* **Note 15**).
5. If tissue processing cannot be performed straight after 24 h of fixation, the samples can be stored in 70% ethanol at room temperature.

3.4 Embedding

1. Following fixation, the samples are placed into tissue cassettes indelibly labeled with a unique sample identifier.
2. The samples are then immersed in sufficient volume of 20% formic acid decalcification solution for 1 week at room temperature (replacement with fresh solution is not necessary) or in 0.5 M EDTA at room temperature for 2 weeks on a plate shaker, changing the solution every 2 days (*see* **Note 16**).

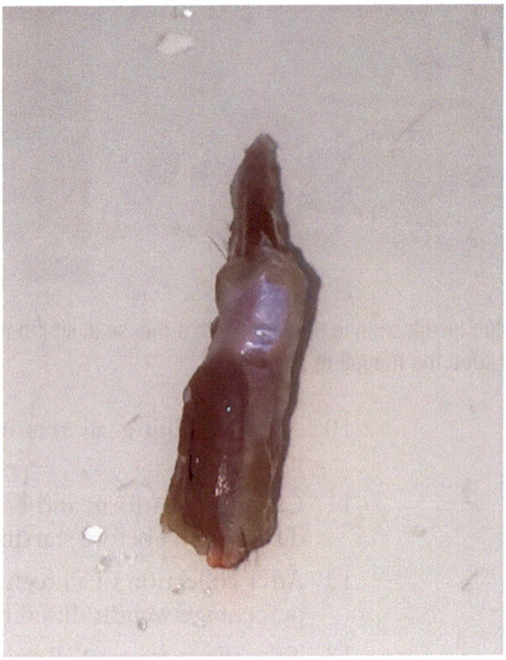

Fig. 7 The dissected limb with the majority of muscle removed, prior to processing for histology

3. The fully decalcified mouse knees can be processed for histological analysis. The samples can be frozen and cryosectioned, but for this purpose, paraffin embedding is preferred because of the good tissue morphology preservation and cost-effectiveness without need of specialized instrumentation.
4. Tissue cassettes are then transferred to a tissue processor for standard overnight tissue processing as follows (*see* **Note 17**):
   (a) 1 h and 30 min in formal saline at 40 °C.
   (b) 1 h and 30 min in 70% ethanol at 40 °C.
   (c) Five changes of 100% ethanol for 1 h and 30 min each at 40 °C.
   (d) Three changes of xylene for 1 h and 30 min at 40 °C.
   (e) Four changes of paraffin for 1 h and 30 min at 63 °C.
5. Fill a mold of appropriate depth and size for the tissue embedding station with molten paraffin wax, and transfer the sample from the tissue cassette to the metallic mold on the hot plate, with heated forceps.
6. To embed the joints coronally, the patella tendon should be positioned at the bottom of the metallic mold along its long axis (*see* **Note 18**).
7. After orienting the sample, move the mold to the cold plate carefully to avoid changes in the position of the sample and let the paraffin wax to solidify.

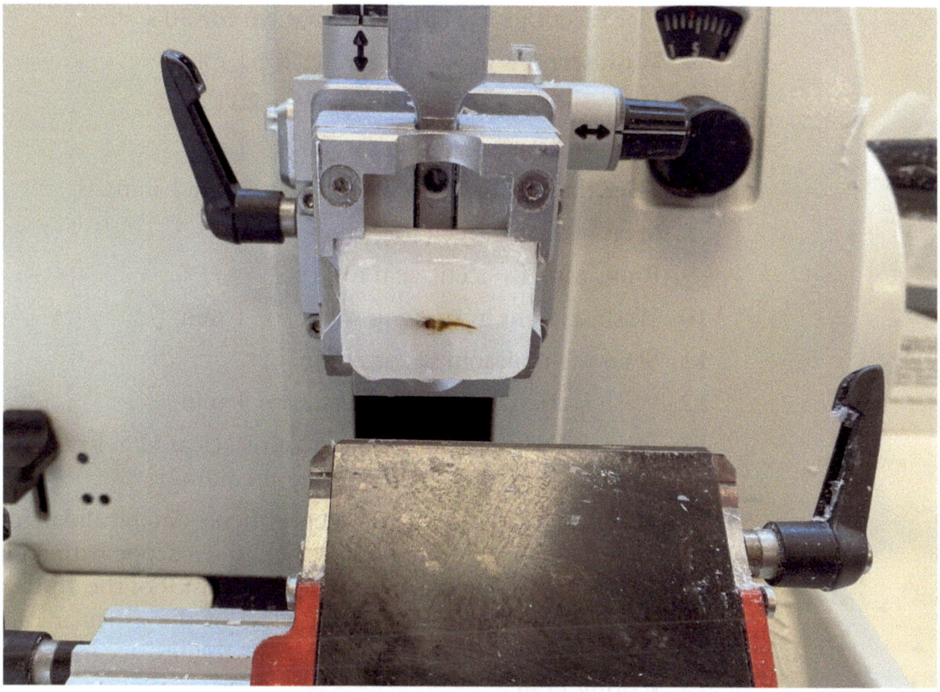

Fig. 8 The correct orientation of the embedded limb for sectioning

3.5 Sectioning

1. Cool paraffin blocks to set the paraffin wax before sectioning.
2. Place the paraffin block in the microtome block holder so that the orientation of the cartilage interface is parallel to the blade in order to avoid folding of the articular cartilage during sectioning, which can hinder appropriate disease scoring (Fig. 8).
3. Trim the block until menisci on the medial and lateral compartments of the joint appear distinct. This can be monitored in unstained sections (*see* **Note 19**).
4. Start collecting 4 μm tissue sections. Keep the block cool and hydrated using an ice box as it facilitates the sectioning process.
5. 15 sections should be obtained for each block, with 80 μm intervals between sections, allowing the examination of the articular cartilage throughout the joint.
6. Float sections onto adhesive microscope glass slides in the water bath.
7. To minimize the risk of having unscorable sections due to histological artifacts, two sections can be placed on each slide.
8. Bake slides overnight at 60 °C before staining.

3.6 Safranin O/Fast Green Staining of Sections

1. Dewax sections in two changes of xylene 5 min each (*see* **Note 20**).
2. Hydrate sections (three changes of 100% ethanol 1 min each followed by running tap water for 20 s).

3. Place sections in Harris hematoxilyn for 30 s.
4. Wash sections in running water for 1 min.
5. Differentiate with acid alcohol for 20 s.
6. Wash sections in running water for 1 min.
7. Place the sections in blueing solution for 1 min.
8. Stain with fast green for 6 min.
9. Rinse with 1% acetic acid for 15 s.
10. Wash sections in running water for 1 min.
11. Stain with safranin O for 2 min.
12. Wash sections in running water for 1 min.
13. Dehydrate (3 × 1 min 100% ethanol), clear (2 × 1 min xylene), and mount with DPX mounting medium.
14. Take images of both joint compartments (lateral and medial) at 10× magnification on a bright-field microscope with an image acquisition software, repeating for at least 9 of the 15 sections. This will give a representative view of the articular cartilage throughout the entire joint, enabling accurate disease scoring [12].

4 Notes

1. The following surgical instruments (or equivalent instrumentation) should be autoclaved and kept sterile prior to use.
2. Ensure the oxygen cylinder is secured and cannot be knocked over.
3. As an additional level of aseptic technique and protection for the user from allergens, the surgery may be carried out in a laminar airflow cabinet (*see* Fig. 2).
4. The Linton incapacitance meter should be in a quiet room and cleaned between each use. Avoid positioning under an air source that could upset the readings.
5. This is in accordance with UK regulations; local regulations may vary.
6. The suggested concentration for inducing anesthesia in rodents is 3–5%, and typically we found 4% good for induction.
7. There is no particular reason for the right knee to be the operated knee; it is purely for consistency throughout the procedure and analysis.
8. The suggested doses are Vetergesic 0.05–0.1 mg/kg and Torbugesic 1–5 mg/kg, and we use 0.05 mg/kg (assuming mouse is 25 g) and 1 mg/kg (assuming mouse is 25 g), respectively.

9. This is an important welfare consideration as there is a significant risk of eyes drying out under anesthesia. The mice should be monitored throughout the procedure and recovery, and additional Viscotears applied if necessary.
10. Fine jeweler's forceps can get damaged very easily, so use tip guards to protect the points when not in use.
11. Analgesics are only administered once before surgery and the effects last for many hours.
12. We found that due to mice being inherently more active, a 1 s interval is most suitable. This may vary by strain, and when working with other organisms such as rats a longer interval is often used.
13. In addition to the Linton incapacitance, we have employed the Librae incapacitance system (Ugo Basile) which employs a very similar protocol to the Linton system but has the advantage of real-time data with mean standard deviation and maximum weight of each recording. We are also testing the Dynamic Weight Distribution 2.0 (Bioseb in vivo research instruments, France) apparatus which has a video to record movement and a pressure-sensitive floor to the cage, thus having the advantage of reducing stress and allowing the mice to move naturally.
14. Some groups prefer to fix joints in a semi-flexed position as mice weight bear on a bent knee and loaded areas on the tibia and femur are more likely to be directly contacting one another.
15. Cartilage damage can be assessed in tissues that have been fixed for longer than 24 h. However, over-fixation (more than 24 h) will interfere with immunohistochemistry, immunofluorescence, and in situ hybridization as it may have a negative impact on antigen detection.
16. Formic acid results in a rapid decalcification; however the presence of formalin within the solution may cause antigen masking, limiting the use of immunodetection techniques. Therefore, when immunohistochemistry techniques are to be included in the histological analysis, a gentler decalcification method, such as 0.5 M EDTA, is preferred.
17. If an automatized tissue processing system is not available, manual tissue processing can be performed as followed: The samples are placed overnight in 70% ethanol, transferred to 90% ethanol for 2 h, then 100% ethanol for 2 h, perform two changes of xylene 2 h each are performed, and then incubate specimens in melted paraffin in the oven at 60 °C overnight.

18. Coronal embedding allows for simultaneous assessment of the medial and lateral tibiofemoral joints, and so it is essential the orientation of the sample at this stage is correct. This allows the evaluation of the entire joint. Some groups prefer sagittal sections for assessment of synovium.
19. When trimming the block, care is needed to see if the orientation is correct. If it is not, the sample can be rotated using the block adjustment screws in the microtome. However, if the orientation is sagittal rather than coronal, the block must be melted and the sample re-embedded.
20. An automated stainer should be used in order to keep the staining consistent. The following protocol describes the steps to achieve a satisfactory safranin O/fast green staining of knee joint sections.

References

1. Losina E, Paltiel AD, Weinstein AM et al (2015) Lifetime medical costs of knee osteoarthritis management in the United States: impact of extending indications for total knee arthroplasty. Arthritis Care Res (Hoboken) 67(2):203–215. https://doi.org/10.1002/acr.22412
2. Vincent TL (2020) Of mice and men: converging on a common molecular understanding of osteoarthritis. Lancet Rheumatol 2(10):e633–e645. https://doi.org/10.1016/S2665-9913(20)30279-4
3. Bialkowska AB, Ghaleb AM, Nandan MO et al (2016) Improved Swiss-rolling technique for intestinal tissue preparation for immunohistochemical and immunofluorescent analyses. J Vis Exp (113):54161. https://doi.org/10.3791/54161
4. Blease A, Das Neves Borges P, Curtinha M et al (2018) Studying osteoarthritis pathogenesis in mice. Curr Protoc Mouse Biol 8(4):e50. https://doi.org/10.1002/cpmo.50
5. Fu K, Robbins SR, McDougall JJ (2018) Osteoarthritis: the genesis of pain. Rheumatology (Oxford) 57(suppl_4):iv43–iv50. https://doi.org/10.1093/rheumatology/kex419
6. Serra CI, Soler C (2019) Animal models of osteoarthritis in small mammals. Vet Clin North Am Exot Anim Pract 22(2):211–221. https://doi.org/10.1016/j.cvex.2019.01.004
7. McCoy AM (2015) Animal models of osteoarthritis: comparisons and key considerations. Vet Pathol 52(5):803–818. https://doi.org/10.1177/0300985815588611
8. Johnson CI, Argyle DJ, Clements DN (2016) In vitro models for the study of osteoarthritis. Vet J 209:40–49. https://doi.org/10.1016/j.tvjl.2015.07.011
9. von Loga IS, El-Turabi A, Jostins L et al (2019) Active immunisation targeting nerve growth factor attenuates chronic pain behaviour in murine osteoarthritis. Ann Rheum Dis 78(5):672–675. https://doi.org/10.1136/annrheumdis-2018-214489
10. Butterfield NC, Curry KF, Steinberg J et al (2021) Accelerating functional gene discovery in osteoarthritis. Nat Commun 12(1):467. https://doi.org/10.1038/s41467-020-20761-5
11. Das Neves Borges P, Forte AE, Vincent TL et al (2014) Rapid, automated imaging of mouse articular cartilage by microCT for early detection of osteoarthritis and finite element modelling of joint mechanics. Osteoarthr Cartil 22(10):1419–1428. https://doi.org/10.1016/j.joca.2014.07.014
12. Glasson SS, Chambers MG, Van Den Berg WB et al (2010) The OARSI histopathology initiative – recommendations for histological assessments of osteoarthritis in the mouse. Osteoarthr Cartil 18(Suppl 3):S17–S23. https://doi.org/10.1016/j.joca.2010.05.025
13. Chinzei N, Rai MF, Hashimoto S et al (2019) Evidence for genetic contribution to variation in posttraumatic osteoarthritis in mice. Arthritis Rheumatol 71(3):370–381. https://doi.org/10.1002/art.40730
14. Bygrave AE, Rose KL, Cortes-Hernandez J et al (2004) Spontaneous autoimmunity in 129 and C57BL/6 mice-implications for autoimmunity described in gene-targeted mice. PLoS Biol 2(8):E243

15. Loeser RF, Olex AL, McNulty MA et al (2012) Microarray analysis reveals age-related differences in gene expression during the development of osteoarthritis in mice. Arthritis Rheum 64(3):705–717. https://doi.org/10.1002/art.33388
16. Percie du Sert N, Hurst V, Ahluwalia A et al (2020) The ARRIVE guidelines 2.0: updated guidelines for reporting animal research. J Physiol 598(18):3793–3801. https://doi.org/10.1113/JP280389
17. Ma HL, Blanchet TJ, Peluso D et al (2007) Osteoarthritis severity is sex dependent in a surgical mouse model. Osteoarthr Cartil 15(6):695–700. https://doi.org/10.1016/j.joca.2006.11.005

Chapter 27

Collagenase-Induced Osteoarthritis in Mice

Peter M. van der Kraan

Abstract

The collagenase-induced experimental osteoarthritis model is in general applied in mice but can also be used in other small species. The model is mainly based on the induction of joint laxity but has also a major inflammatory component. In this chapter, the induction is described by two injections of collagenase at the start of the model. Investigators who will use this model have to have ample experience in intra-articular injection in mice.

Key words Collagenase, Joint laxity, Intra-articular injection, Anesthesia, Mouse

1 Introduction

The early pathogenesis of osteoarthritis (OA) cannot be effectively studied in man, and therefore interest has been directed to animal models of osteoarthritis that simulate the human disease. Spontaneous osteoarthritis has been studied among others in mice, rats, guinea pigs, dogs, monkeys, and horses. However, spontaneous OA in animal models is in general age-related and/or dependent on specific mutations, which results in large interindividual variation and unpredictability in starting point of disease development. To overcome these limitations, OA models have been developed that are induced at a specific moment in time, either by surgical methods or by intra-articular injection of chemicals interfering with joint stability or cartilage metabolism. Recently, murine OA models have been developed based on mechanical damage induced by noninvasive loading [1]. These models can be regarded as models that most reliably mimic post-traumatic OA.

This collagenase-induced experimental OA model in mice has been first described by us in 1989 [2, 3]. In addition to mice, collagenase-induced OA has been carried out in a number of other animal species, for instance, rats and rabbits [4–7]. Collagenase-induced OA is a model of post-traumatic OA that is mainly

Martin J. Stoddart et al. (eds.), *Cartilage Tissue Engineering*,
Methods in Molecular Biology, vol. 2598, https://doi.org/10.1007/978-1-0716-2839-3_27,

based on induction of joint laxity. A strong correlation between early ligament damage and cartilage destruction and osteophyte formation has been found in this animal model [8]. The collagenase injection has little direct effects on the articular cartilage but induces, as a result of soft tissue damage, a profound early inflammatory reaction in the joint, characterized by an influx of neutrophils [2, 3]. Collagenase injection results in a short, 1 day, depression of articular cartilage proteoglycan synthesis, which is followed by a prolonged period of elevated synthesis [3]. Cartilage damage, as measured by chondrocyte death, fibrillation, or erosion, is not seen before day 21. Furthermore, also osteophyte formation is a rather late phenomenon [3].

Collagenase-induced OA can be induced in various mouse strains and both sexes. We have compared disease development in male and female mice of two closely related strains, C57BL6 and C57BL10 [9]. Differences in prevalence of cartilage damage between strains and sexes were observed. Prevalence was higher in C57BL10 than in C57BL6, and the prevalence was twice as high in males as in females in both strains. Also, the site of the damage appeared to be dose and strain dependent. Male C57BL6 mainly showed damage on the medial tibial plateau, while in male C57BL10 damage almost always appeared on the lateral tibial. Since it is known that male mice are more prone to spontaneous OA than female mice, and C57BL10 are more prone than C57BL6 mice, these results indicate that predisposition to spontaneous OA increases the risk of developing injury-induced osteoarthritis.

This procedure describes how collagenase-induced osteoarthritis is elicited in mice by the intra-articular injection of highly purified bacterial collagenase. The methodology is identical for all mouse strains and gender independent. However, OA severity is in general higher in males than in females, but variability between individuals is higher in males. Very severe cases can result in joint dislocation, and due to severity of the disease process, these animals could be removed from analysis based in the specific research question.

Initially we used a single injection of 5 or 10 units to induce osteoarthritis. However, to reduce variability and to circumvent joint dislocations, our standard protocol is twice an injection with 1 unit collagenase on day 1 and day 3.

2 Materials

2.1 Solutions

1. Anesthetic agent: isoflurane.
2. Collagenase type VII in physiological saline (Sigma, final concentration 165 units/mL).
3. Sterile physiological saline (NaCl 0.9%).
4. Ethanol 70%.
5. 10 mL tubes with 7 mL 4% buffered formalin.

2.2 Supplies

1. Hamilton gas tight syringe without needle (glass syringe to be able to inject 6 μL).
2. Needle (30GA1/2 0.3 × 13).
3. Isoflurane anesthetic setup, including 5 mL syringe to keep mouse sedated during procedure.
4. Sterile gauze pads, crooked tweezers, small scissors.
5. Clean container with towel.
6. Sterile gauze swaps (5 cm × 5 cm).
7. Curved tweezers.
8. Small scissors.
9. Bone cutter.

3 Methods

All animal handling procedures must be carried out by a qualified person according to all local laws and ethical requirements.

3.1 Preparation for Osteoarthritis Induction

1. Prepare the collagenase solution immediately before the start of the experiment.
2. Before injecting the mouse intra-articularly, the mouse is anesthetized with isoflurane vapor.
3. During the procedure, the mouse is placed with its head in the casing of the 5 mL syringe, in such a way that the mouse is kept under complete anesthesia during the entire procedure.
4. Fill the Hamilton syringe with 6 μL collagenase solution (*see* **Note 1**).

3.2 Induction of Osteoarthritis

Day 1:

Fill the Hamilton syringe with 6 μL collagenase solution.

1. Disinfect the skin of the knee joint to be injected with 70% ethanol.
2. Make a small opening through the skin (3–4 mm) at the level where the patella tendon is attached to the patella by means of a small scissor.
3. Inject at the medial site of the patella tendon 6 μL (1 unit) of collagenase into the knee joint cavity (*see* **Note 2**).

Day 3:

1. Fill the Hamilton syringe with 6 μL collagenase solution.
2. Disinfect the skin of the knee joint to be injected with 70% ethanol.

3. Make a small opening through the skin at the level where the patella tendon is attached to the patella by means of a small scissor.
4. Inject at the medial site of the patella tendon 6 μL (1 unit) of collagenase into the knee joint cavity (*see* **Note 2**).

3.3 Osteoarthritis Induction Aftercare

1. The mice are placed on a towel in the cage and kept warm.
2. Animals are housed in group cages according to local regulations.

The typical experiment runs for 42 days. However, this can be adjusted to a shorter or a longer period depending on the specific research questions. A frequently used method to evaluate OA is by histology of the knee joints.

3.4 Knee Joint Dissection

Preparation before you start:

1. Fill the 10 mL tubes with 7 mL 4% formalin.
2. Sacrifice the mice via cervical dislocation.

Dissection procedure (Fig. 1):

1. Disinfect and make the knee joint wet with a sterile gauze swap and ethanol 70%.
2. Remove the skin with tweezers and scissors.
3. Cut with the bone cutter 0.5 cm above the patella tendon and 0.5 cm under the patella tendon through the bone.
4. Hold the knee joint with the tweezers and cut it out with the scissors.

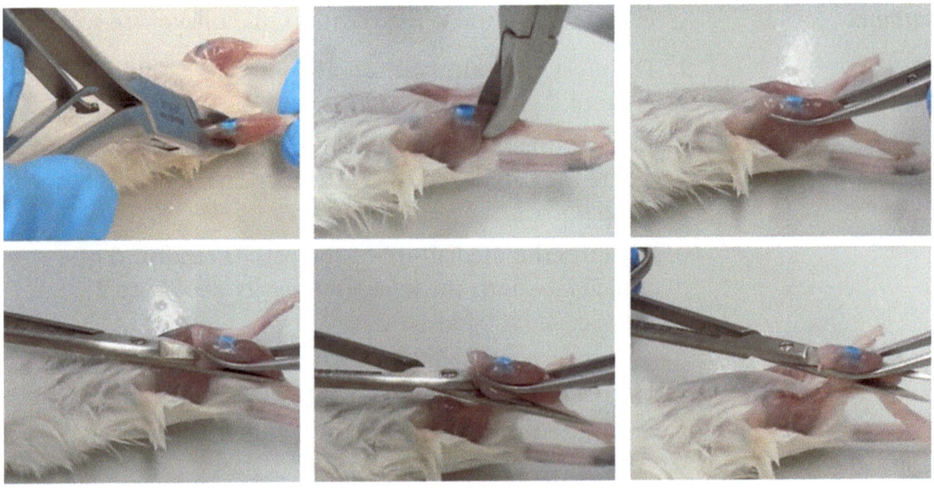

Fig. 1 Dissection of the murine knee joint for histology. Procedure is from top left to bottom right

5. Put the knee joint gently in the tube of formalin, leaving the natural positioning of the knee joint intact.
6. Process the sample for histology using standardized methods for the staining required.

4 Notes

1. Be aware that due to the dead volume of the needle you need much more 6 μL to fill the syringe and needle.
2. Intra-articular injection has to be performed by trained investigator. Intra-articular injection can be trained by use of injection of trypan blue in death overcomplete mice.

References

1. Poulet B (2016) Non-invasive loading model of murine osteoarthritis. Curr Rheumatol Rep 18(7):40. https://doi.org/10.1007/s11926-016-0590-z
2. van der Kraan PM, Vitters EL, van Beuningen HM et al (1990) Degenerative knee joint lesions in mice after a single intra-articular collagenase injection. A new model of osteoarthritis. J Exp Pathol (Oxford) 71(1):19–31
3. van der Kraan PM, Vitters EL, van de Putte LB et al (1989) Development of osteoarthritic lesions in mice by "metabolic" and "mechanical" alterations in the knee joints. Am J Pathol 135(6):1001–1014
4. Yeh TT, Wu SS, Lee CH et al (2007) The short-term therapeutic effect of recombinant human bone morphogenetic protein-2 on collagenase-induced lumbar facet joint osteoarthritis in rats. Osteoarthr Cartil 15(12):1357–1366. https://doi.org/10.1016/j.joca.2007.04.019
5. Adaes S, Almeida L, Potes CS et al (2017) Glial activation in the collagenase model of nociception associated with osteoarthritis. Mol Pain 13: 1744806916688219. https://doi.org/10.1177/1744806916688219
6. Choi JH, Choi JH, Kim DY et al (2002) Effects of SKI 306X, a new herbal agent, on proteoglycan degradation in cartilage explant culture and collagenase-induced rabbit osteoarthritis model. Osteoarthr Cartil 10(6):471–478. https://doi.org/10.1053/joca.2002.0526
7. Cheng NT, Guo A, Cui YP (2016) Intra-articular injection of Torin 1 reduces degeneration of articular cartilage in a rabbit osteoarthritis model. Bone Joint Res 5(6):218–224. https://doi.org/10.1302/2046-3758.56.BJR-2015-0001
8. van Osch GJ, van der Kraan PM, Blankevoort L et al (1996) Relation of ligament damage with site specific cartilage loss and osteophyte formation in collagenase induced osteoarthritis in mice. J Rheumatol 23(7):1227–1232
9. van Osch GJ, van der Kraan PM, Vitters EL et al (1993) Induction of osteoarthritis by intra-articular injection of collagenase in mice. Strain and sex related differences. Osteoarthr Cartil 1(3):171–177. https://doi.org/10.1016/s1063-4584(05)80088-3

Chapter 28

Platelet-Rich Plasma for Intra-articular Injections: Preclinical and Clinical Evidence

Angelo Boffa and Giuseppe Filardo

Abstract

The use of platelet-rich plasma (PRP) has been supported by encouraging data from in vitro and preclinical in vivo studies, both in terms of safety and efficacy. This led to the wide use of PRP injections in the clinical practice, with promising results especially as a minimally invasive treatment for cartilage degeneration and osteoarthritis (OA). While many controversies remain on the best PRP formulation, the overall available clinical studies support the benefits of PRP, with functional improvement and reduction of pain-related symptoms up to 12 months, especially in young patients and early OA stages.

Key words Platelet-rich plasma, PRP, Intra-articular, Injection, Cartilage, Osteoarthritis

1 Introduction

Intra-articular injective therapies are considered a suitable option for the management of patients with symptomatic degenerative cartilage lesions and osteoarthritis (OA), especially when the other conservative treatments are ineffective. Traditional injections with corticosteroids and hyaluronic acid are largely used in the clinical practice [1]. These injectables showed to be safe and able to provide positive effects with patient satisfaction, although results present a significant variability among different patients, and the effectiveness generally decreases over time, leading in the end to the need for more invasive surgical procedures [2]. Moreover, the evidence on corticosteroids and hyaluronic acid is often questioned, and the recommendations are still inconsistent among the international scientific societies [3–5].

To overcome the limits of these options, autologous therapies with intra-articular orthobiologic approaches have been introduced to address joints affected by OA, exploiting their disease-modifying potential effects [6, 7]. Among these, platelet-rich plasma (PRP) is gaining a large interest in the clinical practice as a minimally invasive

Martin J. Stoddart et al. (eds.), *Cartilage Tissue Engineering*,
Methods in Molecular Biology, vol. 2598, https://doi.org/10.1007/978-1-0716-2839-3_28,

injective treatment of OA, thanks to its safety, the low costs, and the simple preparation technique to obtain its biologically active content [6]. PRP consists of a volume of autologous plasma with a higher concentration of platelets compared to the whole blood [8]. It exploits the high concentration of growth factors, cytokines, and bioactive molecules stored in platelet α-granules, which showed to take part in the homeostasis of joint tissues, with anabolic effects and even more immunomodulatory properties able to counteract inflammatory and catabolic molecules characterizing the OA joint environment [9, 10].

This chapter addresses the overall evidence supporting this biological approach, describing the main preclinical and clinical results on the intra-articular use of PRP for the treatment of patients with OA.

2 Preparation Method and Administration

Different preparation methods are available to obtain PRP, and several types of kits are available in the market by various manufacturers, although most of these procedures follow a common line (Fig. 1). Generally, all procedures entail a peripheral venous blood sample harvested from the patient arm. Blood is then processed using a centrifuge to separate the blood components according to their density, obtaining three layers: the acellular one, platelet poor plasma (PPP, 55% of whole blood); the "buffy coat" (<1% of whole blood) which contains concentrated platelets and white blood cells; and the third one with erythrocytes (45% of whole blood). PPP and erythrocytes are discarded to obtain the intermediate layer concentrated in platelets (with or without leukocytes). The final product can have various concentrations of platelets, although some authors suggest a platelet concentration of at least 1,000,000 platelets/μL

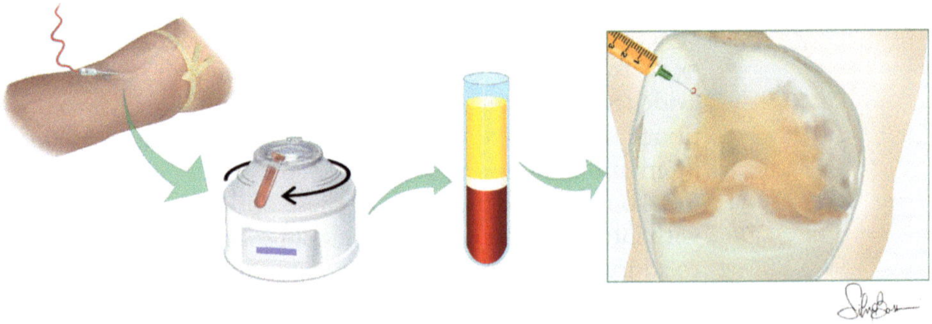

Fig. 1 Example of a blood preparation procedure. Blood is first collected from the patient and immediately centrifuged. The initial centrifugation separates red blood cells (RBCs) from platelet-poor plasma (PPP) and the "buffy coat." Then, the PPP and RBC layers are discarded, while the platelet concentrate is collected. The obtained platelet-rich plasma (PRP) is ready to be applied into the knee

[8, 11]. Greater platelet concentrations have not been shown to further improve healing, although several variables interact with the platelet concentrations affecting the biologic activity of PRP preparations [11]. Procedures can differ in terms of volume of blood collected, force and duration of centrifugation, single-step or two-step centrifugation protocols, and the resulting volume of platelets [12]. Moreover, the platelets contained in PRP may be activated using thrombin, calcium gluconate, or calcium chloride before being applied to the injury site as a liquid injection, and some authors prefer not to use any activator and to inject PRP relying on the in situ activation by local environmental factors.

Different application modalities have been proposed in terms of number of injections, mostly ranging from one to three injections, and recently some authors explored PRP combination with hyaluronic acid. Regardless of the administration strategy, before the injection, the skin is sterilely dressed, and the injection is generally performed through a classic superlateral approach using a 22-g needle. At the end of the procedure, the patient is encouraged to bend and extend the knee a few times to allow PRP to spread throughout the joint. After the injection, patients are sent home with instructions to restrict the use of the leg for at least 24 h and to use ice or other cold therapy modalities on the affected area to relieve pain. Rest or mild activities are permitted, and subsequently a gradual resumption of normal sport or recreational activities is allowed as tolerated. More specific indications on the post-injective management after PRP have been not clarified, yet.

3 In Vitro Evidence

Platelet activation triggers the degranulation and release of growth factors including platelet-derived growth factor (PDGF), epidermal growth factor (EGF), insulin-like growth factor (IGF-1), fibroblast growth factor (FGF), transforming growth factor-β (TGF-β), and vascular endothelial growth factor (VEGF) [11, 13]. These biologically active molecules seem to influence and promote a favorable joint environment, fostering the restoration of a homeostatic balance in degenerative joints by targeting not only cartilage but also other tissues [13]. In particular, in vitro studies investigating the effect of PRP on chondrocytes showed various and heterogeneous mechanisms of action, including the increase of the chondrocyte proliferation rate, the inflammation modulation, and the matrix production stimulation [14]. However, recent in vitro results showed conflicting findings, demonstrating that PRP was unable to inhibit the inflammatory response in OA chondrocytes and did not promote their regeneration despite an increased proliferation [15]. On the other hand, PRP may significantly enhance synoviocytes' hyaluronic acid secretion and switch synovial angiogenesis to a more balanced status [16]. Moreover, PRP may upregulate the viability of meniscal cells in a dose-dependent manner, as well as their mRNA expression of biglycan and decorin [17].

Current evidence further supports the role of PRP in modulating the intra-articular environment by affecting inflammation processes in degenerative joint diseases: after an initial pro-inflammatory action, with the stimulation of synoviocytes to release metalloproteinases and cytokines, a following phase of modulation and reduction of the inflammatory response has been demonstrated, with a decrease of pro-inflammatory cytokines, contrasting the chemotaxis of monocyte-like cells [18]. PRP also showed analgesic effects, possibly by modulating cannabinoid receptors in chondrocytes, and antimicrobial activity due to the release of platelet microbicidal proteins and multiple chemokines [14, 19]. Based on the available evidence, PRP may not directly promote a long-lasting tissue regenerative effect, but it may rather act through its different bioactive molecules by affecting tissue homeostasis and offering a benefit in terms of symptom relief and functional improvement.

4 Preclinical Evidence

The evaluation of PRP potential in counteracting OA progression largely relies on animal models, which play a crucial role in understanding the structural effects of novel therapeutic interventions [20]. Animals treated with PRP were reported to sustain a marked reduction in the severity of cartilage destruction and surface loss, as well as less fibrillation and irregularity compared to control groups [21–23]. These findings observed at macroscopic level were confirmed by phase-contrast computed tomography, which revealed greater cartilage volume and better surface in mice treated with PRP [24]. Histologically, better cellularity was seen following PRP administration, which prevented chondrocyte apoptosis and increased their viability with a higher proliferation rate, resulting in higher chondrocyte numbers [25–27]. Chondrocytes also showed better ultrastructural characteristics with a relatively intact cell membrane, abundant cellular organelles, relatively uniform cytoplasmic staining, and roughly normal nuclear morphology [28]. The cartilage matrix also presented better characteristics, showing no proteoglycan loss and collagen fibers uniformly arranged and a strong type II collagen expression with some fractures, suggesting the repair of damaged articular cartilage by hyaline cartilage [29, 30].

The positive effects on cartilage could derive from the direct anabolic action of PRP, as well as from the indirect effects in reducing synovial inflammation, thus positively modulating the overall joint environment. In fact, the histological evaluation of animals receiving intra-articular PRP injections showed less synovial hyperplasia with thinner synovial membrane [24, 31]. PRP counteracted the inflammatory reactions caused by OA, reducing inflammatory cell infiltration, synovial vascularity, and fibrosis and leading to less edema [32, 33]. Moreover, the macrophage subtype evaluation showed a reduction of inflammatory macrophages ($iNOS^+$) [33]. Synovium specimens revealed a strong expression of PDGF subunit A in synoviocytes, vascular endothelium, and stromal cells. Likewise, there was a higher percentage of immunostained area for VEGF in the synovial membrane from PRP specimens compared to controls [31].

The measurement of synovial fluid (SF) or serum biomarkers related to cartilage metabolism or inflammation was also investigated to assess PRP benefits in animal models. SF cartilage oligomeric matrix protein (COMP) was reported to be significantly lower after PRP administration, supporting its potential to slow down OA processes [22]. PRP also decreased the serum and SF concentrations of interleukin-1 beta (IL-1β), tumor necrosis factor-α (TNF-α), and prostaglandin E2 (PGE2) [34, 35]. Injections also provided a significant increase in serum levels of VEGF and PDGF-A, further suggesting a possible role of PRP in modulating angiogenesis and cartilage healing [31].

5 Clinical Evidence

Since the first clinical applications, safety and effectiveness of PRP for the treatment of knee OA have been supported by several studies, mostly showing satisfactory results in terms of functional improvement and reduction of pain-related symptoms up to 12 months, especially in young patients and early OA stages [36–38]. Importantly, several series demonstrated that intra-articular PRP injections are safe with a rate of adverse events not higher compared to the other intra-articular injectable products [39]. Despite the large placebo effect of any intra-articular injections, and in particular when dealing with new products such as PRP [40], a recent meta-analysis on 34 RCTs showed that PRP injections provided a statistically and clinically significant improvement in patients with knee OA compared to saline injections at 6 and 12 months of follow-up, without increasing the risk of adverse events [41]. Moreover, PRP injections led to better clinical results than other commonly used injective strategies such as hyaluronic acid or corticosteroids at 6 and 12 months of follow-up [41]. Some evidence suggest that the clinical improvement provided by PRP can be perceived by some patients also beyond 24 months, with a subsequent gradual reduction over time [42]. A recent retrospective study and survival analysis also showed that PRP may delay, with a median of 4 years, the need for knee arthroplasty, with a survival rate of 85% at 5 years of follow-up [43].

Clinical studies currently available in the literature support the intra-articular injection of PRP for the treatment of knee OA. Nevertheless, its ability to affect OA progression has not been demonstrated with radiographic evaluation [44]. On the other side, some studies reported possible effects of PRP injections on cartilage and synovium tissues through ultrasound or magnetic resonance imaging (MRI) evaluation. At ultrasound assessment, Ahmad et al. reported less effusion, reduced synovial hypertrophy, and improved vascularity scores in knee OA patients treated with three intra-articular PRP injections compared to hyaluronic acid at 6 months of follow-up [45]. A recent small trial documented better MRI findings in terms of patellofemoral cartilage volume and synovitis in patients with grade 1–2 knee OA treated with two intra-articular PRP injections compared with patients treated with hyaluronic acid at 12 months of follow-up [46]. However, these preliminary findings need to be confirmed by larger trials investigating if PRP can actually provide structural effects beside the demonstrated clinical improvement.

Besides the promising clinical results, no consensus or guidelines exist among scientific societies of orthopedic surgeons, rheumatologists, and physiatrists, on the most suitable indications for the use of PRP in the treatment of OA [3–5]. This is likely due to the lack of a standardized protocol and to the availability of different PRP preparation methods which can yield products with different compositions and characteristics in terms of number of platelets, presence of leukocytes, and red blood cells [47]. In particular, the presence of leukocytes is one of the most debated aspects regarding PRP effects, considering the pro-inflammatory properties ascribed to leukocytes. Several in vitro studies suggested their detrimental effects on joint tissues, reporting a greater synoviocytes and chondrocytes' death, a larger production of pro-inflammatory mediators, and the promotion of catabolic pathways after leukocyte-rich PRP (LR-PRP) application compared to leukocyte-poor PRP (LP-PRP) [16, 48, 49]. On the other side, PRP without leukocytes often presents a limited number of platelets, and in vitro evidence showed that LP-PRP did not provide superior results on chondrocyte growth and anabolism compared to the simple PPP, thus suggesting that the lower concentrations of platelets in LP-PRP may lead to a significantly lower anabolic potential compared to LR-PRP [16].

In the animal model with experimental knee OA, the increase of pro-inflammatory factors such as IL-1β and PGE2 after LR-PRP injections has been demonstrated, which underlines the possible pro-inflammatory effect of leukocytes [35]. Still, PRP is a composite concentrate with a complex balance of several factors that could influence the final effects, which questions the simplistic assumption of a direct correlation of leukocytes and results, with LP-PRP supposedly being a "good" PRP and LR-PRP being a "bad" PRP for knee OA. In fact, controversial findings were also observed in animal studies, with some authors reporting overall better results for LP-PRP in reducing the progression of cartilage degeneration and preventing synovitis, while other authors did not show differences between the two types of PRP at histological evaluation and even reported a better protection from thermal hyperalgesia when applying LR-PRP [24, 32].

Another issue is the translation from the in vitro and small animal model findings to the clinical scenario in humans. How does the higher "cell death" in terms of clinical outcome? Unfortunately, there is sparse high-level clinical evidence on this issue. The potential detrimental effects of leukocytes were documented in a comparative clinical study which evaluated 144 patients treated with a single-spinning procedure (PRP without leukocytes and low platelets) or a double-spinning approach (PRP rich in platelets and leukocytes). In fact, while the two treatments provided a similar clinical improvement up to 12 months of follow-up, more swelling and pain reaction were found after LR-PRP injections, supporting the negative role played by leukocytes [50]. On the other side, these negative findings could be considered minor or even negligible, as they spontaneously resolved after a few days, with the trend of clinical improvement being overlapping between the two PRPs up to 1 year. Moreover, a biomarker analysis performed in 36 patients with knee OA showed inconsistent findings compared to animal models, reporting no increase in pro-inflammatory cytokine levels before and after 1 week from LR-PRP injections [51]. Therefore, the role of leukocytes appears less detrimental in the clinical practice compared to in vitro studies, and further studies should investigate the effects of leukocytes on both safety and effectiveness of PRP for the treatment of knee OA.

There are additional variables to consider regarding PRP efficacy including the volume of blood to harvest, which is related to the volume of PRP to inject, the number and speed of centrifugations, the use of fresh or freeze-thawed product, the activation by different substances and the timing of activation with respect to the intra-articular PRP injection, and the number and the time interval between injections [52]. Unresolved issues related to each of these parameters must be addressed to understand if different products may provide different results and thus to optimize the therapeutic strategy in the use of platelet concentrates. In this scenario, a new coding system has been recently proposed as a tool to help basic researchers as well as clinicians in the difficult process of delineating more precise information about PRP [52]. The use by researchers and clinicians of a common reference system is needed to convey all the essential data to compare results of different trials and orient future investigations on the intra-articular use of PRP for the treatment of knee OA.

References

1. Ayhan E, Kesmezacar H, Akgun I (2014) Intraarticular injections (corticosteroid, hyaluronic acid, platelet rich plasma) for the knee osteoarthritis. World J Orthop 5(3):351–361. https://doi.org/10.5312/wjo.v5.i3.351
2. Crawford DC, Miller LE, Block JE (2013) Conservative management of symptomatic knee osteoarthritis: a flawed strategy? Orthop Rev 5(1):e2. https://doi.org/10.4081/or.2013.e2
3. Jevsevar DS (2013) Treatment of osteoarthritis of the knee: evidence-based guideline, 2nd edition. J Am Acad Orthop Surg 21(9):571–576. https://doi.org/10.5435/JAAOS-21-09-571
4. Bannuru RR, Osani MC, Vaysbrot EE et al (2019) OARSI guidelines for the non-surgical management of knee, hip, and polyarticular osteoarthritis. Osteoarthr Cartil 27(11):1578–1589. https://doi.org/10.1016/j.joca.2019.06.011
5. Bruyere O, Honvo G, Veronese N et al (2019) An updated algorithm recommendation for the management of knee osteoarthritis from the European Society for Clinical and Economic Aspects of Osteoporosis, Osteoarthritis and Musculoskeletal Diseases (ESCEO). Semin Arthritis Rheum 49(3):337–350. https://doi.org/10.1016/j.semarthrit.2019.04.008
6. Filardo G, Kon E, Longo UG et al (2016) Non-surgical treatments for the management of early osteoarthritis. Knee Surg Sports Traumatol Arthrosc 24(6):1775–1785. https://doi.org/10.1007/s00167-016-4089-y
7. Kon E, Filardo G, Drobnic M et al (2012) Non-surgical management of early knee osteoarthritis. Knee Surg Sports Traumatol Arthrosc 20(3):436–449. https://doi.org/10.1007/s00167-011-1713-8
8. Marx RE (2001) Platelet-rich plasma (PRP): what is PRP and what is not PRP? Implant Dent 10(4):225–228. https://doi.org/10.1097/00008505-200110000-00002
9. Sundman EA, Cole BJ, Fortier LA (2011) Growth factor and catabolic cytokine concentrations are influenced by the cellular composition of platelet-rich plasma. Am J Sports Med 39(10):2135–2140. https://doi.org/10.1177/0363546511417792
10. Di Matteo B, Kon E, Filardo G (2016) Intra-articular platelet-rich plasma for the treatment of osteoarthritis. Ann Transl Med 4(3):63. https://doi.org/10.3978/j.issn.2305-5839.2016.01.18
11. Foster TE, Puskas BL, Mandelbaum BR et al (2009) Platelet-rich plasma: from basic science to clinical applications. Am J Sports Med 37(11):2259–2272. https://doi.org/10.1177/0363546509349921
12. Dhurat R, Sukesh M (2014) Principles and methods of preparation of platelet-rich plasma: a review and author's perspective. J Cutan Aesthet Surg 7(4):189–197. https://doi.org/10.4103/0974-2077.150734
13. Filardo G, Kon E, Roffi A et al (2015) Platelet-rich plasma: why intra-articular? A systematic review of preclinical studies and clinical evidence on PRP for joint degeneration. Knee Surg Sports Traumatol Arthrosc 23(9):2459–2474. https://doi.org/10.1007/s00167-013-2743-1
14. Park SI, Lee HR, Kim S et al (2012) Time-sequential modulation in expression of growth factors from platelet-rich plasma (PRP) on the chondrocyte cultures. Mol Cell Biochem 361(1–2):9–17. https://doi.org/10.1007/s11010-011-1081-1
15. Rikkers M, Dijkstra K, Terhaard BF et al (2021) Platelet-rich plasma does not inhibit inflammation or promote regeneration in human osteoarthritic chondrocytes in vitro despite increased proliferation. Cartilage 13-(2_suppl):991S–1003S. https://doi.org/10.1177/1947603520961162
16. Assirelli E, Filardo G, Mariani E et al (2015) Effect of two different preparations of platelet-rich plasma on synoviocytes. Knee Surg Sports Traumatol Arthrosc 23(9):2690–2703. https://doi.org/10.1007/s00167-014-3113-3
17. Ishida K, Kuroda R, Miwa M et al (2007) The regenerative effects of platelet-rich plasma on meniscal cells in vitro and its in vivo application with biodegradable gelatin hydrogel. Tissue Eng 13(5):1103–1112. https://doi.org/10.1089/ten.2006.0193
18. Pereira RC, Scaranari M, Benelli R et al (2013) Dual effect of platelet lysate on human articular cartilage: a maintenance of chondrogenic potential and a transient proinflammatory activity followed by an inflammation resolution. Tissue Eng A 19(11–12):1476–1488. https://doi.org/10.1089/ten.TEA.2012.0225
19. Mariani E, Canella V, Berlingeri A et al (2015) Leukocyte presence does not increase microbicidal activity of Platelet-rich Plasma in vitro. BMC Microbiol 15:149. https://doi.org/10.1186/s12866-015-0482-9
20. Kuyinu EL, Narayanan G, Nair LS et al (2016) Animal models of osteoarthritis: classification,

update, and measurement of outcomes. J Orthop Surg Res 11:19. https://doi.org/10.1186/s13018-016-0346-5
21. Boffa A, Salerno M, Merli G et al (2021) Platelet-rich plasma injections induce disease-modifying effects in the treatment of osteoarthritis in animal models. Knee Surg Sports Traumatol Arthrosc 29(12):4100–4121. https://doi.org/10.1007/s00167-021-06659-9
22. Kanwat H, Singh DM, Kumar CD et al (2017) The effect of intra-articular allogenic platelet rich plasma in Dunkin-Hartley guinea pig model of knee osteoarthritis. Muscles Ligaments Tendons J 7(3):426–434. https://doi.org/10.11138/mltj/2017.7.3.426
23. Olifirenko O, Savosko S, Movchan O (2020) Knee joint structural changes in osteoarthritis and injections of platelet rich plasma and bone marrow aspirate concentrate. Georgian Med News (303):184–188
24. Jayaram P, Liu C, Dawson B et al (2020) Leukocyte-dependent effects of platelet-rich plasma on cartilage loss and thermal hyperalgesia in a mouse model of post-traumatic osteoarthritis. Osteoarthr Cartil 28(10): 1385–1393. https://doi.org/10.1016/j.joca.2020.06.004
25. Asjid R, Faisal T, Qamar K et al (2019) Platelet-rich plasma-induced inhibition of chondrocyte apoptosis directly affects cartilage thickness in osteoarthritis. Cureus 11(11): e6050. https://doi.org/10.7759/cureus.6050
26. Yan L, Zhou L, Xie D et al (2019) Chondroprotective effects of platelet lysate towards monoiodoacetate-induced arthritis by suppression of TNF-alpha-induced activation of NF-kB pathway in chondrocytes. Aging 11(9):2797–2811. https://doi.org/10.18632/aging.101952
27. Yun S, Ku SK, Kwon YS (2016) Adipose-derived mesenchymal stem cells and platelet-rich plasma synergistically ameliorate the surgical-induced osteoarthritis in Beagle dogs. J Orthop Surg Res 11:9. https://doi.org/10.1186/s13018-016-0342-9
28. Wu JZZ, Qin X, Cai X (2016) Platelet-rich-plasma alleviates pathological symptoms in a rabbit model of osteoarthritis. Int J Clin Exp Med 9(11):21038–21047
29. Chouhan DK, Dhillon MS, Patel S et al (2019) Multiple platelet-rich plasma injections versus single platelet-rich plasma injection in early osteoarthritis of the knee: an experimental study in a guinea pig model of early knee osteoarthritis. Am J Sports Med 47(10): 2300–2307. https://doi.org/10.1177/0363546519856605
30. Yokoyama M, Sato M, Tani Y et al (2017) Platelet-activated serum might have a therapeutic effect on damaged articular cartilage. J Tissue Eng Regen Med 11(12):3305–3312. https://doi.org/10.1002/term.2238
31. Almasry SM, Soliman HM, El-Tarhouny SA et al (2015) Platelet rich plasma enhances the immunohistochemical expression of platelet derived growth factor and vascular endothelial growth factor in the synovium of the meniscectomized rat models of osteoarthritis. Ann Anat 197:38–49. https://doi.org/10.1016/j.aanat.2014.10.006
32. Araya N, Miyatake K, Tsuji K et al (2020) Intra-articular injection of pure platelet-rich plasma is the most effective treatment for joint pain by modulating synovial inflammation and calcitonin gene-related peptide expression in a rat arthritis model. Am J Sports Med 48(8):2004–2012. https://doi.org/10.1177/0363546520924011
33. Khatab S, van Buul GM, Kops N et al (2018) Intra-articular injections of platelet-rich plasma releasate reduce pain and synovial inflammation in a mouse model of osteoarthritis. Am J Sports Med 46(4):977–986. https://doi.org/10.1177/0363546517750635
34. Ahmad MR, Badar W, Ullah Khan MA et al (2020) Combination of preconditioned adipose-derived mesenchymal stem cells and platelet-rich plasma improves the repair of osteoarthritis in rat. Regen Med 15(11): 2285–2295. https://doi.org/10.2217/rme-2020-0040
35. Yin WJ, Xu HT, Sheng JG et al (2016) Advantages of pure platelet-rich plasma compared with leukocyte- and platelet-rich plasma in treating rabbit knee osteoarthritis. Med Sci Monit 22:1280–1290. https://doi.org/10.12659/msm.898218
36. Filardo G, Kon E, Buda R et al (2011) Platelet-rich plasma intra-articular knee injections for the treatment of degenerative cartilage lesions and osteoarthritis. Knee Surg Sports Traumatol Arthrosc 19(4):528–535. https://doi.org/10.1007/s00167 010-1238-6
37. Southworth TM, Naveen NB, Tauro TM et al (2019) The use of platelet-rich plasma in symptomatic knee osteoarthritis. J Knee Surg 32(1): 37–45. https://doi.org/10.1055/s-0038-1675170
38. Altamura SA, Di Martino A, Andriolo L et al (2020) Platelet-rich plasma for sport-active patients with knee osteoarthritis: limited return

to sport. Biomed Res Int 2020:8243865. https://doi.org/10.1155/2020/8243865
39. Campbell KA, Saltzman BM, Mascarenhas R et al (2015) Does intra-articular platelet-rich plasma injection provide clinically superior outcomes compared with other therapies in the treatment of knee osteoarthritis? A systematic review of overlapping meta-analyses. Arthroscopy 31(11):2213–2221. https://doi.org/10.1016/j.arthro.2015.03.041
40. Previtali D, Merli G, Di Laura Frattura G et al (2021) The long-lasting effects of "placebo injections" in knee osteoarthritis: a meta-analysis. Cartilage 13(1_suppl):185S–196S. https://doi.org/10.1177/1947603520906597
41. Filardo G, Previtali D, Napoli F et al (2021) PRP injections for the treatment of knee osteoarthritis: a meta-analysis of randomized controlled trials. Cartilage 13(1_suppl):364S–375S. https://doi.org/10.1177/1947603520931170
42. Di Martino A, Di Matteo B, Papio T et al (2019) Platelet-rich plasma versus hyaluronic acid injections for the treatment of knee osteoarthritis: results at 5 years of a double-blind, randomized controlled trial. Am J Sports Med 47(2):347–354. https://doi.org/10.1177/0363546518814532
43. Sanchez M, Jorquera C, Sanchez P et al (2021) Platelet-rich plasma injections delay the need for knee arthroplasty: a retrospective study and survival analysis. Int Orthop 45(2):401–410. https://doi.org/10.1007/s00264-020-04669-9
44. Buendia-Lopez D, Medina-Quiros M, Fernandez-Villacanas Marin MA (2018) Clinical and radiographic comparison of a single LP-PRP injection, a single hyaluronic acid injection and daily NSAID administration with a 52-week follow-up: a randomized controlled trial. J Orthop Traumatol 19(1):3. https://doi.org/10.1186/s10195-018-0501-3
45. Ahmad HS, Farrag SE, Okasha AE et al (2018) Clinical outcomes are associated with changes in ultrasonographic structural appearance after platelet-rich plasma treatment for knee osteoarthritis. Int J Rheum Dis 21(5):960–966. https://doi.org/10.1111/1756-185X.13315
46. Raeissadat SA, Ghorbani E, Sanei Taheri M et al (2020) MRI changes after platelet rich plasma injection in knee osteoarthritis (randomized clinical trial). J Pain Res 13:65–73. https://doi.org/10.2147/JPR.S204788
47. Mautner K, Malanga GA, Smith J et al (2015) A call for a standard classification system for future biologic research: the rationale for new PRP nomenclature. PM R 7(4 Suppl):S53–S59. https://doi.org/10.1016/j.pmrj.2015.02.005
48. Braun HJ, Kim HJ, Chu CR et al (2014) The effect of platelet-rich plasma formulations and blood products on human synoviocytes: implications for intra-articular injury and therapy. Am J Sports Med 42(5):1204–1210. https://doi.org/10.1177/0363546514525593
49. Cavallo C, Filardo G, Mariani E et al (2014) Comparison of platelet-rich plasma formulations for cartilage healing: an in vitro study. J Bone Joint Surg Am 96(5):423–429. https://doi.org/10.2106/JBJS.M.00726
50. Filardo G, Kon E, Pereira Ruiz MT et al (2012) Platelet-rich plasma intra-articular injections for cartilage degeneration and osteoarthritis: single- versus double-spinning approach. Knee Surg Sports Traumatol Arthrosc 20(10):2082–2091. https://doi.org/10.1007/s00167-011-1837-x
51. Mariani E, Canella V, Cattini L et al (2016) Leukocyte-rich platelet-rich plasma injections do not up-modulate intra-articular pro-inflammatory cytokines in the osteoarthritic knee. PLoS One 11(6):e0156137. https://doi.org/10.1371/journal.pone.0156137
52. Kon E, Di Matteo B, Delgado D et al (2020) Platelet-rich plasma for the treatment of knee osteoarthritis: an expert opinion and proposal for a novel classification and coding system. Expert Opin Biol Ther 20(12):1447–1460. https://doi.org/10.1080/14712598.2020.1798925

INDEX

Martin J. Stoddart et al. (eds.), *Cartilage Tissue Engineering*,
Methods in Molecular Biology, vol. 2598, https://doi.org/10.1007/978-1-0716-2839-3,

Q

R

S

T

U

MIX
Papier aus verantwortungsvollen Quellen
Paper from responsible sources
FSC® C105338

If you have any concerns about our products,
you can contact us on
ProductSafety@springernature.com

In case Publisher is established outside the EU,
the EU authorized representative is:
Springer Nature Customer Service Center GmbH
Europaplatz 3, 69115 Heidelberg, Germany

Printed by Libri Plureos GmbH
in Hamburg, Germany